SOME COMMON MOSSES

OF

BRITISH COLUMBIA

W.B. SCHOFIELD
Department of Botany
University of British Columbia

Illustrations by Patricia Drukker-Brammall

Published by the Royal British Columbia Museum, 675 Belleville Street, Victoria, British Columbia, V8V 1X4, Canada.

Printing History:

First edition 1969
Reprinted 1973

Second edition 1992

Cover photograph of *Isothecium stoloniferum*
by Andrew Niemann, RBCM.
Cover and general design by Chris Tyrrell, RBCM.
Production co-ordinated by Gerry Truscott, RBCM.
Typeset 11/13 in Garamond No. 3 by The Typeworks, Vancouver.

Printed in Canada by Queen's Printer.

This edition is published with financial assistance from the Ministry of Environment.

Canadian Cataloguing in Publication Data

Schofield, W.B.
Some common mosses of British Columbia

(Royal British Columbia Museum handbook, ISSN 1188-5114)

Previously published; British Columbia Provincial Museum, 1969. (Handbook ; no. 28)

ISBN 0-7718-9165-2

1. Mosses—British Columbia. I. Royal British Columbia Museum. II. Title. III. Series.

QK541.7.B7S36 1992 588′.2′09711 C92-092080-2

PREFACE

The first edition of this book endeavoured to introduce an intriguing but neglected group of plants to the curious naturalist; it proved popular with students and amateurs, but had its limitations in the clarity of the keys, in particular, and in the number of species included. The present edition has been completely rewritten and is constructed on different assumptions. Detailed descriptions have been eliminated, with the beautiful drawings of Patricia Drukker-Brammall serving to show the essential appearance. The scale bar always refers to the magnification of the habit sketch of the whole plant; sizes of enlarged leaves, shoots or sporangia can be determined readily through comparison with this habit sketch. The information provided with each main species treated greatly expands the number of additional species that can be determined in spite of lack of illustrations. The derivation of the name of the moss as well as information concerning features that distinguish it and species that resemble it are presented to add interest and to better understand the species. The distribution maps are based on specimens examined, and those preserved at the University of British Columbia Herbarium. It is hoped that greater familiarity with the identity of mosses will lead to increased awareness of their importance in the landscape as well as the great beauty they exhibit.

ACKNOWLEDGEMENTS

The illustrations for this book were financed by the University of British Columbia Faculty of Graduate Studies Committee on Research. Much of the field work in acquiring the specimens has been supported by Natural Sciences and Engineering Research Council of Canada. The Nippon Paint Company of Japan subsidized improvement of the typed manuscript.

Debra Wadsworth typed the manuscript and Tami Chappell typed the index; my grateful thanks to them. Gratitude is also due to the editors at the Royal British Columbia Museum: Harold Hosford and Gerry Truscott. I am indebted to Richard Hebda and Michael Ryan for initial selection and identification of the mosses in the photographs.

The idea for this book came from John Steuart Erskine of Wolfville, Nova Scotia. It was he who led the author to become fascinated by mosses. My debt to him is incalculable. The book is dedicated to his memory.

R.R. Ireland and D.H. Vitt reviewed the manuscript of this book and offered valuable corrections and recommendations; I gratefully acknowledge this assistance. If errors persist, they are my responsibility.

Contents

PREFACE iii

ACKNOWLEDGEMENTS iv

INTRODUCTION 1

- Mosses in British Columbia 1
- What are Mosses? 2
- Life Cycle 8
- How to Collect 10
- Where to Collect 12
- When to Collect 12
- History of Collectors 13
- Ecology 14
- Distribution in British Columbia 16
- Habitats in British Columbia 19
- Critical Determination 26
- Names of Mosses 27
- Moss Gardens 29
- Uses of Mosses 30
- Use of the Hand Lens 31
- Format of this Book 31
- Some Uncommon Mosses of British Columbia 32

KEY TO MOSSES ILLUSTRATED IN THIS BOOK 35

SOME COMMON MOSSES OF BRITISH COLUMBIA 57

CHECKLIST OF THE MOSSES OF BRITISH COLUMBIA 307

DISTRIBUTION MAPS 319

GLOSSARY 383

INDEX 387

INTRODUCTION

Mosses in British Columbia

Mosses are nearly everywhere in British Columbia. Floors of coniferous forests are heavily blanketed by soft green carpets; the trees of the foggy coast often carry filmy banners that festoon their branches, and their trunks are sheathed by thick sleeves of mosses. Peatland is often composed entirely of wet spongy carpets of moss varying in colour from vivid red-purple to pale yellow and many shades of green. Tufts and mats of mosses soften the outline of boulders and cliffs, and enrich the landscape with colours that glow vividly after rain.

Mosses invade lawns, particularly when lawns are shaded and moist much of the year; even roofs and walls have tufts of moss squeezing among the shingles, between the tiles and into neglected rain gutters. Among the paving-stones and cracks in sidewalks mosses establish themselves; even on unpainted concrete or stone steps mosses find a niche and flourish.

The luxuriance of mosses as well as their diversity in form inevitably capture the interest of a naturalist. Once a moss can be recognized and identified, its species name leads the naturalist into the literature concerning its biology. If the naturalist creates a filing system within which personal observations can be placed, each species name becomes a magnet that attracts and holds information.

Technical manuals tend to be somewhat daunting to a beginner, and extremely simplified guides can frustrate a serious naturalist by inadvertently misleading the reader with over-simplified information. This guide attempts to avoid these pitfalls, but a limited amount of specialized vocabulary is necessary to escape the repetition of cumbersome descriptive phrases. These specialized terms are defined either in the text or in the Glossary (p. 383).

This guide treats only a fraction of the mosses that occur in British Columbia. It includes most species that are common in areas that are most accessible.

There is great satisfaction in learning more about the organisms that surround us. Such knowledge not only enriches our intellectual life but also enhances our pleasure of the world around us. This guide endeavours to lead the user into a deeper understanding of the mosses, their biology and beauty. A rich by-product of this new knowledge is a greater appreciation of all aspects of the natural world. The more we are captivated by this appreciation, the more likely we are to endeavour to live in harmony with that world.

What Are Mosses?

Many people, including far too many biologists, do not recognize what a moss is. Mosses belong to their own class of the plant kingdom, the *Musci* or *Bryopsida*. They are plants that have a leafy green shoot that lacks both a complex vascular system and roots. A moss, if it reproduces sexually, produces a sporophyte (spore-producing plant) that has a sporangium (spore-producing organ) containing spores. The sporangium generally opens by means of a lid (*operculum*). The sporophyte is attached to the leafy green plant (gametophyte) and is parasitic on it.

Many plants that are not mosses but resemble them are also called mosses; the so-called "club-mosses" are actually leafy green plants that have a complex vascular system and roots, and the green plant is a sporophyte rather than a gametophyte. The spores of club-mosses actually produce a gametophyte that is never leafy and is rarely seen except by an experienced student who knows where to look.

"Reindeer mosses" are lichens; lichens are fungi that have trapped algae in their plant body and which depend on the algae for sustenance. Other lichens that festoon tree branches are sometimes popularly called mosses - their lack of leaves and their fungal structure soon identifies them as lichens.

"Spanish moss" is a flowering plant related to pineapple. It produces small flowers. Spanish moss is not found in British Columbia.

"Sea mosses" are marine algae or seaweeds. These plants are aquatic, never leafy, and are confined to salt water. No true mosses live in salt water. Sometimes freshwater algae are popularly called mosses. These aquatic plants are not leafy and form slimy masses in water bodies.

"Scale mosses", the leafy liverworts (class *Hepaticae*), are closely related to mosses and can be mistaken for them. When they possess sporophytes, leafy liverworts can be distinguished readily from mosses; in liverworts the seta is very soft and collapses soon after the sporangium sheds its spores (in mosses the seta is usually rigid and persists for an extended period); in leafy liverworts the sporangium usually opens by means of four longitudinal slits so that it resembles a flower with four petals (in mosses the sporangium opens by an operculum); among the spores of leafy liverworts are corkscrew twisted threads (*elaters*) that are never present in mosses; these are visible at 12X. The gametophytes of leafy liverworts usually have the leaves in three rows, two lateral and one ventral (on the undersurface), while in mosses leaves are usually in more than three rows. The lateral leaves of leafy liverworts frequently have a lobe (sometimes several lobes) and the underleaves are usually smaller and different in shape from the lateral leaves; this is rare in mosses. Leaves of leafy liverworts never have a midrib, while a midrib is common in mosses. The leaf arrangement in most leafy liverworts is complanate (the leafy shoot appears compressed). In leafy liverworts the ventral leaves are usually flattened against the surface of the stem and the lateral leaves are flattened in the same plane, parallel to the stem; this means that they extend outward like wings from the lateral surface of the stem. Such a leaf arrangement is uncommon in mosses, where the leaf arrangement is usually spiral and the leafy shoot is generally not flattened. A further difference, apparent only under the microscope, is that the rhizoids of leafy liverworts are unicellular and unbranched while they are multicellular and branched in mosses. The female sex organs of leafy liverworts are often sheathed by a special sleeve of fused leaves, the perianth. Such a structure is absent in mosses, where the female sex organs are surrounded by many specialized leaves that constitute the perichaetium.

Liverworts are rather difficult to determine without the use of a microscope, but these beautiful, small plants offer the same fascination as the mosses. British Columbia has a rich flora of more than 220 species of liverworts.

The following key should separate most mosses in the province from other groups of organisms that might be confused with them:

1. Plants with many leaves borne on a stem ______________________ 2
1. Plants not composed of leaves on a stem ______________ not a moss.

2. Sporangia borne in flowers, cones, or on leaves, each sporophyte producing many sporangia ______________________ not a moss.
2. Sporangium parasitic on leafy plant, each sporophyte producing one sporangium ______________________ 3

3. Leaves lobed ______________________ not a moss.
3. Leaves not lobed ______________________ 4

4. Leaves with midrib ______________________ moss.
4. Leaves lacking midrib ______________________ 5

5. Leaves with sharply pointed tip ______________________ moss.
5. Leaves with rounded tip ______________________ 6

6. Sporangium with operculum ______________________ moss.
6. Sporangium lacking operculum ______________________ not a moss.

Figure 1. Structure of a moss. A moss consists of two distinct parts, the *gametophyte* and the *sporophyte*. The gametophyte is the leafy green plant, attached to the substratum by branched *rhizoids*. If the plant produces female sex organs, these are usually protected by leaves, termed *perichaetial leaves*. A perichaetial leaf consists of several distinct regions: at the base, near the margin, are the *alar regions*. The cells of this region are sometimes different in size, shape and colour from the remainder of the leaf. The edge of the leaf is termed its *margin*; the midrib is usually termed the *costa*, but for the sake of simplicity, the term midrib is used in this book. At the apex of the leaf are the *gemmae*. Gemmae are small masses of cells that are produced on the gametophyte and which fall off readily. They can grow into a new leafy gametophyte if they reach a suitable environment.

The sporophyte grows partly as a parasite on the gametophyte. In most mosses it is made up of a rigid stalk or *seta* and a *sporangium* at the apex of the seta. The sporangium is usually covered by a little cap, the *calyptra*, and often has a swollen base, the *apophysis*, and a lid, the *operculum*. When the calyptra and operculum are shed, the *peristome* is revealed; this is the opening at the apex of the sporangium, which usually has a ring of *peristome teeth* around it. These teeth respond to moisture changes by pulsating in and out of the mouth of the sporangium. The jagged inner surfaces of the teeth catch spores and thus pull them out to be carried away by the moving air or water, or even by small invertebrates.

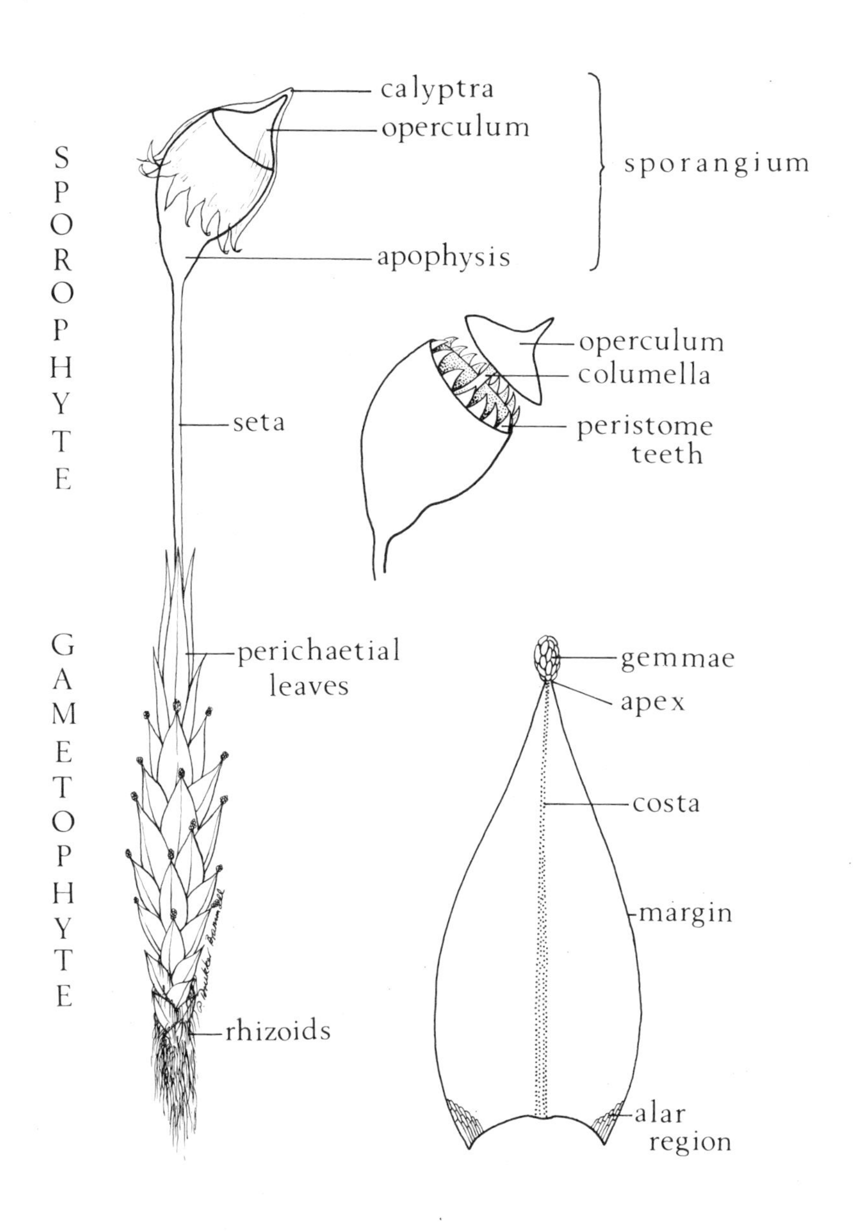

Structure of a moss.

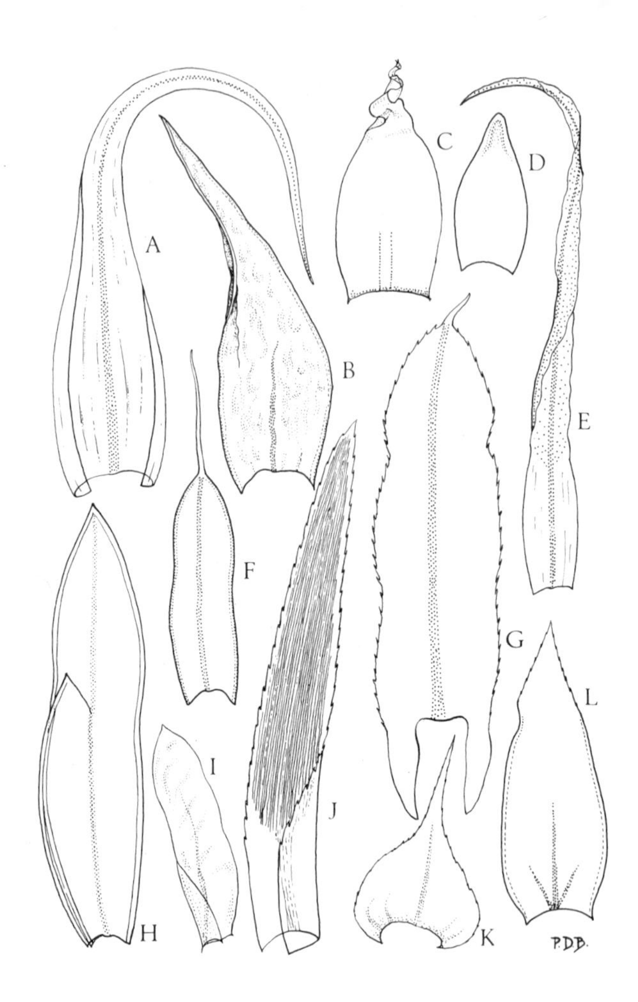

Figure 2. Shapes and other features of the leaves of mosses. Constancy of shape, presence or absence of a midrib, nature of margins, as well as other features within a species are useful for identification. A. *Drepanocladus uncinatus*: ovate-lanceoate and falcate, with revolute margins and longitudinal pleats. B. *Rhytidium rugosum*: ovate with acute apex and irregular wrinkles and undulations. C. *Hylocomium splendens*: ovate with a sinuous apex and double midrib. D. *Andreaea rupestris*: ovate with a blunt apex. E. *Tortella tortuosa*: lanceolate, with somewhat undulate margins. F. *Tortula muralis*: tongue-shaped, with an abrupt hair point. G. *Plagiomnium insigne*: oblong ovate to tongue-shaped with differentiated and toothed margin and a decurrent base. H. *Fissidens limbatus*: oblong, with differentiated margin and a sheathing flap emerging from the midrib. I. *Metaneckera menziesii*: oblong-ovate with undulate surface. J. *Pogonatum contortum*: broadly lanceolate to oblong, with abundant parallel lamellae and toothed margins. K. *Kindbergia praelonga*: broadly heart-shaped. L. *Antitrichia curtipendula*: ovate, toothed toward apex, midribs flaring.

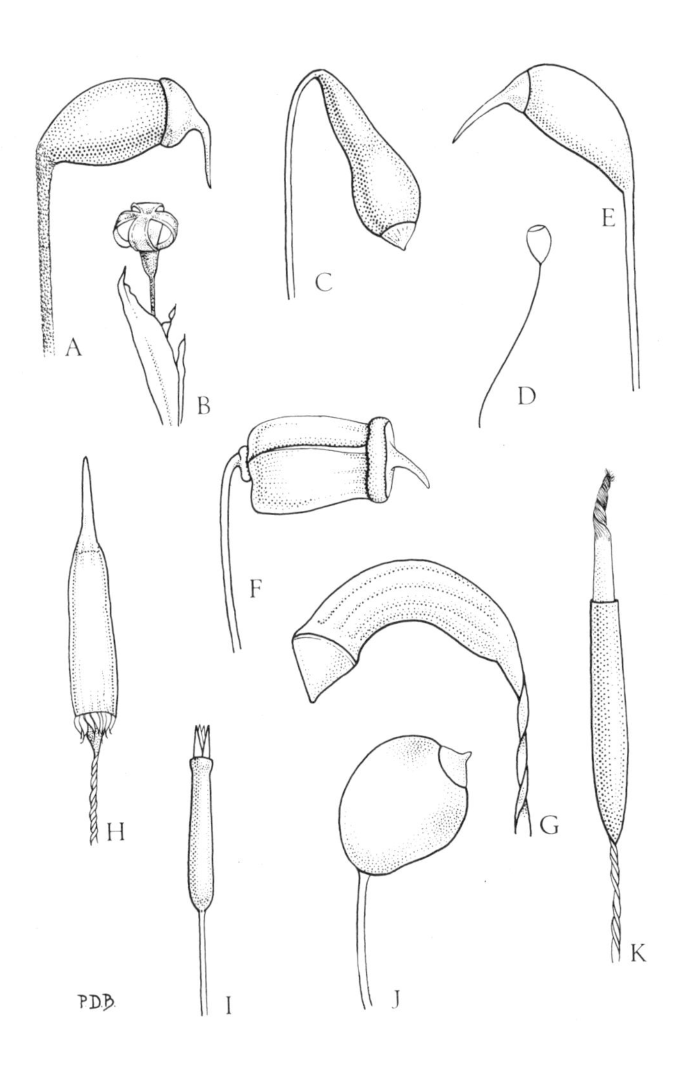

Figure 3. Shapes and orientations of the sporangia of mosses. Shape and orientation are consistent features of each moss species and can be used to help identify the species. In the examples illustrated shape is noted first, followed by orientation of the sporangium. A. *Kindbergia praelonga*: ovoid, inclined; B. *Andreaea rupestris*: elliptic (the gaping portion subspherical), erect; C. *Leptobryum pyriforme*: pear-shaped, nodding; D. *Schistostega pennata*: subspherical, erect; E. *Dicranum fuscescens*: cylindric, curved, inclined; F. *Poytrichum juniperinum*: rectangular, inclined; G. *Aulacomnium palustre*: cylindric, curved, inclined; H. *Encalypta ciliata* (with calyptra): cylindric, erect; I. *Tetraphis pellucida*: cylindric, erect; J. *Philonotis fontana*: subspherical, sub-erect to inclined; K. *Tortula princeps*: cylindric, erect.

Life Cycle

The simplest stage in the life of a moss is the *spore*. The spore is generally borne by wind or water. If deposited on a humid substratum, it usually germinates to produce a filament. This filament (the *protonema*) branches and creeps over the substratum. It differentiates into a green cobwebby mass attached by colourless (or brownish), multicellular, filamentous branches, the *rhizoids*. In time, the protonema produces "buds" of cell masses. Each bud differentiates an apical cell that gives rise to a stem with leaves.

The leafy *gametophyte* ultimately produces sex organs, on either a terminal or lateral branch. The female sex organ, the *archegonium*, is essentially microscopic and shaped like a tiny flask with an *egg* inside the flask. Each female sexual branch (*perichaetium*) has a mass of protective leaves (*perichaetial leaves*) surrounding several archegonia with filaments among them. When mature and moist, each flask-shaped archegonium has cells within the neck that disintegrate to produce a fluid that is exuded from the mouth of the flask.

Meanwhile, male sexual branches (*perigonia*), also composed of specialized leaves (*perigonial leaves*), have produced within this sheath of leaves many male sex organs (*antheridia*) and filaments among them. These many-celled antheridia are irregularly elliptic or spherical and surrounded by a single layer of sterile cells that make up the jacket; each contains numerous sperms. When mature and moist, the antheridial wall ruptures and the many sperms are released into the water. Each sperm is microscopic and swims about by means of two terminal, hair-like flagella that are in constant motion carrying the sperm in a corkscrew pathway through the water.

The fluid that has diffused from the archegonial necks into the water dissipates into the water. When a sperm encounters archegonial fluid, it swims toward the stronger concentration of the fluid, down the neck of an archegonium to the egg. When a sperm touches the egg wall, it burrows through the wall and its nucleus unites with the nucleus of the egg. This cell is now the *zygote* and is the first stage of the *sporophyte*. All the growth up to the production of the zygote has been in the gametophyte. The gametophyte is the complex leafy plant in the moss. A structural equivalent to the gametophyte, in flowering plants, would be a microscopic pollen grain or the embryo sac.

The zygote undergoes cell division, in time differentiating a stalk or *seta* and a foot that burrows into the gametophyte. The archegonium also grows to form a sheath that caps the growing sporophyte; this sheath is the *calyptra*. As the seta elongates, the calyptra is torn from the apex of the gametophyte and persists as a protective cap at the apex of the sporophyte. Beneath the calyptra and at the apex of the sporophyte, enlargement occurs as differentiation of the sporangium proceeds. The sporangium is now wider in diameter than the seta.

Further differentiation of the sporangium usually results in a sterile jacket of cells forming the wall of the sporangium; the differentiation of a weakened ring of cells cuts out the lid (*operculum*) around the apex of the sporangium. Within the sporangium there is generally a cylinder of numerous spore mother cells that surround a central cylinder (*columella*) of sterile cells. The many spore mother cells undergo meiosis, each producing four spores. These spores represent the first stage of the gametophyte. Throughout these youthful stages the sporophyte is green, producing some of its own organic material through photosynthesis, but also extracting water and nutrients from the gametophyte by way of the foot.

When the sporangium is mature, the calyptra usually dries up and is carried away by a slight breeze. Further drying of the sporangium causes it to shrink and the operculum breaks loose. Beneath the operculum is the peristome, usually ringed with peristome teeth. When moist, these teeth curve into the sporangium and pick up spores on their irregular surfaces. When the teeth dry out, they flick outward, throwing spores out of the sporangium with each flick. This process of expelling spores continues as the peristome teeth moisten and dry out again and again. The expelled spores are then carried away by any slight air movement. Some spores may ultimately reach a suitable site to germinate.

This process is described as *sexual reproduction*, but mosses are not completely dependent on sexual reproduction for effective propagation. Since the leafy plant of a moss is structurally very simple, almost any living cell is capable of becoming the progenitor of an entire leafy plant. Thus, if a plant is fragmented into many pieces, each fragment in which there are some whole living cells can produce a protonema if the substratum is humid enough and suitable illumination and temperature are available; the protonema differentiates to produce a bud and finally a leafy plant. Some mosses reproduce entirely by this method of *asexual reproduction*, since no sex organs or sporophytes have been discovered in these mosses. This is a method widely used by all mosses, even those that reproduce sexually.

Sexual reproduction is dependent on the presence of moisture during the many stages that lead to the spore landing in a suitable habitat; this makes the process a chancy one, particularly in an area where suitable substratum may be limited and is subject to invasion by other organisms. Fortunately some moss spores can remain alive for a relatively long period of time, and when suitable conditions finally appear, they germinate. Another advantage for many mosses is that the protonema can also tolerate drying for extended periods. Of course, not all mosses are so tolerant; those that are not have more restricted habitats.

Many mosses produce special masses of cells that break from the shoot readily. These are *gemmae*. Mosses like *Tetraphis* and *Aulacomnium* commonly produce shoots terminated by masses of gemmae. These gemmae are carried by wind or water to a new site where they may germinate to produce a protonema and, in time, a new leafy plant that may produce gemma-bearing, sexual or asexual shoots. Some mosses have fragile leaves or stems. Asexual reproduction is a highly effective process in producing fragments of the gametophyte, some of which may end up in an environment suitable for growth.

How to Collect

Collecting mosses is extremely simple. Most mosses are easily removed from the substratum. Frequently, a knife is needed to pry the moss loose. After the specimen has been cleaned of extraneous matter (soil, detritus, etc.) it can be air-dried. No pressing is necessary. Normally the specimen is teased apart so that its growth appearance is clearly shown and it fits neatly into a folded packet.

When collecting, the collector should endeavour to find sporangium-bearing material. In some mosses, sporangia are needed for an accurate determination; in most mosses, the presence of sporophytes make identification easier. Some mosses rarely produce sporophytes, therefore even specimens lacking sporophytes should be collected. If the specimen is undeterminable, it can be discarded. A number of endemic or rare species in the province, including *Wijkia carlottae*, *Gebeebia gigantea* and *Leptodontium recurvifolium*, have never been collected with sporophytes; even some common mosses, like *Pleurozium schreberi* and *Hylocomium splendens*, usually lack sporophytes.

When collecting mosses, the following data should be recorded and placed on the label of the packet containing the specimen:

(1) Habitat—the substratum (rock, rotten wood, etc.), exposure (sun-

ny, shaded), moisture (wet, submerged, dry), nature of surrounding vegetation (tundra, coniferous forest), elevation (if possible).

(2) Collector's name.

(3) Locality collected (example: Grafton Bay, Bowen Island, Howe Sound, British Columbia) with approximate latitude and longitude.

(4) Date collected (example: 12 August 1956).

(5) When the specimen is determined, the scientific name can be placed on the label.

Specimens can be collected in improvised paper packets. To make a collecting packet begin with a letter-size sheet of paper (8½ x 11 inches or use standard metric size letter sheet 210 x 300 mm). First, fold the sheet so that the bottom edge is 3 inches (or 80 mm) from the top edge. Then fold the top 3-inch (80-mm) portion downward to make a flap over the lower portion. Fold the two sides of the paper under to make 1-inch (or 20-mm) flaps. This produces a closed packet. Later, paper packets of the same design can be used to store the dried specimens. Plastic bags should not be used; specimens kept in plastic bags tend to decompose or discolour before they dry.

Usually a collector assigns a collection number to each specimen. This number identifies the specimen for reference purposes. It is useful to record localities and collection numbers in a record book. This record serves as a useful future reference of what has been found at particular localities. If a specimen is divided and placed in several packets, each packet should be assigned the same number.

Specimens can be stored easily and effectively in shoe boxes (men's shoe boxes are a convenient size). The specimens are placed in thin packets, treating them like file cards, in the shoe boxes. Initially, a collection can be filed alphabetically by the name of the genus and species. Possession of an accurately determined collection enables a student to compare undetermined specimens with named material. The collection placed in a permanent herbarium represents a valid record of the presence of the species from a given area and, if the vegetation of an area changes, is ultimately a fragment of the record of a past vegetation. Many mosses and other plants have been eliminated from many areas of the earth's surface. Collections made by both amateur and professional botanists of the past are the only surviving evidence of this flora.

Amateurs can make valuable contributions to an understanding of the flora by assembling well-documented and processed specimens for a particular region. The distribution maps included in this book show that some areas are very poorly documented. Such collections, or replicates, should be donated to an institution that preserves the material for future reference and study. In British Columbia, the herbaria of the Botany Department of the University of B.C. in Vancouver and Royal British Columbia Museum in Victoria are appropriate. In Ottawa, the herbarium of the National Museum of Nature assembles such specimens.

Where to Collect

The discussion under Ecology (p. 14) and Distribution of Mosses in British Columbia (p. 16) gives a good idea where mosses are most abundant and also what species might be found there. As mentioned earlier, mosses can be collected nearly anywhere in the province. Humid forests and canyons and open cliffs where moisture persists give the richest floras.

Where humidity is constantly high and where habitats are not occupied by flowering plants, mosses tend to be abundant and varied. In parts of the province where there is a distinct and extended dry season or in which precipitation is low (as in the Interior), mosses tend to be less abundant and more restricted in distribution. In these localities, mosses are found in habitats where moisture and illumination are suitable. Surprisingly, boulders in sagebrush and the earth under the scattered shrubs have a number of mosses peculiar to them. These are usually inconspicuous during the dry season, but become more apparent when wet.

When to Collect

Mosses can be collected at any time of the year. In most areas of the province the best time is in early spring when the temperatures begin to rise and moisture is still available either from melting snow or from continuing precipitation. At this time the mosses appear most verdant and also are producing sporophytes. Some mosses delay sporophyte production until the cooler, humid autumn. Many mosses can be collected in summer, however, since the sporophytes persist from the spring or from the preceding autumn, although both calyptra and operculum are likely to be gone and, in many cases, the peristome teeth will have broken away. In areas that have little snowfall in winter or in which the ground is not frozen, mosses can be collected in winter.

Of course, at higher elevations on mountains, collecting must be delayed until the mosses are exposed by the melting of the snow. Mountain habitats have numerous different mosses in the vegetation. In some mountains, late summer and early autumn yield great riches in moss-collecting.

History of Collectors

Until recent years, collection of mosses in the province has not been great, but casual collections have been made since the late 18th century, when Dr Archibald Menzies, surgeon with the fur-trading vessel *Prince of Wales* in 1786 and 1789, collected from what is now British Columbia. During the expeditions of Capt. George Vancouver, Menzies returned and collected mosses between 1791 and 1795, some certainly from southern Vancouver Island.

The next important collection was made by Thomas Drummond during 1826 in the Rocky Mountains; these specimens were distributed to various herbaria in 1828.

Dr David Lyall collected mosses in the province between 1858 and 1861. His collections comprise the richest representation of the moss flora up to that time. They were made during his tenure as naturalist for the survey party marking out the 49th parallel, the boundary between the United States and Canada.

John Macoun, the Dominion Botanist, made numerous highly important collections of mosses in the Province, the first in 1872 and the last during his retirement on Vancouver Island in 1916. His contribution to the knowledge of mosses of the province is probably the most important until recent time.

A significant collection was made by R.S. Williams in 1898 while on his way across the mountains of the White Pass route to Dawson, Yukon, during the gold-rush. A number of amateurs have made significant collections in the province in the first half of the 20th century; among these are A.H. Brinkman, Faye MacFadden and Elizabeth MacKenzie.

Three Scandinavian botanists have made significant collections in British Columbia: T. Ahti (largely in Wells Gray Park), V. Kujala (in various localities) and H. Persson (in the Queen Charlotte Islands).

Dr V.J. Krajina has made significant collections of mosses since 1949 and his students have contributed smaller collections. F.M. Boas has also made many important collections.

Since 1960, the author, financed by the Natural Sciences and Engineering Research Council of Canada, has made extensive collections of mosses in the province. These, as well as duplicates of many earlier collections made by collectors discussed above, are housed in the herbarium of the University of British Columbia, the most comprehensive collection of mosses of the province. Since 1960, too, there has been a considerable enrichment of collections from the province from other sources; among the major contributors are R.J. Belland, C.C. Chuang, N. Djan-Chekar, R. Halbert, D.G. Horton, A. von Hubschmann, R.R. Ireland, D.W. Jamieson, T.T. McIntosh, J.R. Spence, B.C. Tan, T. Taylor and D.H. Vitt. Other amateurs and professionals have made limited collections of mosses in the province.

Mosses have attracted the interest of a rather limited number of amateurs. The need for costly and extensive literature and a microscope for critical identification of material may discourage many would-be enthusiasts. An additional problem is the false notion that all mosses look alike. A quick look at the illustrations in this book should effectively dispel this idea.

The coastal portion and the humid and dry interior of the province have been relatively well collected, but even there, especially in topographically complex terrain, the moss flora is not well documented. Extensive regions of the province, including a large proportion of the northern half, have few or no mosses documented; even the Rocky Mountains and adjacent mountain ranges need careful exploration, which would likely yield species of particular interest. Only a few moss collections have been taken from much of Vancouver Island and the fiord-lands of the mainland and from the islands that flank the mainland.

Ecology

While mosses are widely distributed over the earth's surface, many mosses have special requirements. Some, for example, grow only on decayed animal waste; these include many species of the family Splachnaceae. Interestingly enough, many of these mosses also produce a foul smell that attracts flies. Flies, when they alight on the moss sporangium, receive a dusting of moss spores. When the flies move to other decayed matter, some of these spores reach this new substratum. *Splachnum rubrum* and *Tetraplodon mnioides* are Splachnaceous mosses.

Some mosses grow best in habitats with high illumination and humidity. Wetlands have these conditions; there, many mosses flourish, especially *Sphagnum*. Each of the many species of *Sphagnum* has a particular type of site in which it thrives. Some are confined to wet depressions in the bog, others to the drier margins. Still others tolerate only a certain amount of acidity, and die if this is exceeded, to be replaced by mosses that can tolerate the altered conditions.

Sphagnum, the peat moss, forms extensive quaking carpets in bogs and around lakes and ponds, especially in the boreal coniferous forest. Some species of *Sphagnum* are aquatic and float in the marginal waters of quiet bodies of water. These mosses, as they grow, absorb chemical substances from the water, leaving it highly acid. This acid water inhibits decay and permits few organisms to survive in it. Thus, any organic material added to the water, either by the growth of the *Sphagnum* or by litter that blows into the water, leads to a deposition of this material on the lake bottom. In time, the lake is slowly filled in; the depth of the marginal shallows decreases and a quaking mat of *Sphagnum* is formed, growing over the accumulating organic muck. As the lake fills with organic material, other plants move in on the *Sphagnum* mat, among them, ultimately, forest tree species. Initially there is an open bog, which is later replaced by an open bog forest and finally by a closed forest.

Sphagnum growth, however, can also inhibit the growth of forest trees. In some areas, the *Sphagnum* expands from an open bog into the floor of the adjacent forest. The expansion of the *Sphagnum* population means that the growing moss absorbs water from adjacent *Sphagnum* in the water. The whole population acts as an immense absorptive sponge as water moves from the water-body or wet bog outward to the perimeter of the colony. If the perimeter invades a forest, the water brought to the forest floor can drown the roots of the trees and kill them. The death of the trees increases illumination at ground level and improves moss growth—the *Sphagnum* population expands further into the forest. The destruction of forests by the encroachment of *Sphagnum* bogs can be significant, particularly in northern British Columbia.

Many mosses grow well on exposed rock faces. Since these mosses can tolerate periods of complete drying out, they are well suited to such extreme environments. As the moss colony grows on the rock surface, its decaying remains produce various acids and hold moisture—conditions suitable for the chemical breakdown of the rock surface itself. In time, a primitive soil is formed, containing minerals from both the chemical breakdown of the rock and decay of the moss itself. This soil retains moisture and serves as a substratum for seed plants. Thus mosses act as pioneers on rock surfaces and lead to colonization by seed plants. These seed plants further alter the substratum and cause shading, creating an unfavourable environment for the pioneer moss but a favourable environment for some woody plants. Slowly, then, the rock surface is colonized by scattered trees and, as the rock surface itself is broken down and covered by an accumulating soil, the forest moves in.

With such a forest environment, new habitats are created that favour still other mosses. Many mosses thrive under relatively low illumination and cannot tolerate high illumination, particularly when such habitats dry rapidly. The humid forest frequently has luxuriant carpets of mosses, some growing on decaying logs, others on the trunks and the branches of living trees, and still others on the humus of the forest floor. The forest floor usually has several kinds of mosses, many restricted to a specific substratum but all influenced by the shade and humidity of the forest itself.

Many mosses are very specific to substratum; for example, *Andreaea* is found only on acidic rocks, usually in exposed sites, while some species of *Tortula* are confined to sites rich in lime. Other mosses are common on cliffs in some parts of their range, while in other parts they grow mainly on tree trunks. *Metaneckera menziesii*, for example, is a common species on tree trunks in coastal British Columbia; in the Interior it is generally found on cliffs.

Distribution in British Columbia

It is useful to know the specific mosses found in a geographic region and the particular habitat each occupies. This allows a naturalist to narrow the choices of possibilities by eliminating the many species not found in a particular area. The following lists include only species treated in this book. Many other species also occur in these areas and sites, but are generally less abundant.

1. Throughout the province

Some mosses are found throughout the province. Most are particularly common in the northern and the mountain portions of the province. Considering only the mosses treated in this book, the following species are widely distributed throughout British Columbia:

Amblystegium serpens
Amphidium lapponicum
Andreaea rupestris
Aulacomnium palustre
Blindia acuta
Bryum argenteum
Calliergon giganteum
Calliergon stramineum
Campylium stellatum
Ceratodon purpureus
Climacium dendroides
Cratoneuron commutatum
Cratoneuron filicinum
Dichodontium pellucidum
Dicranum fuscescens
Dicranum scoparium
Distichium capillaceum
Ditrichum flexicaule
Drepanocladus exannulatus
Drepanocladus uncinatus
Eurhynchium pulchellum
Encalypta ciliata
Fissidens adianthoides
Funaria hygrometrica
Hedwigia ciliata
Hylocomium splendens
Leptobryum pyriforme
Mnium spinulosum
Oncophorus wahlenbergii
Philonotis fontana
Plagiopus oederi
Pleurozium schreberi
Pohlia cruda
Pohlia nutans
Pohlia wahlenbergii
Polytrichum alpinum
Polytrichum commune
Polytrichum juniperinum
Polytrichum piliferum
Ptilium crista-castrensis
Racomitrium lanuginosum
Rhytidiadelphus triquetrus
Schistidium apocarpum
Sphagnum capillifolium
Sphagnum palustre
Sphagnum squarrosum
Tetraphis pellucida
Tetraplodon mnioides
Timmia austriaca
Tortella tortuosa
Tortula ruralis

Each of these species is restricted in its habitat, but the populations are very large for those with extensive habitat.

2. Coastal Forests and Interior Humid Forest

The following species are confined largely to the coastal coniferous forests or, in some cases, re-appear in the cedar-hemlock forests of the Interior:

Antitrichia curtipendula
Atrichum selwynii
Aulacomnium androgynum
Bryum capillare
Bryum miniatum
Buxbaumia piperi
Claopodium crispifolium
Dicranella heteromalla
Dicranella rufescens
Ditrichum heteromallum
Fissidens limbatus
Fontinalis antipyretica
Grimmia pulvinata
Heterocladium macounii
Homalothecium fulgescens
Hookeria lucens
Hypnum circinale
Hypnum subimponens
Isothecium stoloniferum
Kiaeria starkei
Kindbergia oregana
Kindbergia praelonga
Leucolepis acanthoneura
Neckera douglasii
Oligotrichum aligerum
Orthotrichum lyellii
Plagiomnium insigne
Plagiothecium undulatum
Pogonatum contortum
Pogonatum urnigerum
Pohlia annotina
Pseudotaxiphyllum elegans
Rhizomnium glabrescens
Rhytidiadelphus loreus
Rhytidiopsis robusta
Roellia roellii
Schistostega pennata
Scleropodium obtusifolium
Scouleria aquatica
Tortula muralis
Tortula princeps
Ulota obtusiuscula

3. More frequent away from the coast

The following are species absent or rare near the coast and found predominantly on calcium-rich substrata:

Abietinella abietina
Catoscopium nigritum
Dicranum polysetum
Hypnum revolutum
Meesia triquetra
Paludella squarrosa
Rhytidium rugosum
Thuidium recognitum
Tomentypnum nitens
Tortula mucronifolia

4. Widespread, but absent in the north

Several species are widespread in the province, but absent in the northwestern corner where boreal forest predominates. Some of these apparent absences may result from inadequate collections from this part of the province.

Bartramia pomiformis
Dicranum tauricum
Didymodon insulanus
Encalypta ciliata
Hygrohypnum ochraceum
Plagiothecium denticulatum

Racomitrium aciculare
Racomitrium canescens
Racomitrium heterostichum
Racomitrium sudeticum

5. Confined to the coastal region
Several species are confined to the coastal region and do not extend beyond the mountain ranges along the coastal mainland. Of these, some are confined to the summer-dry climatic region of the southwestern coast.

Brachythecium asperrimum
Claopodium crispifolium
Dendroalsia abietina
Dicranoweisia cirrata
Homalothecium nuttallii
Tortula muralis
Tortula princeps
Ulota obtusiuscula

6. Semi-arid Interior
One moss, *Grimmia pulvinata*, although predominatly coastal, is also found locally in the semi-arid Interior. Another, *Homalothecium aeneum*, is predominantly an Interior species of the southeastern part of the province, where it is common, but extends rarely to dry open sites along the coast.

Two mosses treated in this book are confined to the semi-arid Interior where they are common:

Coscinodon calyptratus
Encalypta rhaptocarpa

7. More northern distribution
Two species treated here represent those of more northern distribution in the province:

Aulacomnium turgidum
Splachnum rubrum

Habitats in British Columbia

Considered on the basis of habitat, the mosses treated in this book may be divided as follows:

1. These mosses grow on rock surfaces that are subject to drying, such as boulders and outcrop knobs:

Abietinella abietina (occasionally)
Andreaea rupestris
Antitrichia curtipendula
Coscinodon calyptratus
Dicranum scoparium
Didymodon insulanus
Drepanocladus uncinatus (occasionally)
Grimmia pulvinata
Hedwigia ciliata
Heterocladium macounii
Homalothecium aeneum
Homalothecium fulgescens
Hylocomium splendens
Hypnum circinale
Hypnum revolutum
Hypnum subimponens
Isothecium stoloniferum
Kindbergia oregana
Kindbergia praelonga
Kiaeria starkei
Pleurozium schreberi
Pogonatum urnigerum
Polytrichum juniperinum
Polytrichum piliferum
Racomitrium canescens
Racomitrium heterostichum
Racomitrium lanuginosum
Rhizomnium glabrescens
Rhytidiadelphus triquetrus
Rhytidium rugosum
Schistidium apocarpum
Tortula muralis
Tortula princeps

2. These mosses grow on rock surfaces that are subject to splashing or seepage:

Aulacomnium palustre
Blindia acuta
Bryum miniatum
Cratoneuron commutatum
Cratoneuron filicinum
Dichodontium pellucidum
Ditrichum flexicaule (occasionally)
Drepanocladus exannulatus
Fissidens adianthoides
Fontinalis antipyretica (submerged)
Hygrohypnum ochraceum
Kindbergia praelonga
Oncophorus wahlenbergii
Philonotis fontana
Racomitrium aciculare
Rhytidiadelphus squarrosus
Scleropodium obtusifolium
Scouleria aquatica

3. The ledges and crevices of cliffs on the shaded faces of water courses have a rich moss flora. Indirect light and relatively high humidity make good habitats for the following species:

Antitrichia curtipendula
Bartramia pomiformis
Claopodium crispifolium
Dicranum fuscescens
Dicranum scoparium
Dicranum tauricum (rarely)
Didymodon insulanus
Ditrichium capillaceum (crevices)
Ditrichum flexicaule
Encalypta ciliata (shelves)
Eurhynchium pulchellum
Fissidens adianthoides

Fissidens limbatus (crevices)
Heterocladium macounii
Homalothecium aeneum
Homalothecium fulgescens
Homalothecium nuttallii
Hylocomium splendens
Hypnum circinale
Hypnum revolutum
Hypnum subimponens
Isothecium stoloniferum
Kindbergia oregana
Kindbergia praelonga
Leucolepis acanthoneuron (shelves)
Metaneckera menziesii
Mnium spinulosum (shelves)
Philonotis fontana
Plagiopus oederi
Plagiothecium denticulatum
Plagiothecium undulatum
Pleurozium schreberi
Pogonatum urnigerum
Pohlia cruda (crevices)
Polytrichum alpinum
Polytrichum juniperinum
Pseudotaxiphyllum elegans
Racomitirum canescens
Racomitrium heterostichum
Rhizomnium glabrescens
Rhytidiadelphus loreus
Rhytidiadelphus triquetrus
Roellia roellii
Schistidium apocarpum
Tortella tortuosa
Tortula mucronifolia (shelves)
Tortula princeps
Tortula ruralis

Among other habitats frequented by these mosses are the boulders and cliff bases banking rapid streams and rivers as well as cliff faces subject to perennial wash by erratic or permanent seepage.

4. These mosses grow in dry, sunny, exposed sites, usually on earth:

Abietinella abietina
Aulacomnium androgynum
Aulacomnium palustre
Aulacomnium turgidium
Bryum argenteum
Bryum capillare
Ceratodon purpureus
Dicranum scoparium
Didymodon insulanus
Distichium capillaceum
Ditrichum heteromallum
Ditrichum flexicaule
Drepanocladus uncinatus
Encalypta rhaptocarpa
Funaria hygrometrica
Hypnum revolutum
Kindbergia oregana
Leptobryum pyriforme
Oligotrichum aligerum
Philonotis fontana
Pleurozium schreberi
Pogonatum urnigerum
Pohlia annotina
Pohlia nutans
Pohlia wahlenbergii
Polytrichum alpinum
Polytrichum commune
Polytrichum juniperinum
Polytrichum piliferum
Racomitrium canescens
Racomitrium lanuginosum
Rhytidiadelphus squarrosus

Rhytidiadelphus triquetrus
Rhytidium rugosum
Tortula princeps
Tortula ruralis

5. Several mosses grow on the shaded forest floor, mostly in rather dark, coniferous forests:

Abietinella abietina
Amblystegium serpens (logs and rocks)
Dicranella heteromalla (trails)
Dicranum fuscescens (stumps and logs)
Dicranum polysetum
Dicranum scoparium
Dicranum tauricum (logs and trees)
Drepanocladus uncinatus
Eurhynchium pulchellum
Homalothecium aeneum
Hookeria lucens (relatively moist areas)
Hylocomium splendens
Isothecium stoloniferum
Leptobryum pyriforme (raw earth)
Leucolepis acanthoneuron
Mnium spinulosum
Oncophorus wahlenbergii (logs)
Plagiomnium insigne
Plagiothecium denticulatum
Plagiothecium undulatum
Pleurozium schreberi
Pogonatum contortum (raw earth)
Pohlia nutans
Polytrichum alpinum
Polytrichum juniperinum
Pseudotaxiphyllum elegans
Ptilium crista-castrensis
Racomitrium canescens
Rhytidiadelphus loreus
Rhytidiadelphus triquetrus
Rhytidiadelphus squarrosus
Rhytidiopsis robusta
Rhytidium rugosum (dry areas)
Thuidium recognitum
Timmia austriaca

6. These mosses grow in wet, swampy or boggy areas:

Amblystegium serpens
Aulacomnium palustre
Calliergon giganteum
Calliergon stramineum
Campylium stellatum
Catoscopium nigritum
Climacium dendroides
Cratoneuron commutatum
Cratoneuron filicinum
Fissidens adianthoides
Fontinalis antipyretica
Hookeria lucens
Kindbergia praelonga
Leucolepis acanthoneuron
Meesia triquetra
Paludella squarrosa
Philonotis fontana
Plagiomnium insigne
Pohlia nutans
Pohlia wahlenbergii
Polytrichum commune
Racomitrium lanuginosum
Rhytidiadelphus squarrosus
Sphagnum capillifolium
Sphagnum palustre
Sphagnum squarrosum

Splachnum rubrum
Tetraplodon mnioides
Tomentypnum nitens

7. These mosses are most common on disturbed earth banks, either along trails or roadsides, or sometimes on sandy or silty cliffs:

Atrichum selwynii
Bryum capillare
Ceratodon purpureus
Dichodontium pellucidum
Dicranella heteromalla
Dicranella rufescens
Didymodon insulanus
Ditrichum heteromallum
Fissidens limbatus
Funaria hygrometrica
Leptobryum pyriforme
Oligotrichum aligerum
Philonotis fontana
Pogonatum contortum
Pogonatum urnigerum
Pohlia annotina
Pohlia nutans
Pohlia wahlenbergii
Polytrichum juniperinum
Racomitrium canescens

8. Some mosses, including many of those mentioned earlier, are common garden weeds, growing on the soil of shaded sites, among the rocks of rock gardens and on soil left free from cultivation for a few months (near the edge of flower beds, for example):

Atrichum selwynii
Bryum argenteum
Bryum capillare
Ceratodon purpureus
Dichodontium pellucidum
Dicranella heteromalla
Dicranella rufescens
Didymodon insulanus
Ditrichum heteromallum
Fissidens limbatus
Funaria hygrometrica
Leptobryum pyriforme
Oligotrichum aligerum
Philonotis fontana
Pogonatum contortum
Pogonatum urnigerum
Pohlia annotina
Pohlia nutans
Pohlia wahlenbergii
Polytrichum juniperinum
Racomitrium canescens
Rhytidiadelphus squarrosus

9. Lawns are often overtaken by mosses, particularly in patches that are shaded and humid in winter. Lawns that have acidic, nutrient-poor soils are also favourable habitats for mosses. An application of ferrous aluminum sulphate (fertilizer grade) at the rate of about ¼ kg per 12 m^2 is usually sufficient to destroy the mosses in the lawn and fertilize the lawn as well. This chemical is relatively inexpensive and no special

"moss killers" are necessary. Probably the most weedy of the mosses are *Rhytidiadelphus squarrosus* and several species of *Brachythecium*. These mosses often take over immense tracts of lawn near houses and in golf courses. Other common lawn mosses are:

Aulacomnium palustre
Ceratodon purpureus
Dicranum scoparium
Kindbergia oregana
Kindbergia praelonga
Plagiothecium undulatum
Plagiomnium insigne
Polytrichum commune
Polytrichum juniperinum
Polytrichum piliferum
Pseudotaxiphyllum elegans
Racomitrium canescens
Rhizommium glabrescens
Rhytidiadelphus loreus
Rhytidiadelphus triquetrus

Another moss, *Pseudoscleropodium purum*, rare in North America and probably inadvertently introduced from Europe, is locally abundant in some lawns in south coastal British Columbia. *P. purum* resembles *Pleurozium schreberi*, discussed in this book.

10. These mosses commonly grow on the mortar between bricks and stones, the stucco on buildings, or on concrete retaining walls and sidewalks:

Aulacomnium androgynum
Bryum argenteum
Bryum capillare
Ceratodon purpureus
Dicranoweisia cirrata
Didymodon insulanus
Funaria hygrometrica
Grimmia pulvinata
Orthotrichum lyellii
Racomitrium canescens
Racomitrium heterostichum
Schistidium apocarpum
Tortula muralis
Tortula princeps
Tortula ruralis

11. Several mosses are common on tree trunks. *Dicranoweisia cirrata*, for example, is a common moss on trees that border streets. It is fertilized by road dust and rain washing down the trunk and often forms rich green turfs on the bark. Other epiphytic mosses are:

Amblystegium serpens(tree bases)
Antitrichia curtipendula
Brachythecium asperrimum
Claopodium crispifolium
Dendroalsia abietina
Dicranella heteromalla (tree bases)
Dicranum fuscescens
Dicranum scoparium

Dicranum tauricum (occasionally)
Didymodon insulanus (tree bases)
Drepanocladus uncinatus (occasionally)
Homalothecium nuttallii
Hylocomium splendens (rarely)
Heterocladium macounii
Homalothecium fulgescens
Hypnum circinale
Hypnum subimponens
Isothecium stoloniferum
Kindbergia oregana
Kindbergia praelonga
Leucolepis acanthoneuron (occasionally)
Metaneckera menziesii
Mnium spinulosum
Racomitrium lanuginosum (rarely)
Eurhynchium pulchellum
Neckera douglasii
Orthotrichum lyellii
Plagiomnium insigne
Plagiothecium denticulatum
Plagiothecium undulatum
Pohlia cruda (tree bases)
Polytrichum alpinum (rarely)
Pseudotaxiphyllum elegans
Rhizomnium glabrescens (occasionally)
Rhytidiadelphus loreus
Rhytidiadelphus triquetrus
Tortella tortuosa (occasionally)
Tortula princeps
Ulota obtusiuscula

Most of these mosses are not restricted to epiphytic habitats.

12. When trees fall, they expose the root system with soil persisting on it and form a somewhat shaded site where the sterile soil can be colonized by these mosses:

Atrichum selwynii
Aulacomnium androgynum
Dicranella heteromalla
Ditrichum heteromallum
Pohlia nutans
Polytrichum alpinum
Polytrichum juniperinum
Pseudotaxiphyllum elegans
Schistostega pennata

13. Rotten stumps and logs on the forest floor often bear luxuriant growths of mosses. Many mosses on tree trunks persist on the trunk when the tree has fallen, but other mosses replace the epiphytes as the wood decomposes. Stumps and logs, where the wood becomes dark brown and remains moist much of the year, harbour the following mosses:

Aulacomnium androgynum
Buxbaumia piperi
Dicranoweisia cirrata
Dicranum fuscescens
Dicranum scoparium
Hypnum circinale
Kindbergia oregana
Kindbergia praelonga
Plagiomnium insigne
Plagiothecium undulatum

Rhizomnium glabrescens
Rhytidiadelphus loreus
Rhytidiadelphus triquetrus
Rhytidiopsis robusta
Tetraphis pellucida

14. Open tundra contains several mosses in abundance. Of those included in this book, the following are characteristic:

Abietinella abietina
Aulacomnium turgidum
Dicranum scoparium
Hypnum revolutum
Philonotis fontana
Pohlia wahlenbergii
Pohlia nutans
Polytrichum alpinum
Polytrichum juniperinum
Polytrichum piliferum
Racomitrium lanuginosum
Rhytidiadelphus squarrosus
Rhytidium rugosum

Critical Determination

A student of mosses determines specimens with the use of a compound microscope for detailed examination and a lower-powered dissecting microscope for both dissection and observation of gross structures. For careful study of mosses there are two other vital tools: a collection of accurately determined, dried specimens (herbarium) and a library.

The easiest way to obtain the herbarium is by collection and determining specimens yourself. Determination involves using keys, descriptions and illustrations in the appropriate literature. To become confident that the specimens are accurately determined, the student compares a specimen with an accurately named specimen in a herbarium, such as that at the University of British Columbia (Vancouver), the Royal British Columbia Museum (Victoria), the University of Alberta (Edmonton), the National Museum of Nature (Ottawa) or the University of Washington (Seattle). Specimens can be sent to professional bryologists for an opinion at any of these institutions. It is wise to write to the expert before sending specimens to ask if she or he can offer an opinion. Most bryologists are willing to examine specimens, but sometimes their schedules are completely occupied with other duties and they must reluctantly refuse.

The acquisition of a library is more complicated and can be rather costly. The most useful book for the area is Elva Lawton's (1971) *Moss Flora of the Pacific Northwest*. It contains keys, descriptions and illustrations of most of the mosses known from southern British Columbia. Other useful books for the province are:

Mosses, Lichens and Ferns of Northwest North America by D.H. Vitt, Janet E. Marsh and Robin B. Bovey (1988)—this attractive and popular guide contains colour photographs of many of the common species as well as useful comments and general keys for identification.

Mosses of Eastern North America (two volumes) by H.A. Crum and L.E. Anderson (1981) also includes many mosses found in British Columbia with detailed descriptions, keys and illustrations.

Moss Flora of the Maritime Provinces by R.R. Ireland (1982) is a useful guide that contains many species known from British Columbia; the illustrations are especially useful.

Mosses: Utah and the West by Seville Flowers (1973) is lavishly illustrated and includes species from British Columbia, especially from the interior and semi-arid parts of the province.

Unfortunately no book treats all mosses known from the province, so a student must accumulate the periodical literature that treats these species or extract the information from published manuals in which the species are treated.

Besides this literature, a student should have access to a full set of the journal of The American Bryological and Lichenological Society, *The Bryologist*. Anyone can join the society by paying an annual membership fee and receive the journal (issued quarterly) simply by writing to the Business Manager, American Bryological and Lichenological Society. The journal discusses and illustrates most newly discovered mosses. *The Bryologist* also includes careful studies and re-evaluations of moss genera and species and many other articles of interest. Articles are directed at professionals but an enthusiastic amateur can soon develop sufficient expertise to gain considerable information from this source.

Names of Mosses

Most mosses lack popular names in English. Consequently, the only names available are the scientific names in Latin or latinized Greek, the classical languages of science.

In this book, only the genus and species names are given. Occasionally, several species are treated for a single genus. Thus, *Polytrichum* is the genus for the common "hair cap" mosses. If several species of a genus are discussed in a single paragraph, in which the name of the genus is clearly understood, the genus is abbreviated by the first letter of its name; for example, four species of *Polytrichum* are treated in this book: *P. alpinum*, *P. commune*, *P. juniperinum* and *P. piliferum*. The abbreviated names of the author who first described the species are also part of the complete scientific name. If an author's name is in parentheses it means that the author treated that species as belonging to another genus (or perhaps as a variation of a species). The author whose name follows the parentheses is the one who transferred the species to the genus that it is currently credited to by the scientific name. An example is *Metaneckera menziesii* (Hook.) Steere. Hooker, in 1828, named this moss *Neckera menziesii* Hook. In 1967, W.C. Steere proposed the name *Metaneckera* for this species, because it differed from *Neckera* in several significant features. The scientific name, therefore, gives the name of the author who originally provided the species name and also the author who transferred it to another genus, thus the full scientific name is *Metaneckera menziesii* (Hook.) Steere.

The common English names that exist for mosses usually refer to one or several genera. Thus the "hair cap" mosses include all species of *Polytrichum* and *Pogonatum*, the "feather" mosses include many genera, among them *Ptilium*, *Hylocomium*, *Pleurozium*. Peat mosses include all species of the genus *Sphagnum*.

Whenever a moss new to science is discovered, the discoverer must describe it carefully in Latin and, preferably, provide an accurate illustration. This description and illustration must then be published in a scientific journal. To gain wide acceptance, the new species should be accompanied by a detailed discussion noting the relationships of the species and including clear reasons why the author considers the species a valid one. Unfortunately, many thousands of new species of mosses have been described inadequately and without illustrations. This means that, when attempting to clear up confusion, it is often difficult to discover exactly what is meant by the author of a new species, which often results in considerable tedious research in an attempt to clarify his or her intent.

Fortunately, the specimen used by the author to describe a new species is usually kept in a herbarium. This specimen—known as the type specimen—is extremely valuable, serving as the only source of clarifying the nature of some described species. If the type specimen is lost, as is too often the case, the names of some described species are treated as synonyms of species already described until the name is clarified.

Scientific names are often descriptive. The name may be constructed to honour a person or to describe the organism. *Distichium capillaceum*, for example, indicates that the leaves are in two ranks and are hair-like; *Atrichum selwynii* notes that the calyptra is hairless and that the name honours a noted Canadian geologist, A.C. Selwyn.

Scientific names have considerable advantages over common names, at least in theory. Only one name is the correct scientific name for a species. This name is internationally recognized and is the same for all parts of the world. Common names are frequently applied to entirely different kinds of plants in a single area. Hair cap mosses, for example, may be any of more than 40 species of mosses. Furthermore, the same common name may apply to one species in one geographic area and to an entirely different species in another area. As in all studies, however, researchers may disagree concerning the placement of a given species in a particular genus. Thus, some people consider *Kindbergia* not to be a genus independent from *Eurhynchium*, and would treat *Kindbergia oregana* under the name *Eurhynchium*. In this case, the name reflects the judgement of the user.

Users of the first edition of this book may be exasperated by the name changes of some of the mosses described in this edition. These changes reflect a refinement of concepts in many cases, resulting from more thorough research. Other name changes correct errors in application of names.

Moss Gardens

The Japanese have constructed beautiful gardens in which mosses are carefully cultured for ornamental purposes. Japanese gardeners endeavour to re-create those aspects of nature that exhilarate and inspire people; each component that gives a natural vegetation restful beauty is placed in the garden so that its beauty is accentuated.

Since many mosses invade gardens in coastal British Columbia, attempts to encourage them to occupy given portions of a garden can add considerably to the beauty of a garden. The gardener can enhance the beauty of shaded humid areas by introducing moss-covered boulders. If these boulders are attractive in shape and the mosses are colourful, they can become objects of great beauty.

Moss carpets must also be tended carefully to remain attractive. A mixture of lawn grass and moss is often considered unsightly. A gardener who uses mosses in the garden is likely to select the most attractive available mosses. These must be chosen as carefully as any other garden plant and their environmental requirements must be satisfied in the garden. The best way to satisfy these requirements is to understand the environment in which the moss grows in nature and to re-create that environment in the garden. The process is likely to be one of trial and error, but the resulting garden can be most rewarding and especially attractive in winter and spring. It should be noted that such gardens are likely to require as much care as a flower garden; the problem of invading seed plants can be particularly hazardous for mosses.

Uses of Mosses

Sphagnum (peat moss) has been used as insulating material both by aboriginal and pioneering people. It traps much air within cells and among the plants and is, consequently, an excellent insulator. Aboriginal people once used dried *Sphagnum* as diaper material. The Inuit used *Sphagnum* mixed with fat as a salve.

The use of *Sphagnum* as surgical dressing, particularly during the First World War, initiated pleas for patriotic citizens to collect and dry the species that were most effective. *Sphagnum* has absorbent qualities superior to cotton (which was difficult to acquire at that time) and also features some antibiotic qualities.

Other mosses have yielded substances of presumed curative properties. Some mosses are sold in Chinese herb shops for their medicinal qualities—the moss is frequently boiled to produce a tea.

Sphagnum peat is the only moss with an important commercial use. The peat, resulting from centuries of accumulated growth of *Sphagnum*, is surface mined and dried, then sold as gardening material. Its water-holding and antiseptic qualities make it admirable as a mixture used in seed beds and gardens. Locally, the feather moss, *Rhytidiadelphus loreus*, is used decoratively in shop windows as attractive material around plants in hanging baskets and in the construction of whimsical sculptures of animals.

Peat, cut into blocks and dried, has long been used as heating fuel, particularly in eastern European countries. It has also been used in some European countries as fuel to generate power. Other uses have been as packing material, mattress stuffing and bedding for animals.

Moss sporangia are eaten by various animals, including birds. Some birds consistently use mosses as nesting material.

No mosses are known to be poisonous, but attempts at eating mosses seem to have been rather limited. The food value of the leafy plants is rather low.

Use of the Hand Lens

A relatively inexpensive hand lens (10X or 15X) can be purchased at most university bookstores.

The hand lens, to be used effectively, should be held relatively close to the eye and the examined object held in the other hand so that light is reflected from the object through the lens to the eye. If the hand lens is held far from the eye, much less detail can be seen.

Format of this Book

This edition is a complete revision of the original Handbook first published in 1968. Only the introductory material remains similar to the first edition, although it is considerably expanded. The key is entirely new, attempting to reduce specialized terminology, and defining terms as they are introduced.

The body of the text is designed to present the illustrations as an accurate representation of each species. The bar scale always denotes the habit sketch. All other structures are drawn to scale large enough to show detail. The scale of these can be determined by simple comparison to their size in the habit sketch. **Name** explains the derivation of the name, which is often a useful way to remember the name and often reflects an interesting aspect of the history or structure of the plant. **Habit** gives the appearance of the plant including colour. **Habitat** denotes the environments where the plant grows in the province. **Reproduction** presents sporophytic features and vegetative reproductive means. **World Distribution** indicates the general pattern and a more explicit outline of its North American distribution. The maps show the known distribution in British Columbia, based mainly on specimens in the herbarium of the University of British Columbia. To accommodate large-format illustrations, the maps are located in a separate section (p. 319). **B.C. Distribution** shows the location of each map. **Distinguishing Features** are presented to note useful features that make the species distinctive while **Similar Species** point out comparisons with other species that might be mistaken for the species under discussion. **Comments** are occasionally given to note particular information about the species that might be of interest.

Some Uncommon Mosses of British Columbia

The three mosses illustrated here are not common, but are so distinctive that they are likely to attract the attention of the novice.

(A) *Schistostega pennata* (or "luminous moss") gives such a startling effect when it glows in caverns and the shaded earth of tree roots where it grows that it will undoubtedly attract the interest of a curious amateur. It is discussed in more detail on page 270.

(B) *Buxbaumia piperi* ("bug moss") has an unusual shape: its lack of a leafy green gametophyte, thus the seeming emergence of the sporophyte directly from the substratum will arouse questions in the person who discovers this moss. It is discussed in detail on page 86.

(C) *Splachnum rubrum* ("umbrella moss") is likely to remind a collector of a mushroom with an attenuated stalk. The brilliant red-purple umbrella will attract the attention of any collector. The leafy gametophyte, however, will immediately inform the collector that the plant is indeed a moss. A detailed discussion is on page 282.

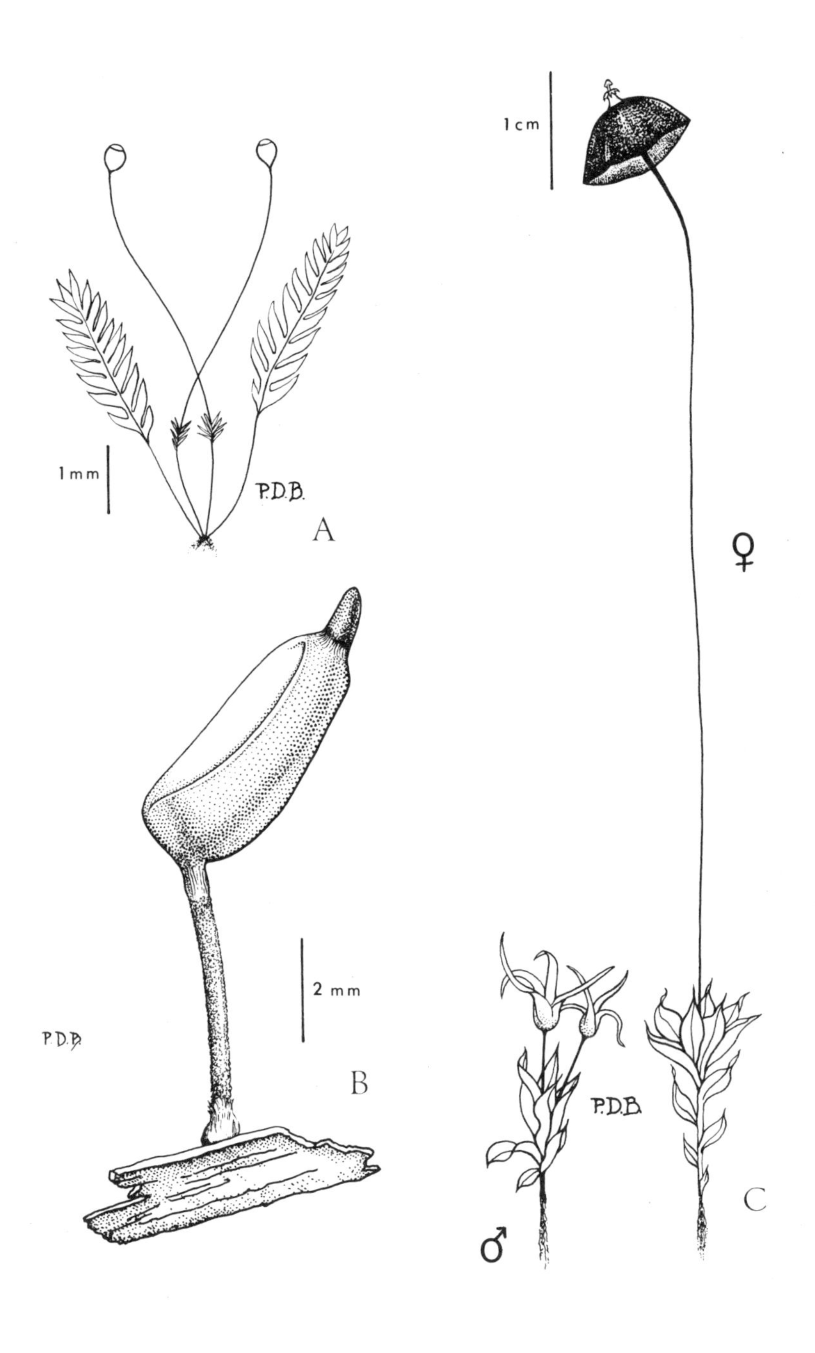

Three mosses that are uncommon in British Columbia.

KEY TO THE MOSSES ILLUSTRATED IN THIS BOOK

The following key has been constructed so that most of the species can be determined using gametophytic features. Gametophytes are always present, while sporophytes are seasonal. In spite of this, however, specimens bearing sporophytes can be determined more reliably than those without. The illustrations show features that should be useful in naming a specimen accurately.

Use the hand lens to discern whether the leaf has a midrib or is pleated, whether its margins are entire or toothed, etc. For more effective interpretation of leaves, proceed as follows:

Strip leaves from the main stem. This is most easily done with forceps, but leaves can be scraped off with a razor blade. Normally, the leaves are stripped or scraped downward. The moss should be moist to remove leaves intact. Then place the leaves in a drop of water on a piece of clear glass. If light is transmitted through the glass (and thus through the leaves), the features of the leaf can be seen more clearly with the aid of a hand lens. Most leaves are transparent when damp. If leaves cannot be removed easily, most details can be observed while the leaf is attached to the shoot, although the shape may not be easily discerned, and some important features, such as an obscure or a double midrib, may not be noticed at all.

For some mosses, this book describes features of changes in the leaf shape from moist to dry conditions. Freshly collected mosses air-dry rather rapidly; the addition of moisture to a dried moss (usually by submerging it in water for a few minutes) will permit the moss to appear fresh again.

The key is constructed using pairs of contrasting features. If the specimen does not fit one of the pair of features, proceed to the number as directed by the second alternative. This process is followed until a name is discovered. Reference to the species that has been keyed out leads to a comparison of the specimen with the illustration and to the discussion. The notes under *Similar Species* in each species treatment may guide the student to the correct species if the specimen matches neither the illustration nor the description. One should be particularly careful in following the correct alternative in each pair of choices; a wrong choice can lead to an entirely unrelated species. To ensure that no mistake has been made, it is often wise to key the specimen out twice to discover whether the same name results each time. If it fits both the description and the illustration, the specimen is determined.

It should be emphasized, however, that less that 20% of the mosses known in the province (there are more than 700 species) are treated in this book. Many species not included in the book strongly resemble those that are. Consequently, it is possible to key out such a moss specimen as one of the species treated here. Some of these similar species (not illustrated) are noted under the species that they resemble.

If a specimen remains puzzling, place it in a carefully labelled packet and send it to the author of this book for identification. Send only dried, labelled packets; moist specimens in plastic bags are subject to decay and are extremely difficult to determine in such a state. Do not send large numbers of specimens for an opinion. If large numbers are sent, they are likely to be neglected because they take a long time for determination. If a few specimens are sent each time, a prompt answer is more likely.

1. Sporangium resembling a miniature open umbrella, red-purple, on a long slender seta ________ *Splachnum rubrum*
1. Sporangium not umbrella-like, or absent ________ 2

2. Sporangium obliquely oriented with a flat diagonal upper surface, leafy plant not visible ________ *Buxbaumia piperi*
2. Sporangium not oblique, leafy plant always apparent, or sporangium absent ________ 3

3. Leaves in two rows, the leafy shoots appearing flattened ________ 4
3. Leaves in more than two rows, the leafy shoots usually rounded, occasionally flattened ________ 7

4. Leaves with bases fused and longitudinally attached on shoots,

making each tiny leafy shoot resemble a pinnately lobed leaf ___ *Schistostega pennata*

4. Leaves with bases sheathing the shoot, not having bases fused ___ 5

5. Leaf bases glossy, the leaves very narrow and with the upper part not flattened, with points diverging outward ___ *Distichium capillaceum*

5. Leaf bases not glossy, the blade obviously very flattened, with the sheathing portion of the upper face of the leaf fused with the midrib ___ 6

6. Leaves 3 mm long or longer, the leafy shoots usually more than 10 mm long ___ *Fissidens adianthoides*

6. Leaves less than 2 mm long, the leafy shoots rarely exceeding 5 mm ___ *Fissidens limbatus*

7. Leafy shoots and branches flattened as though pressed ___ 8

7. Leafy shoots and branches not conspicuously flattened ___ 14

8. Leaves never with undulations on surface ___ 9

8. Leaves with undulations on surface ___ 11

9. Leaf cell network clearly visible at 10X, leaf apex blunt ___ *Hookeria lucens*

9. Leaf cell network not clearly visible at 10X, leaf apex sharply pointed ___ 10

10. Leaves usually more than 2 mm long, abruptly tapering to point, leafy shoots flattened dry or moist ___ *Plagiothecium denticulatum*

10. Leaves usually 1 mm long or less, gradually tapering to point, only leafy shoots with leaves curling downward toward substratum when dry ___ *Pseudotaxiphyllum elegans*

11. Midrib single; plants often with many slender brittle branchlets; sporophytes immersed on underside of leafy shoot ___ *Metaneckera menziesii*

11. Midrib double or obscure; plants lacking slender brittle branchlets; sporophytes on elongate seta ___ 12

12. Plants with regular pinnate branching, usually epiphytic or on cliffs ________ *Neckera douglasii*
12. Plants irregularly branched; usually terrestrial or on cliffs or occasionally epiphytic ________ 13

13. Plants pale yellow-green to whitish green, undulations of leaves very regular, leaves not glossy ________ *Plagiothecium undulatum*
13. Plants dark green, not yellow-green or whitish green, undulations of leaves obscure, leaves glossy ________ *Plagiothecium denticulatum*

14. Branches forming a cluster at the apex; lateral branches on the main stem in tufts of two or three diverging branches and one or two branches that hang downward on the stem ________ 15
14. Branches never in lateral tufts on main stem ________ 17

15. Branch leaves conspicuously squarrose (=with the tips bending out abruptly) ________ *Sphagnum squarrosum*
15. Branch leaves not squarrose ________ 16

16. Branch leaves narrow, with narrowed tips (at 10X) ________ *Sphagnum capillifolium*
16. Branch leaves broadly ovate, with rounded tips ________ *Sphagnum palustre*

17. Leafy plants resembling a tiny tree with trunk and a mass of radiating apical branches ________ 18
17. Leafy plants not resembling a tiny tree with a radiating mass of apical branches ________ 19

18. Main stem leaves white, narrowly triangular and with sharply toothed margins visible at 10X ________ *Leucolepis acanthoneuron*
18. Main stem leaves green, broadly heart-shaped and with margins without obvious teeth at 10X ________ *Climacium dendroides*

19. Leaves in three regular rows, giving the leafy shoot a triangular outline ________ 20
19. Leaves in more than three rows ________ 22

20. Leaves spirally arranged in three rows, sometimes not strongly;

spherical sporangia on elongate setae usually present ________ *Plagiopus oederi*

20. Leaves arranged in vertical rows, usually very conspicuously; sporangia never spherical ________ 21

21. Plants submerged aquatics, reclining, leaves imbricate and keeled, wet or dry ________ *Fontinalis antipyretica*
21. Plants of marshes, erect, leaves strongly divergent when humid, contorted when dry ________ *Meesia triquetra*

22. Plants with gemma producing shoots ________ 23
22. Plants lacking gemma producing shoots ________ 28

23. Gemmae borne in an exposed cluster at the tip of a slender leafless shoot, or on a shoot with reduced leaves ________ 24
23. Gemmae not in an exposed terminal cluster ________ 27

24. Gemmae surrounded by a cup-like whorl of blunt leaves ________ *Tetraphis pellucida*
24. Gemmae lacking surrounding leaves ________ 25

25. Gemmae nearly circular and flattened ________ *Tetraphis pellucida*
25. Gemmae narrowly elliptic and pointed ________ 26

26. Gemmae in obvious spherical clusters; plants usually on wood, especially dryish rotten wood of stumps, or humus ________ *Aulacomnium androgynum*
26. Gemmae in irregular clusters; plants of damp sites ________ *Aulacomnium palustre*

27. Plants somewhat flattened and compressed on substratum; gemmae often forming pale yellow-green masses, especially among apical leaves ________ *Pseudotaxiphyllum elegans*
27. Plants not flattened, erect from substratum; gemmae few at leaf axils ________ *Pohlia annotina*

28. Sporangia with four peristome teeth ________ *Tetraphis pellucida*
28. Sporangia with peristome teeth absent or with more than four peristome teeth, or sporangia absent ________ 29

29. Leaf margin with a border of cells differing from those of the rest of the leaf ______ 30
29. Leaves not bordered by differentiated cells ______ 33

30. Leaves tapering to a sharp apex, always with marginal teeth ______ 31
30. Leaves with a blunt apex, leaf shape elliptic to nearly circular, marginal teeth absent ______ *Rhizomnium glabrescens*

31. Leaves with diagonal rows of spines on back, leaves undulate when humid; ripe sporangia erect with membrane across mouth, attached to the incurved teeth ______ *Atrichum selwynii*
31. Leaves lacking spines on back on leaves; sporangia nodding, lacking membrane on mouth ______ 32

32. Teeth of leaves in pairs; leaves weakly contorted when dry; producing a single sporophyte on each leafy shoot ______ *Mnium spinulosum*
32. Teeth of leaves solitary (not in pairs); leaves strongly contorted when dry; producing two or more sporophytes on most leafy shoots ______ *Plagiomnium insigne*

33. Plants nearly black; sometimes dark red-brown or dark green ______ 34
33. Plants green, yellow-green or pale brownish, never black ______ 43

34. Leaves lacking midrib; plants usually 10–20 mm tall; sporangium opening by four longitudinal lines, without an operculum ______ *Andreaea rupestris*
34. Leaves with midrib; plants usually more than 50 mm tall; sporangium opening by operculum ______ 35

35. Leafy plants usually less than 10 mm tall; leaves with lamella ridges on both upper surface and under surface of leaf ______ *Oligotrichum aligerum*
35. Leafy plants 20 mm or taller; leaves without lamellae, or lamellae confined to upper surface ______ 36

36. Plants wine red; lamellae absent; leaves blunt tipped; sporangia nodding ______ *Bryum miniatum*
36. Plants blackish to dark green, sometimes rusty brown; lamellae present or absent ______ 37

37. Leaves with lamellae; leaf margins with sharp marginal teeth (at 10X) ________ 38
37. Leaves lacking lamellae; leaf margins with blunt teeth, or teeth not clearly visible at 10X ________ 40

38. Leaves strongly contorted and twisted when dry; leaf base not strongly sheathing; marginal teeth with chlorophyll ________ *Pogonatum contortum*
38. Leaves not contorted when dry, or only slightly contorted; leaf base sheathing; marginal teeth lacking chlorophyll ________ 39

39. Lower portion of the stem with tiny brown leaves that lack chlorophyll; sporangia circular in cross-section ________ *Polytrichum alpinum*
39. Lower portion of the stem with normal or broken leaves; sporangia 4-angled in cross-section ________ *Polytrichum commune*

40. Leaves tapering gradually to sharp apex; teeth on leaf margins not apparent at 10X ________ 41
40. Leaves tapering abruptly to bluntish apex; teeth on leaf margins blunt and distant ________ 42

41. Sporangia immersed among perichaetial leaves; red-brown peristome teeth flaring outward when dry ________ *Schistidium apocarpum*
41. Sporangia on elongate setae; dark brown peristome teeth erect when dry ________ *Racomitrium sudeticum*

42. Leaves strongly spreading to squarrose when moist; leaves often with rhizoids on lower midrib; sporangia immersed, compressed-subspherical; operculum persistent after sporangium opens ________ *Scouleria aquatica*
42. Leaves spreading but not squarrose when moist; leaves lacking rhizoids; elongate sporangia on long seta; operculum shed to expose long peristome teeth ________ *Racomitrium aciculare*

43. Leaves with white hair points ________ 44
43. Leaves lacking white hair points ________ 56

44. Leaves lacking midrib; sporangia immersed and without peristome teeth; perichaetial leaves with ciliate marginal teeth ________

_____ *Hedwigia ciliata*

44. Leaves with midrib; sporangia immersed or with conspicuous seta and with peristome teeth; perichaetial leaves lacking cilia _____ 45

45. Leaves rigid, with margins incurved over lamellae; sporangia square in cross-section _____ *Polytrichum piliferum*
45. Leaves soft, margins not incurved, lamellae absent; sporangia circular in cross-section or absent _____ 46

46. Plants bearing many short lateral branches off the main shoot _____ 47
46. Plants unbranched, or branches few and identical in size to the main branch _____ 50

47. Plants suberect, white leaf apices with teeth visible at 10X _ 48
47. Plants creeping, white leaf apices with teeth not visible at 10X _____ 49

48. Leaf apex teeth pointed toward apex or obscure at 10X, plants pale yellow-green when damp _____ *Racomitrium canescens*
48. Leaf apex teeth pointed outward and obvious at 10X, plants greyish to dark green or brown when damp _____ *Racomitrium lanuginosum*

49. Sporangium immersed among perichaetial leaves; peristome teeth red and flaring when dry _____ *Schistidium apocarpum*
49. Sporangium on elongate seta; peristome teeth erect when dry _____ *Racomitrium heterostichum*

50. Sporangium nodding at tip of seta when mature _____ *Bryum capillare*
50. Sporangium erect when mature _____ 51

51. Plants forming greyish tufts when wet or dry _____ 52
51. Plants forming turfs, greyish when dry but red-brown to dark green when wet _____ 53

52. Calyptra sheathing entire sporangium; seta straight before operculum is shed; sporangium not grooved when mature _____ *Coscinodon calyptratus*

52. Calyptra half covering upper ½–⅓ of sporangium; seta curved before operculum is shed; sporangium grooved when mature — *Grimmia pulvinata*

53. Hair points with sharp teeth visible at 10X — 54
53. Hair points with teeth not visible at 10X — 55

54. Leaves strongly squarrose when moist; all leaves of similar size along shoots — *Totula ruralis*
54. Leaves not squarrose when moist; leaves forming whorls of larger leaves above areas of smaller leaves, marking annual growth increments along shoots — *Tortula princeps*

55. Peristome teeth corkscrew twisted around each other; sporangium not longitudinally grooved; calyptra obliquely oriented and sheathing only one side of the sporangium — *Tortula muralis*
55. Peristome teeth straight; sporangium longitudinally grooved; calyptra completely sheathing the sporangium — *Encalypta rhaptocarpa*

56. Leaves with irregular wrinkles on upper surface — 57
56. Leaves lacking wrinkles on upper surface — 58

57. Leaves with single midrib and recurved margins; plants with many short reduced lateral branches — *Rhytidium rugosum*
57. Leaves with two midribs, short lateral branches of plants, if present, of similar form to main shoot — *Rhytidiopsis robusta*

58. Leaves with longitudinal pleats or with lamellae on upper surface — 59
58. Leaves lacking longitudinal pleats and lacking lamellae — 75

59. Leaves with lamellae on upper surface of leaf — 60
59. Leaves lacking lamellae, but with surface pleated — 67

60. Lamellae forming three to five lines on costa surface; leaves with diagonal undulations and teeth on leaf back — *Atrichum selwynii*
60. Lamellae forming more than fifteen lines on upper leaf surface — 61

61. Leaf margins (excluding apex) flat, with teeth apparent at 10X ________________ 63
61. Leaf margins incurved, lacking teeth ________ 62

62. Leaf tip reddish, with teeth; leaf blade colourless and overlapping green lamellae of upper surface; lamellae absent on under surface of leaf ________ *Polytrichum juniperinum*
62. Leaf tip green, lacking teeth; leaf blade green and only partly overlapping green lamellae; lamellae on undersurface as well as upper surface of leaf ________ *Oligotrichum aligerum*

63. Leaves with lamellae on both surfaces ________ *Oligotrichum aligerum*
63. Leaves with lamellae confined to upper surface ________ 64

64. Lower portion of the leafy stem (often ⅓–½ the length of the leafy shoot) with reduced, reddish-purple leaves lacking lamellae, making this part of the shoot appear leafless ________ 65
64. Lower portion of the leafy stem with broken leaves or normal leaves ________ *Polytrichum commune*

65. Leaves, when dry, strongly twisted and contorted ________________ *Pogonatum contortum*
65. Leaves, when dry, not strongly twisted and contorted ________ 66

66. Plants appearing bluish-green when fresh; leaves flat when dry ________ *Pogonatum urnigerum*
66. Plants dark green when fresh, leaves somewhat contracted and incurved when dry ________ *Polytrichum alpinum*

67. Leaves falcate-secund (curved and points facing outward on one side of stem) ________ 68
67. Leaves straight, not falcate-secund ________ 70

68. Falcate-secund leaves confined to main stem or branch tips; pleats confined to leaf base; stems red-brown ________________ *Rhytidiadelphus loreus*
68. Falcate-secund leaves throughout stems and branches ________ 69

69. Leaves of branches falcate-secund with points facing base of main shoot; plants densely pinnate ________ *Ptilium crista-castrensis*

69. Leaves of branches falcate-secund with points facing substratum beneath; plants usually loosely pinnate ________ *Drepanocladus uncinatus*

70. Leaf margins recurved; leaves ovate and with central midrib plus two or more radiating midribs ________ *Antitrichia curtipendula*
70. Leaf margins flat; leaves wide and triangular; midrib double with no single central midrib ________ 71

71. Leaves of stem (and often branch) tip strongly and untidily wide-divergent ________ *Rhytidiadelphus triquetrus*
71. Leaves of stem and branch tips tidily imbricate ________ 72

72. Main shoot clothed in dark rhizoids; plants of wet sites ________ *Tomentypnum nitens*
72. Main shoot lacking abundant rhizoids; plants of well-drained sites ________ 73

73. Stem leaves less than 2 mm long, thus stems and branches very slender; sporangia suberect when mature ________ *Homalothecium nuttallii*
73. Stem leaves 2–4 mm long, thus stems and branches appearing coarse; sporangia erect or strongly inclined when mature ________ 74

74. Sporangia erect to suberect, usually much longer than their width; plants usually irregularly branched ________ *Homalothecium fulgescens*
74. Sporangia inclined, usually less than twice as long as their width; plants usually with many short lateral branches from main shoot ________ *Homalothecium aeneum*

75. Leaves falcate-secund (curved and points facing outward on one side of stem) ________ 76
75. Leaves not falcate-secund ________ 93

76. Leaves usually more than ten times longer than the stem width; plants forming turfs of usually unbranched plants with erect shoots ________ 77
76. Leaves usually less than four times longer than the stem width; plants forming mats of usually reclining or partly reclining shoots ________ 84

77. Leaf base with marginal cluster of coloured cells forming a pronounced brownish to red-brown group ________ 81
77. Leaf base lacking coloured basal marginal cluster of cells ____ 78

78. Leaves strongly to somewhat contorted when dry; leafy plants usually more than 40 mm tall ________ 79
78. Leaves little altered wet or dry; leafy plants usually less than 20 mm tall ________ 80

79. Plants forming glossy turfs; sporophytes absent; confined to calcium-rich cliffs ________ *Ditrichum flexicaule*
79. Plants forming non-glossy turfs; sporophytes present, with long seta and curved sporangium ________ *Kiaeria starkei*

80. Sporangia curved, grooved when dry ____ *Dicranella heteronmalla*
80. Sporangia cylindric, straight and erect, ungrooved when dry ________ *Ditrichum heteromallum*

81. Plants of irrigated rock surfaces; sporophytes with curved seta; sporangium short-cylindric (nearly as broad as long) ________ *Blindia acuta*
81. Plants of rotten logs, humus or cliff shelves, not of irrigated rock; sporophytes with straight seta; sporangium long-cylindric, (three to five times longer than their width (sometimes even longer) ________ 82

82. Sporophytes single from each shoot; stems not heavily clothed with rhizoids ________ 83
82. Sporophytes several from each shoot; stems heavily clothed in pale rhizoids ________ *Dicranum polysetum*

83. Leaves glossy, pale green, not strongly contorted when dry ________ *Dicranum scoparium*
83. Leaves not glossy, dark green, strongly contorted when dry ________ *Dicranum fuscescens*

84. Leaves with single midrib ________ 85
84. Leaves with double or obscure midrib ________ 88

85. Plants of wet or seepy habitat ________ 86
85. Plants of dry or well-drained habitat ____ *Drepanocladus uncinatus*

86. Leaves with longitudinal pleats; stem with short hairy structures (paraphyllia) on stem among leaf bases ______ *Cratoneuron commutatum*

86. Leaves lacking pleats; stem lacking obvious paraphyllia ______ 87

87. Plants relatively stiff and regularly pinnately branched ______ *Cratoneuron filicinum*

87. Plants soft and usually distant and irregularly branched ______ *Drepanocladus exannulatus*

88. Plants of wet or seepy habitats, or aquatic ______ *Hygrohypnum ochraceum*

88. Plants of well-drained habitats, never aquatic or in seepage areas ______ 89

89. Leaves pleated, at least near base ______ 90

89. Leaves not pleated ______ 91

90. Plants very densely and regularly pinnate-branched, leaves falcate-secund throughout; stems pale ______ *Ptilium crista-castrensis*

90. Plants distantly branched; falcate leaves mainly confined to near stem tips; stems red-brown ______ *Rhytidiadelphus loreus*

91. Leaf margins strongly revolute; plants dark green to rusty green ______ *Hypnum revolutum*

91. Leaf margins not revolute; plants pale green to golden green ______ 92

92. Sporophytes long-cylindric, erect at maturity ______ *Hypnum subimponens*

92. Sporophytes very short-cylindric to ovoid, inclined ______ *Hypnum circinale*

93. Stem leaves squarrose ______ 94

93. Stem leaves never squarrose ______ 98

94. Stems unbranched, with leaves in 5 regular rows ______ *Paludella squarrosa*

94. Stems branched, leaves in irregular spirals ______ 95

95. Leaves with double midrib, or midribs obscure ________ 96
95. Leaves with single midrib ________ 97

96. Stems red-brown ________ *Rhytidiadelphus squarrosus*
96. Stems yellowish or green ________ *Campylium stellatum*

97. Plants once-pinnate, with branch stems nearly the same diameter as main stem ________ *Kindbergia oregana*
97. Plants twice-to-thrice-pinnate, with the lateral branches much more slender than the main stem ________ *Kindbergia praelonga*

98. Leaves brittle, often with broken tips ________ *Dicranum tauricum*
98. Leaves not brittle; broken tips rare or absent ________ 99

99. Plants forming rounded tufts on trees; calyptrae with erect bristles ________ *Ulota obtusiuscula*
99. Plants not forming rounded tufts on trees; calyptrae usually smooth ________ 100

100. Leaves with hyaline lower portion abruptly differing from opaque upper portion ________ 101
100. Leaves lacking hyaline lower portion abruptly differing from opaque upper portion ________ 106

101. Hyaline area M-shaped at apex, with the sinus of the M extending obliquely down the midrib ________ *Tortella tortuosa*
101. Hyaline area at apex ± at right angles to the midrib ________ 102

102. Hyaline area sheathing with opaque portion abruptly divergent, often coloured; leaf margin toothed ________ 103
102. Hyaline area not sheathing, usually colourless; leaf margin not toothed ________ 104

103. Leaves curled and contorted when dry; plants pale yellow-green; sporangia spherical on elongate seta, and grooved when ripe ________ *Bartramia pomiformis*
103. Leaves imbricated when dry; plants dark green to greyish green; sporangia elongate on long seta, nodding and smooth when ripe ________ *Timmia austriaca*

104. Calyptra asymmetric, sheathing one side of upper portion of sporangium ______ *Tortula mucronifolia*
104. Calyptra symmetric, sheathing whole sporangium ______ 105

105. Calyptra with radiating frill-like fringe at base; sporangium lacking grooves ______ *Encalypta ciliata*
105. Calyptra ± straight at base; sporangium with longitudinal grooves ______ *Encalypta rhaptocarpa*

106. Plants with main stem from which short, lateral branches emerge; plants creeping ______ 141
106. Plants with main stem unbranched or with branches similar to main shoot ______ 107

107. Plants dark brown, red-brown, to nearly black ______ 108
107. Plants green or yellow-green or wine-red ______ 114

108. Plants forming black tufts; leaves without midrib; sporangium opening by four longitudinal lines; lacking operculum ______ *Andreaea rupestris*
108. Plants forming dark turfs; leaves with midrib; sporangium with peristome teeth and operculum ______ 109

109. Leaf apex rounded; plants of irrigated rock surfaces; sporangia nodding and smooth ______ *Bryum miniatum*
109. Leaf apex pointed; plants of dry rock or soil; sporangia erect or, if inclined, with grooves when ripe ______ 110

110. Leaves with small teeth near apex (visible at 10X); sporophytes grooved, rich red-brown and glossy when mature ______ *Ceratodon purpureus*
110. Leaves lacking teeth near apex; sporophytes not grooved when mature ______ 111

111. Sporophytes immersed; plants of rock surfaces or crevices ______ *Schistidium apocarpum*
111. Sporophytes with long seta; plants on soil ______ 112

112. Leaves, except near stem apex, 1 mm long or less, half the size (or less) of the apical leaves, never contorted when dry; plants of humid sandy banks ______ *Dicranella rufescens*

112. Leaves of similar length throughout, contorted when dry; plants of dry open soil ________ 113

113. Sporangia erect and smooth when ripe; peristome teeth cork-screw twisted around each other ________ *Didymodon insulanus*
113. Sporangia inclined and grooved when ripe; peristome teeth straight ________ *Ceratodon purpureus*

114. Plants wine-red; leaves rounded at apex ________ *Bryum miniatum*
114. Plants green, silvery green or yellow-green ________ 115

115. Sporangia on long seta, nodding ________ 116
115. Sporangia on very short seta (shorter than sporangium), or if on long seta, never nodding, or sporophytes absent ________ 124

116. Sporangia black and shiny when mature, nearly spherical ________ *Catoscopium nigritum*
116. Sporangia never black when mature; elongate when moistened ________ 117

117. Sporangia pear-shaped; leaves with long-attenuate upper part ________ *Leptobryum pyriforme*
117. Sporangia not pear-shaped; leaves lacking attenuate upper part ________ 118

118. Plants silvery whitish green, worm-like in form with imbricated leaves ________ *Bryum argenteum*
118. Plants green or yellow-green, never silvery green, leaves somewhat divergent, thus never worm-like in form ________ 119

119. Sporangia 15 mm or more long, leaves usually 5–10 mm long ________ *Roellia roellii*
119. Sporangia less than 8 mm long; leaves usually less than 5 mm long ________ 120

120. Sporangia shrinks to half its length when dry ________ *Pohlia wahlenbergii*
120. Sporangia little changed in length when dry ________ 121

121. Some leaves with hair points ________ *Bryum capillare*
121. Leaves lacking hair points ________ 122

122. Leaves with opalescent sheen (viewed at 10X) ____ *Pohlia cruda*
122. Leaves lacking opalescent sheen ____ 123

123. Leaves ovate, tapering gradually to apex, usually widely separated on stem ____ *Pohlia annotina*
123. Leaves lanceolate, of similar width to ¾ of length, then tapering to point, usually densely covering stem ____ *Pohlia nutans*

124. Sporophytes with seta shorter than sporangia, thus sporangia immersed among leaves ____ 125
124. Sporophytes with elongate seta, or sporophytes absent ____ 126

125. Leaves around sporangium with ciliate marginal hairs; peristome teeth absent ____ *Hedwigia ciliata*
125. Leaves around sporangia lacking marginal hairs; peristome teeth red-brown ____ *Schistidium apocarpum*

126. Lower portion of sporangium swollen, upper portion cylindric; plants in tufts on animal waste ____ *Tetraplodon mnioides*
126. Sporangium absent or sporangium never with swollen lower portion; plants not on animal waste ____ 127

127. Leaf apex blunt ____ 128
127. Leaf apex sharp-pointed ____ 129

128. Plants glossy, slender; leaves 1–2 mm long; rhizoids few to absent on stem ____ *Calliergon stramineum*
128. Plants not glossy, coarse; leaves 3–5 mm long or longer; red rhizoids abundant on stem ____ *Aulacomnium turgidum*

129. Sporangia grooved when dry and mature or sporangia absent ____ 130
129. Sporangia smooth when mature ____ 137

130. Sporangia on elongate seta, at least five times the length of sporangium ____ 132
130. Sporangium with short seta, less than three times the length of sporangium ____ 131

131. Sporangium flaring outward at mouth when dry and mature; peristome teeth absent ____ *Amphidium lapponicum*

131. Sporangium not flaring outward at mouth when dry and mature; two rows of peristome teeth present ________ *Othotrichum lyellii*

132. Sporangium subspherical when mature ________ 133
132. Sporangium cylindric when mature ________ 135

133. Leaves with attenuate tip and sharp marginal teeth, strongly curled, twisted and divergent when dry ________ 134
133. Leaves ovate, lacking attenuate tip and with small marginal teeth, weakly twisted and imbricate when dry ________ *Philonotis fontana*

134. Leaf base glossy; plants pale green to yellow-green; stem circular in cross-section ________ *Bartramia pomiformis*
134. Leaf base not glossy; plants dark green; stem triangular in cross-section ________ *Plagiopus oederi*

135. Sporangium pale brown when ripe; peristome teeth yellowish to pale brown ________ 136
135. Sporangium shiny dark red brown when ripe; peristome teeth dark red brown ________ *Ceratodon purpureus*

136. Leaves 1–2 mm long; plants of well-drained sites, often on rotten wood ________ *Aulacomnium androgynum*
136. Leaves 3–5 mm long; plants of swampy or damp sites, rarely rotten wood except in swamp sites ________ *Aulacomnium palustre*

137. Sporangium with conspicuous swelling in neck ("Adam's apple") ________ *Oncophorus wahlenbergii*
137. Sporangium tapering gradually to seta ________ 138

138. Plants glossy green when dry, usually less than 15 mm tall (excluding sporophytes); leaves imbricate when dry ________ *Ditrichum flexicaule*
138. Plants not glossy when dry, usually 20 mm tall or more; leaves divergent to contorted when dry ________ 139

139. Leaf margins toothed near bluntish apex; plants of damp sites ________ *Dichodontium pellucidum*
139. Leaf margins not toothed near sharp apex; plants of well-drained to dry sites ________ 140

140. Commonly on living trees and on dry logs of open areas at low elevations, often forming tufts or turfs ______ *Dicranoweisia cirrata*
140. Commonly in rock crevices or on rock at higher subalpine to alpine elevations ______ *Dicranoweisia crispula*

141. Plants twice-to-thrice-pinnate ______ 142
141. Plants once-pinnate or irregularly branched ______ 144

142. Stem leaves with sinuous apex; leaves with two midribs; stems red-brown ______ *Hylocomium splendens*
142. Stem leaves lacking sinuous apex; leaves with single midrib; stem green ______ 143

143. Stems clothed in furry paraphyllia at leaf bases; plants not glossy when dry; bright yellow-green ______ *Thuidium recognitum*
143. Stems lacking paraphyllia; plants glossy when dry; light green, but not yellow-green ______ *Kindbergia praelonga*

144. Plants pinnately branched ______ 145
144. Plants very irregularly branched ______ 152

145. Plants resembling flattened trees with "trunk" and many lateral branches in one plane; sporophytes on underside of main shoot and with very short seta; plants coiling downward toward substratum when dry ______ *Dendroalsia abietina*
145. Plants not resembling flattened trees; sporophytes on upper side of shoot and with long seta; plants coiling upward when dry, or not coiling ______ 146

146. Main stem leaves with attenuate points ______ *Claopodium crispifolium*
146. Main stem leaves lacking attenuate points ______ 147

147. Toothed leaf margins apparent at 10X ______ *Isothecium stoloniferum*
147. Toothed leaf margins not apparent at 10X ______ 148

148. Leaf apex blunt or with short apiculus ______ 149
148. Leaf apex sharply pointed ______ 150

149. Midrib single; stem greenish to brown when dry; leaves with swollen enlarged cells at basal corners (alar cells) visible at

10X ______ *Calliergon giganteum*

149. Midrib double; stem reddish; leaves with contracted area of alar cells ______ *Pleurozium schreberi*

150. Plants glossy when dry ______ *Brachythecium asperrimum*
150. Plants not glossy when dry ______ 151

151. Plants not affixed to substratum; lateral branches extremely regular; more attenuate than main stem, emerging nearly at right angles; terrestrial ______ *Abietinella abietina*
151. Plants firmly affixed to substratum; lateral branches of similar dimensions as main stem, emerging at acute angles; on rock or trees ______ *Heterocladium macounii*

152. Plants glossy when dry ______ 153
152. Plants not glossy when dry ______ 163

153. Leaves arranged on shoot to make the leafy shoots and branches appear flattened ______ 154
153. Leaves arranged on shoot to make the leafy shoots appear rounded in sectional view ______ 155

154. Leaves usually 1 mm long or less, curving toward substratum when dry (never undulate); sporangium nodding on seta when ripe ______ *Pseudotaxiphyllum elegans*
154. Leaves usually 2 mm long or more, not curving toward substratum when dry (often undulate); sporangium suberect and curved when ripe ______ *Plagiothecium denticulatum*

155. Leafy shoots with imbricated leaves and a worm-like appearance when dry ______ 156
155. Leafy shoots with leaves somewhat divergent wet or dry ______ 159

156. Leaves lacking teeth at 10X ______ 157
156. Leaves with marginal teeth visible at 10X ______ 158

157. Leaves narrowly triangular; plants slender with leaves 0.5 mm wide or narrower, somewhat widely spaced in lower part of shoot; plants not affixed to substratum ______ *Calliergon stramineum*
157. Leaves broadly ovate; plants swollen and coarse with leaves 1 mm

wide or wider, closely overlapping throughout stem; plants firmly affixed to substratum ______ *Scleropodium obtusifolium*

158. Leaves with central midrib plus two or more lateral ones radiating from leaf base ______ *Antitrichia curtipendula*
158. Leaves with single midrib ______ *Isothecium stoloniferum*

159. Leafy shoots very slender; leaves less than 0.5 mm long ______ 160
159. Leafy shoots coarse; leaves 1 mm or longer ______ 161

160. Leaf margins with teeth visible at 10X ______ *Isothecium stoloniferum*
160. Leaf margins with teeth not visible at 10X ______ *Amblystegium serpens*

161. Leaves with central midrib plus at least two others radiating from leaf base ______ *Antitrichia curtipendula*
161. Leaves with single midrib ______ 162

162. Seta showing visible roughness at 10X ______ *Brachythecium asperrimum*
162. Seta glossy and smooth ______ *Isothecium stoloniferum*

163. Leaves without midrib; sporangium immersed among ciliate-margined leaves; lacking peristome teeth ______ *Hedwigia ciliata*
163. Leaves with midrib; sporangium immersed or on long seta extending from among leaves lacking ciliate margins; peristome teeth present ______ 164

164. Leaves of main stem mostly with attenuate points, strongly contorted when dry ______ *Claopodium crispifolium*
164. Leaves of main stem lacking attenuate points except uppermost ones, not strongly contorted when dry ______ 165

165. Midrib of leaf apparent, extending to leaf apex; sporangia immersed among surrounding leaves ______ *Schistidium apocarpum*
165. Midrib of leaf obscure, confined to lower portion of leaf; sporangium on elongate seta ______ *Heterocladium macounii*

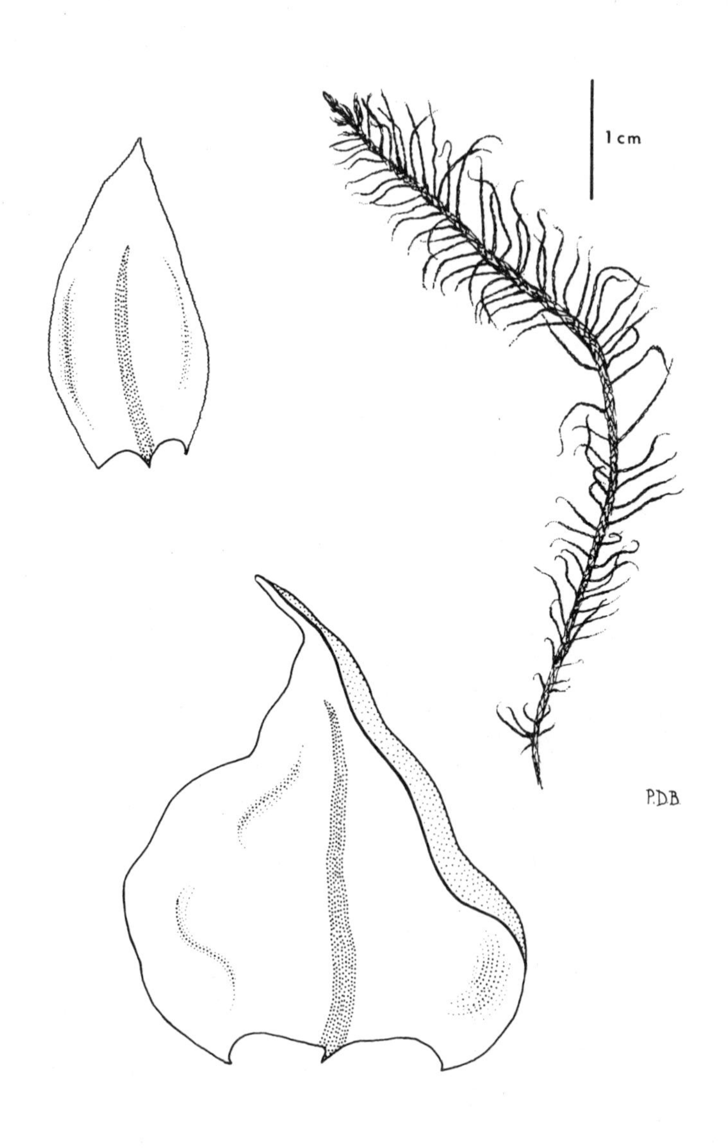

Abietinella abietina

SOME COMMON MOSSES OF BRITISH COLUMBIA

Abietinella abietina (Hedw.) Fleisch.

Name: Genus name derived from *Abies* (fir tree) based on its fancied resemblance to a diminutive fir tree. The species name has the same derivation.
Habit: Reclining or suberect interwoven plants, yellow-green to rusty brown to yellowish; never glossy; when wet the leaves diverge and the plant appears brighter green, upon drying the shoots seem more slender and the leaves sheathe the stem more closely.
Habitat: Dry, usually open, sites in well-drained forest, on cliff shelves and tundra, especially on calcareous substrata; more frequent in interior and northern localities; rare near the coast.
Reproduction: Sporophytes rare, maturing in summer. Sporangium short cylindric and curved, red-brown when ripe; seta smooth, red-brown, elongate. Plants somewhat brittle when dry, probably fragments serve in vegetative reproduction.
World Distribution: Circumboreal in Eurasia and North America; extending, in North America, southward along mountains to Virginia in the east and Arizona in the west.
B.C. Distribution: Map 1, page 319.
Distinguishing Features: The usually dark green to brownish-green, regularly pinnate plants that occur in drier terrestrial sites usually sufficient to distinguish this species.
Similar Species: *Helodium blandowii* is similar in appearance, but this species is in damp sites and the green stems are often covered with paraphyllia readily visible with a hand lens. In *Abietinella*, paraphyllia are small and usually obscured by overlapping leaves, and the stems are

reddish. *Claopodium* may also be confused with *A. abietina* but the two species of similar size (*C. bolanderi* and *C. crispifolium*) have stem leaves with attenuate sharp white points; in *Abietinella* the apices are never attenuate or white. In *Claopodium* the leaves are contorted when dry; in *Abietinella* they are simply imbricate.
Comments: Sometimes named *Thuidium abietinum*, a synonym.

Amblystegium serpens (Hedw.) B.S.G.

Name: Genus name apparently derived from the blunt operculum; the species name referring to the creeping shoots.
Habit: Reclining slender dark green to reddish-brown, thread-like stems forming interwoven mats, somewhat glossy (under lens); not changing markedly from moist to dry conditions, except that the leaves diverge outward when wet and are against the stem when dry.
Habitat: Frequent in swampy areas and on floodplains on soils, logs and tree bases (especially broad-leafed trees), occasionally on cliff shelves, from sea level (upper edge of saltmarsh) to subalpine forest; occasionally, as a weed, on damp, shaded lawns and in greenhouses. Seldom abundant.
Reproduction: Sporophytes frequent and often abundant, maturing in spring to early summer; sporophyte reddish brown or with sporangium green and curved when immature.
World Distribution: Cosmopolitan in both Northern and Southern Hemispheres, especially in temperate to frigid climates. Throughout North America.
B.C. Distribution: Map 2, page 320.
Distinguishing Features: The slender, soft plants with leaves showing an inconspicuous single midrib are usually sufficient to distinguish this species, especially if it occurs on wood.
Similar Species: If on rock or soil, several genera might resemble it. Most require examination of microscopic features for discrimination: *A. compactum* has sharp teeth at the leaf base while *A. serpens* lacks them; *Heterocladium macounii* has papillose leaf cells when viewed under a compound microscope; most *Campylium* species that are the same size as the *Amblystegium* have leaves more spread when wet or dry; *Kindbergia praelonga*, when small, has a strong midrib in the leaves; *Isothecium stoloniferum* has toothed leaves, even in the slender forms of the size of *A. serpens*.

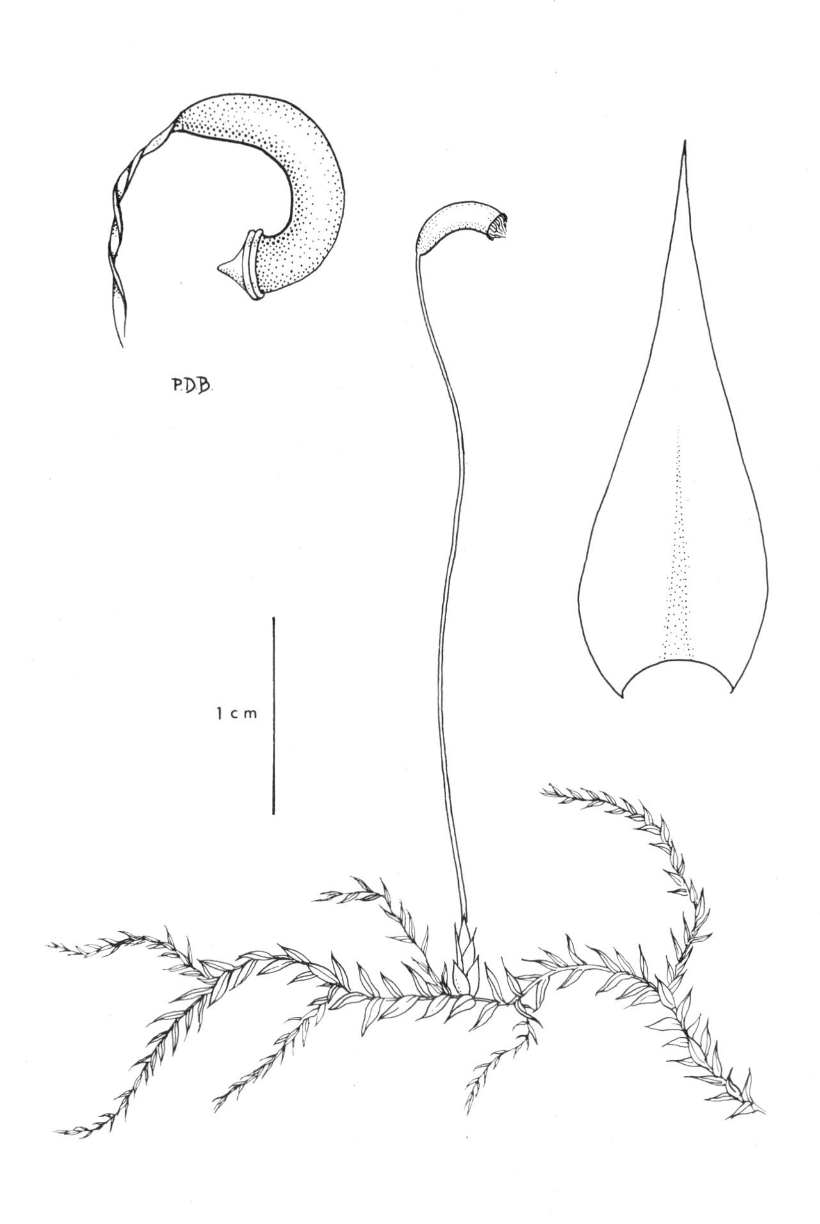

Amblystegium serpens

Amphidium lapponicum (Hedw.) Schimp.

Name: Genus name derived from the sporangium shape (like an urn); the species name, from Lapland, the original place from which the moss was named.
Habit: Forming tight, dark green to light green tufts and cushions; leaves contorted when dry, wide-spreading when moist.
Habitat: Frequent in cliff crevices on siliceous rock, from sea level to alpine sites, usually somewhat shaded.
Reproduction: Sporophytes relatively frequent, maturing in spring to summer; sporangia light brown, grooved when mature and nearly immersed among leaves.
World Distribution: Widespread in temperate and cooler portions of the Northern Hemisphere; in North America extending southward along mountain chains in the east to New Jersey and in the west to Arizona.
B.C. Distribution: Map 3, page 320.
Distinguishing Features: When with sporangia, the grooved urns and lack of peristome teeth distinguish this moss from most others in rock crevices.
Similar Species: *Amphidium mougeotii* cannot be easily distinguished, even on microscopic characters, although the leaves tend to be much narrower and less contorted when dry and blunt marginal teeth are often present (absent in *A. lapponicum*); *Zygodon viridissimus* is somewhat similar and is usually without sporangia (and produces numerous axillary gemmae, absent in *Amphidium*), the leaves are usually sharp toothed near the apex; *Anoectangium aestivum* is superficially similar but the tufts are usually bright yellow-green (dark green in *Amphidium*) and the leaves are not contorted when dry; *Gymnostomum* and *Hymenostylium* are found on calcareous rock and the sporophytes are not grooved and extend on a long seta. *Grimmia torquata* usually forms rounded hard tufts directly on the rock surface and the leaves, when dry, are corkscrew twisted around the stem; in *Amphidium lapponicum* the leaves are contorted, but not spirally twisted around the stem and tufts are in rock crevices. Tiny hyaline tips of the leaves in *Grimmia torquata* and absent in *Amphidium* provide another useful character visible at 10X magnification.

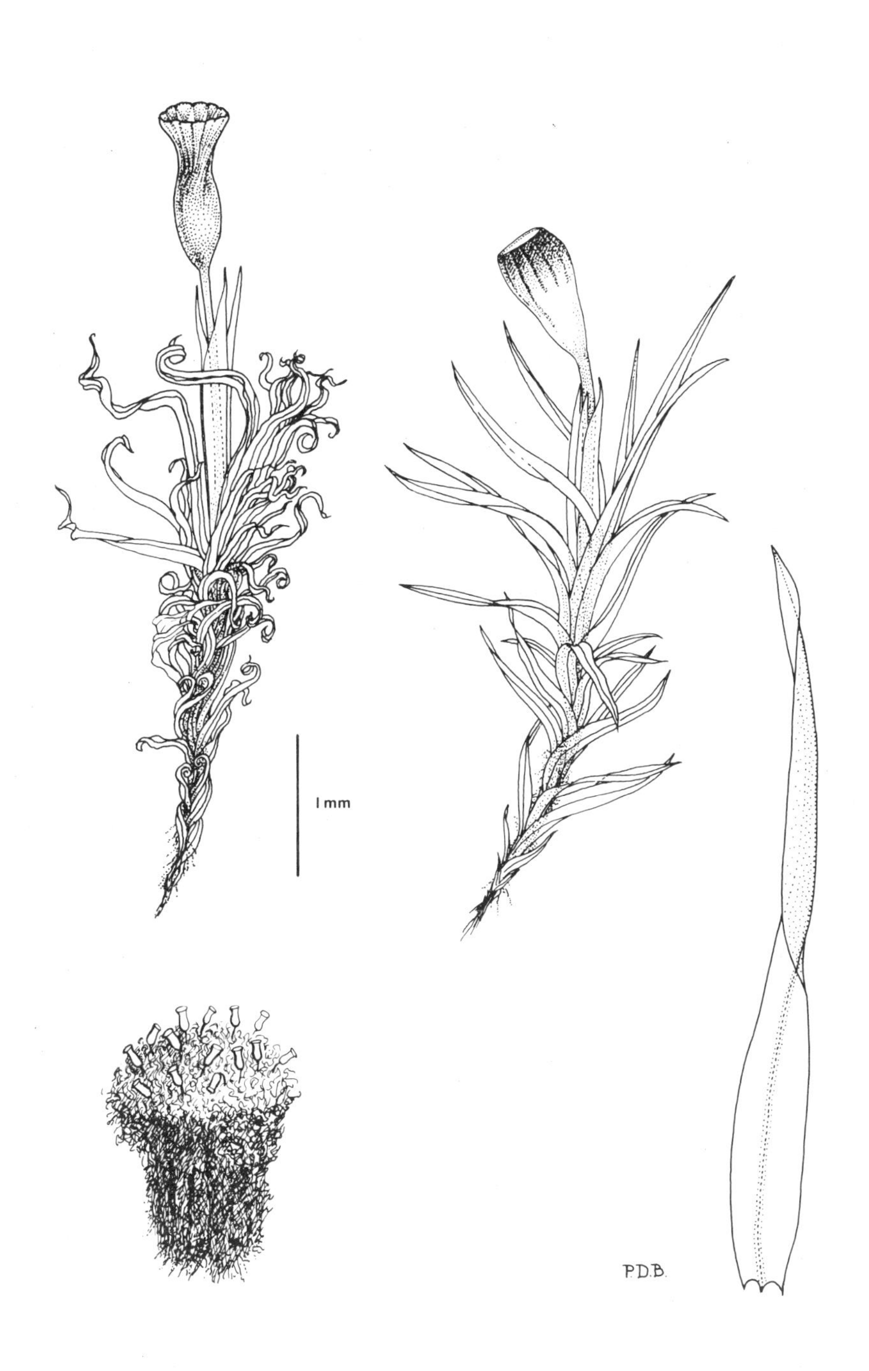

Amphidium lapponicum

Andreaea rupestris Hedw.

Name: Genus named to honour J.G.R. Andreae a German apothecary in the 18th century; the species derived from the habitat: rock.
Habit: Forming black to dark red-brown tufts or cushions.
Habitat: Exposed to somewhat shaded siliceous rock surfaces (cliffs, boulders of talus) from sea level to subalpine and alpine elevations.
Reproduction: Sporophytes black or dark red-brown, relatively frequent, maturing in spring or, at high elevations, after snow melt.
World Distribution: Widespread in cooler climates of the Northern Hemisphere; reported also as scattered in cooler parts of the Southern Hemisphere; in North America, extending across the boreal region and southward along mountain chains of both east and west coasts.
B.C. Distribution: Map 4, page 321.
Distinguishing Features: Any dark brown to black moss that forms tufts of tiny plants on plane rock surfaces (rather than in crevices) is likely to be *Andreaea*. With opened sporangia, when dry, the genus is very characteristic since the four or more longitudinal openings gape to release spores.
Similar Species: There are ten species of *Andreaea* in the province; four resemble *A. rupestris* and are not readily distinguishable without microscopic examination and experience. The other five, *A. blyttii*, *A. nivalis*, *A. megistospora*, *A. rothii*, and *A. schofieldiana* have a distinct midrib. The largest is *A. nivalis* in which the leaves are usually strongly curved and reddish to brown, the plants are usually more than 50 mm long and soft, while *A. rupestris* rarely has stems as long as 10 mm and stems are brittle. *A. nivalis* is usually at higher elevations and most frequent in late snow areas. *A. blyttii* forms short black turfs over rocks in late snow areas.

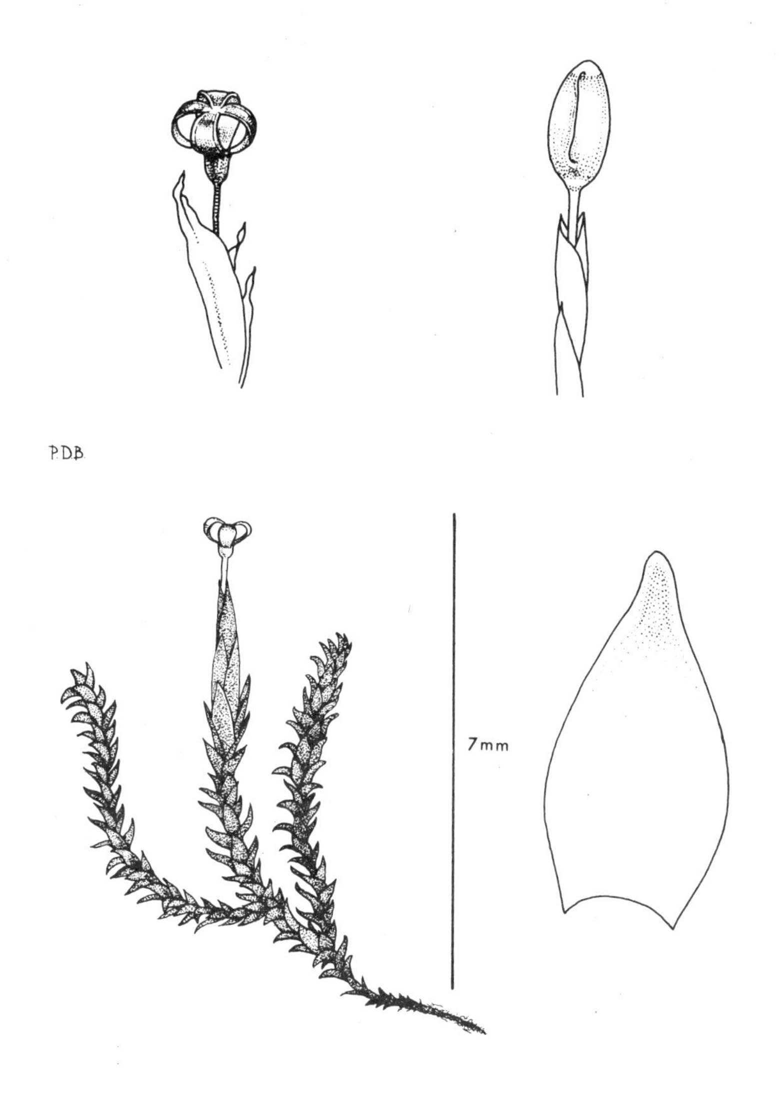

Andreaea rupestris

Antitrichia curtipendula (Hedw.) Brid.

Name: Genus name derived from a reference to the inner peristome segments nearly opposite the outer teeth; the species meaning shortly hanging, in reference to the habit of some plants, but not descriptive of British Columbia specimens that sometimes have very elongate pendent shoots.
Habit: Forming extensive yellowish-green rounded tufts or mats, sometimes with long pendent or hanging shoots.
Habitat: Common on trunks and branches of trees (especially luxuriant on broad-leafed maple), also on cliffs and boulders; occasional on drier hummocks in peatlands. From sea level to subalpine elevations near the coast, where frequent; rare and usually on rock in the interior of the province.
Reproduction: Sporophytes occasional near the coast; sporangia dark brown to red-brown, maturing in spring.
World Distribution: Showing a very interrupted distribution in the Northern Hemisphere: western Europe, western Asia, the mountains of northwestern and northeastern Africa, northeastern North America (Newfoundland) and the west coast of North America south to California.
B.C. Distribution: Map 5, page 321.
Distinguishing Features: The radiating midribs are found in no other local moss.
Similar Species: Sometimes confused with small specimens of *Rhytidiadelphus triquetrus* but the leaves of this species generally form bristly undtidy shoot tips, the leaves are pleated and have two parallel midribs. *A. californica* is usually dark green, the leaves are closely appressed to the stem when dry, making the dry shoots appear very slender, and the midrib is single.
Comments: Sometimes used as packing material and ornamentally in shop windows; has also been used to decorate a ceremonial mask by West Coast First Nations.

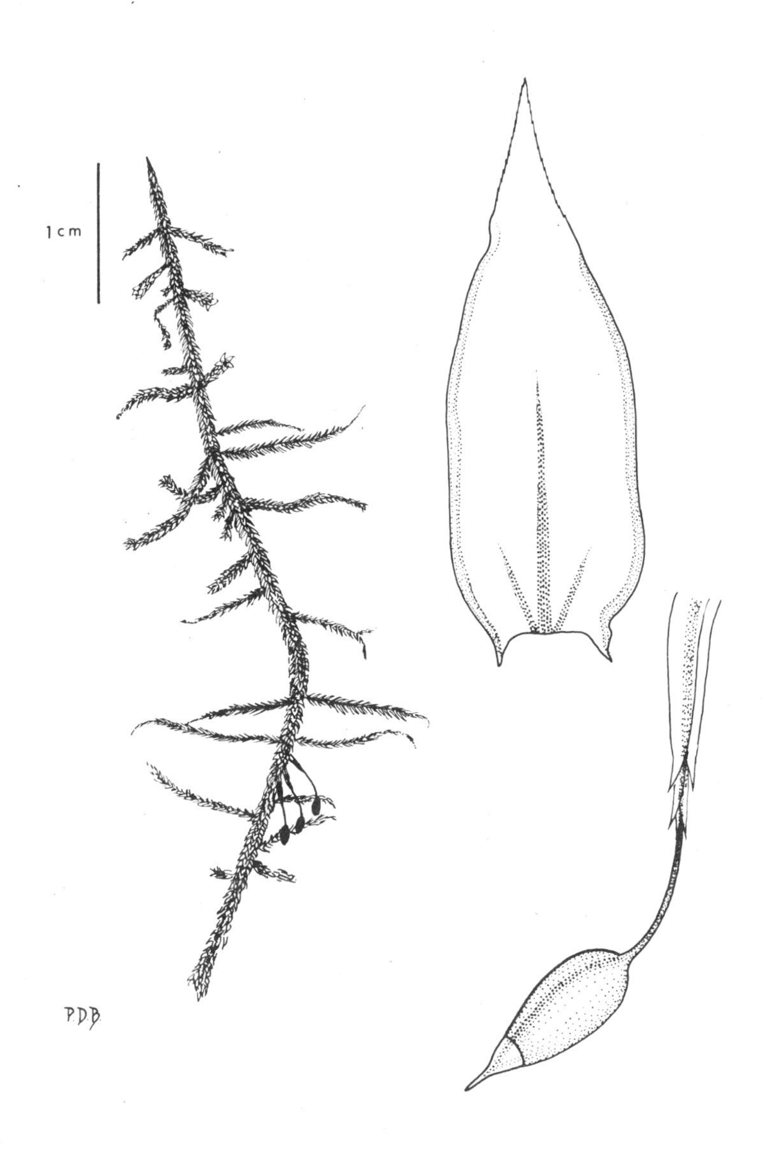

Antitrichia curtipendula

Atrichum selwynii Aust.

Name: Genus name from the hairless calyptra, compared to the hairy calyptra of *Polytrichum*. Species named in honour of A.R.C. Selwyn, director of the Geological Survey of Canada from 1869 to 1895.
Habit: Forming loose turfs of light to dark green plants with sporophyte-bearing plants in colonies usually somewhat separate from male plants.
Habitat: Common on humid mineral soil of disturbed areas, roadsides, banks and near watercourses in open to woodland sites; frequent in flower gardens and on soils disturbed by bulldozing, road cuts, trail margins, and roots of overturned trees. Frequent near the coast; infrequent in the interior. Mainly at lower elevations.
Reproduction: Sporophytes common and abundant; seta and sporangium reddish-brown, maturing in late winter to early spring.
World Distribution: Endemic to western North America from southwestern Alaska southward into California and eastward to Manitoba.
B.C. Distribution: Map 6, page 322.
Distinguishing Features: The extremely low lamellae on the midrib, the teeth in diagonal lines on the lower leaf surface, the markedly contorted leaves when dry, combined with the smooth calyptra distinguish this genus from others related to it.
Similar Species: From *A. undulatum* it is most readily distinguished by the sharper marginal teeth and more sharply acute apices compared to the blunter points of *A. selwynii*. From *A. tenellum* it differs in narrower leaves that are more acute than in *A. selwynii*; the sporangium is usually also shorter in *A. tenellum*. *A. selwynii* is the common species in non-urban sites near the coast while *A. undulatum* is a frequent urban "weed" of gardens and road margins. *Oligotrichum parallelum* of subalpine sites has less acute teeth on the leaf margins and sporophytes are usually light green, rather than dark brown, when mature.
Comments: This genus is sometimes transplanted to Japanese tea gardens as an attractive ground cover.

Atrichum selwynii

Aulacomnium androgynum (Hedw.) Schwaegr.

Name: Genus name derived from the furrowed sporangium; the *mnium* portion of the name being an ancient Greek name for moss. The species name derived from the assumed bisexual condition of the moss, based on a misinterpretation of the gemma-bearing shoot as a male shoot.
Habit: Usually forming bright yellow-green tufts or turfs of erect shoots.
Habitat: Usually on decaying wood of stumps or logs, or peat-like turf, occasionally on rock, tree trunks and disturbed mineral soil. Frequent on fence posts, rail fences and charred stumps. Common near the coast in both woodland and open areas from sea level to subalpine; less common in the interior.
Reproduction: Sporophytes common, maturing in spring, reddish-brown; gemma-bearing shoots common on non-sporophyte-bearing plants, especially conspicuous in winter and late summer. The dusty gemmae fall off readily.
World Distribution: Circumboreal, also in southern South America; in North America across the northern portion of the continent and extending southward to West Virginia in the east and California in the west.
B.C. Distribution: Map 7, page 322.
Distinguishing Features: The spherical masses of gemmae on elongate shoots are unique to this moss in the provincial flora. Any pale yellow-green, tufted moss found on relatively dry wood surfaces is likely to be this species. Sporangia are distinctive with parallel grooves and two rows of peristome teeth.
Similar Species: *Tetraphis pellucida* grows in similar habitats but its peristome has four teeth, the sporangia are not grooved and gemma-bearing shoots are usually terminated by a leaf-fringed cup. *A. palustre* tends to be larger, grows in terrestrial, usually wet, habitats and has gemma-bearing shoots terminated by a small irregular cluster of gemmae. See also notes under *Ceratodon purpureus*.

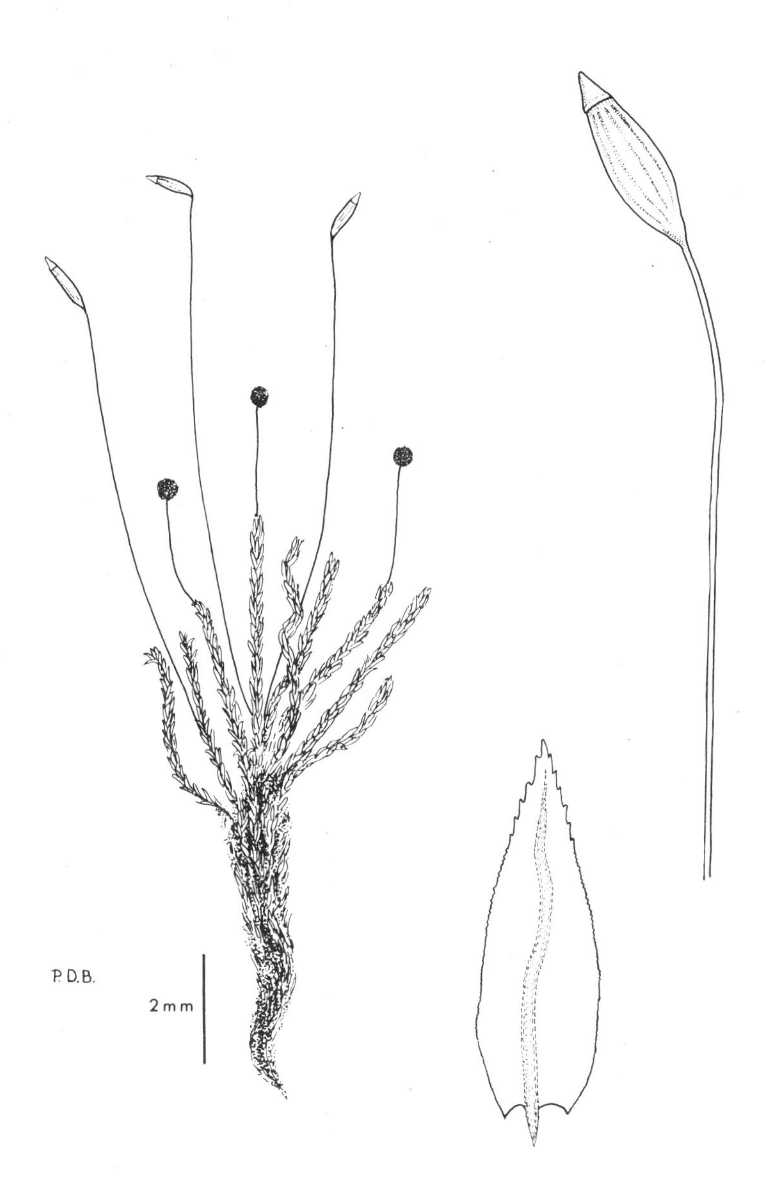

Aulacomnium androgynum

Aulacomnium palustre (Hedw.) Schwaegr

Name: Species name meaning of the marshes, in reference to the usual habitat.
Habit: Forming extensive dense, yellow-green turfs of erect shoots, often interwoven with red rhizoids.
Habitat: Frequent in swampy, boggy, or seepage sites; on lake and pond margins and on cliff shelves; occasionally on rotten logs, from sea level to alpine elevations.
Reproduction: Sporophytes sporadic in occurrence, maturing in spring to summer, reddish-brown and grooved when mature; gemmiferous plants frequent through most of the year.
World Distribution: Cosmopolitan; more frequent in temperate to cool climates.
B.C. Distribution: Map 8, page 323.
Distinguishing Features: The whitish midrib that is shiny on the back of the leaf is distinctive in dried plants; the leaves are somewhat contorted when dry.
Similar Species: Differs from *A. androgynum* in habitat and size. From *A. turgidum*, the contorted leaves and the less turgid shoots are useful distinguishing features; *A. turgidum* also occurs in drier sites. *A. acuminatum* has pointed leaves but the plants are turgid and leaves do not become contorted when dry.

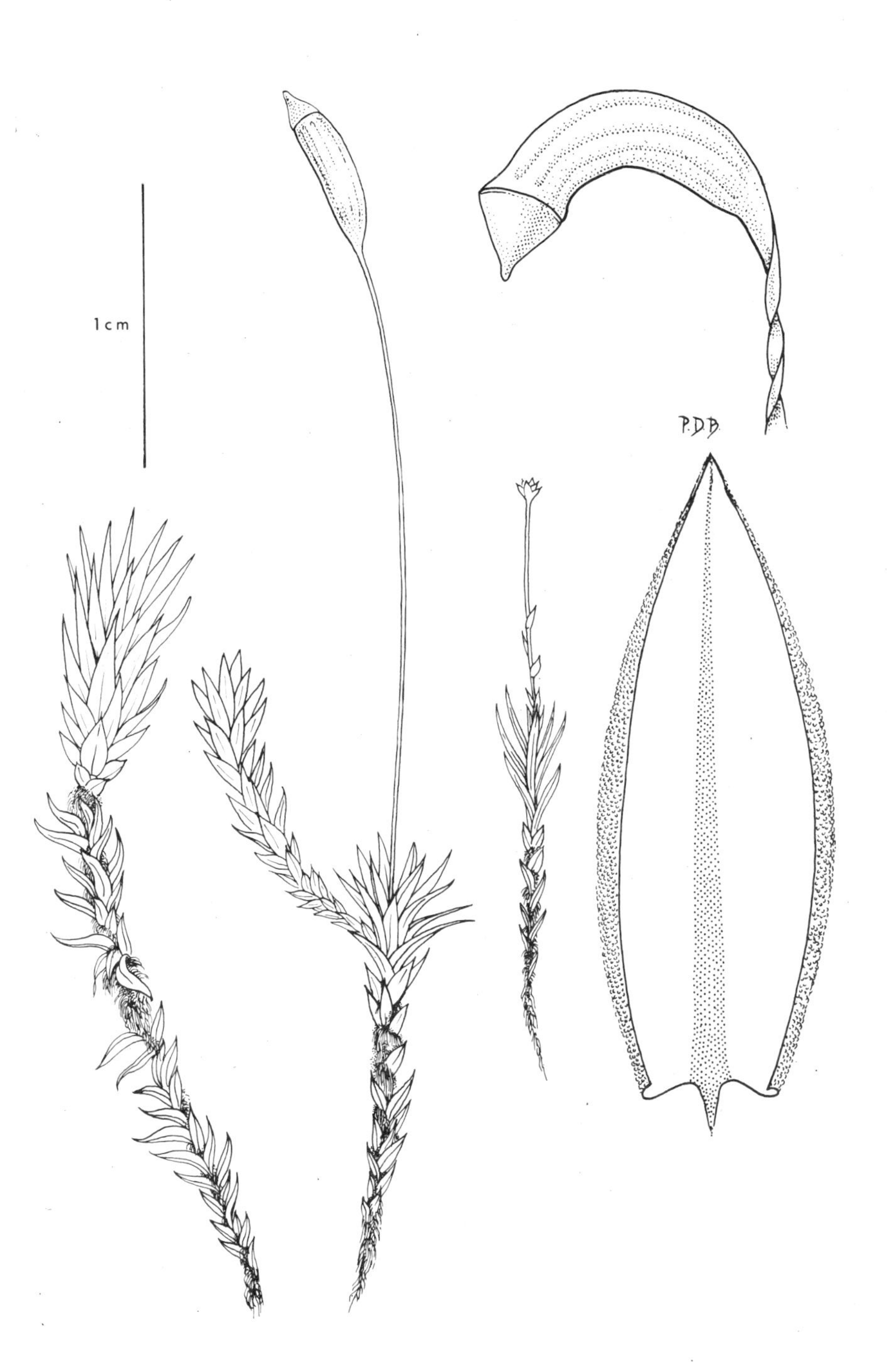

Aulacomnium palustre

Aulacomnium turgidum (Wahlenb.) Schwaegr

Name: Species name meaning turgid, describing the fat leafy shoots.
Habit: Forming yellow-green to golden yellow, tight or loose turfs with abundant red-brown rhizoids among the leaf bases.
Habitat: Usually in tundra in moist to dryish alpine sites, particularly over calcareous to neutral substrata; locally abundant in northern British Columbia.
Reproduction: Sporophytes rare; have never been collected in B.C. The plants fragment readily; presumably this is the main method of propagation.
World Distribution: An arctic, boreal, and alpine species in the Northern Hemisphere, extending as far south as the mountains of East Africa. In North America extending to the Appalachians of Maine and New Hampshire in the east and southward erratically to Mexico in the west.
B.C. Distribution: Map 9, page 323.
Distinguishing Features: The turgid shoots, non-glossy leaves with rounded tips, combined with the yellow-green to golden colour, and open habitat, are distinctive.
Similar Species: From *A. palustre*, which *A. turgidum* sometimes resembles, the absence of gemma-bearing shoots, the leaf shape (especially the rounded rather than pointed tips) and the fact that the leaves do not become twisted when dry are sufficient to separate the two. *A. acuminatum* is similar in form but its leaves are sharply pointed.

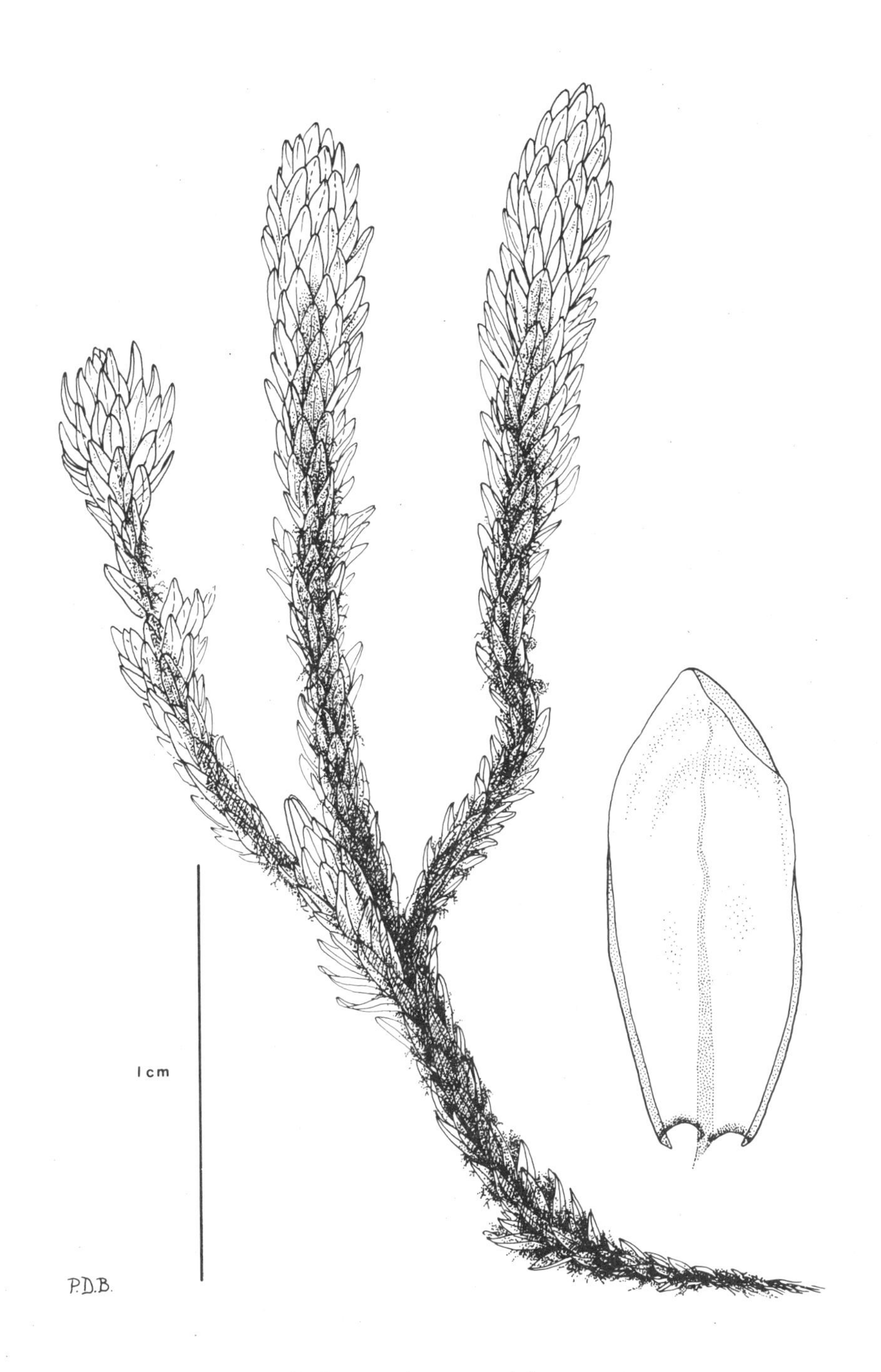

Aulacomnium turgidum

Bartramia pomiformis Hedw.

Name: Genus name in honour of John Bartram, a plant collector and gardener of colonial America. The species name describing the shape of the sporangium: like an apple.
Habit: Forming loose tufts and turfs; leaves much curled and twisted when dry and pale yellowish-green and wide-spreading when moist. Red rhizoids often abundant on the stems.
Habitat: Frequently on cliff shelves, especially in humid sites that are somewhat shaded, rarely epiphytic on lower trunks of trees, from sea level to subalpine elevations. Apparently rare (or absent) in drier climates of the province.
Reproduction: Sporophytes frequent; sporangia pale green and shiny when immature, grooved and brown when ripe.
World Distribution: Circumboreal, extending northward in mountains to arctic latitudes and southward to North Africa. In North America extending southward along the east coast to Georgia and the west coast to California.
B.C. Distribution: Map 10, page 324.
Distinguishing Features: The combination of the spherical sporangia and the long, slender, twisted leaves that diverge strongly when moist, combined with the bluish-green to yellowish-green colour, are usually enough to distinguish the species.
Similar Species: *B. ithyphylla* has leaves with shining clasping leaf bases and the leaves are usually straight, even when dry. *B. halleriana* has the sporangia appearing lateral on relatively short seta, thus nearly buried among the leaves away from the stem apex. Although *Philonotis* has spherical sporangia, the leaves are short and triangular as is the case for *Conostomum tetragonum*. *B. ithyphylla* and *Conostomum* are confined to alpine and subalpine elevations. *Plagiopus oederi* is similar to *Bartramia* but plants tend to be dark green rather than pale green; the leaves appear to be in three irregular rows and lack an expanded base, and the plants grow mainly on calcareous cliffs (*Bartramia* is usually on acidic rock.) *Anacolia menziesii* has sporangia that lack peristome teeth, the sporangium surface is not grooved, and the leaves are not twisted when dry; the plants are usually a rusty, yellowish-green and stems are heavily invested with red-brown rhizoids.
Comments: Commonly called "apple moss", because the unripe sporangium resembles a tiny apple.

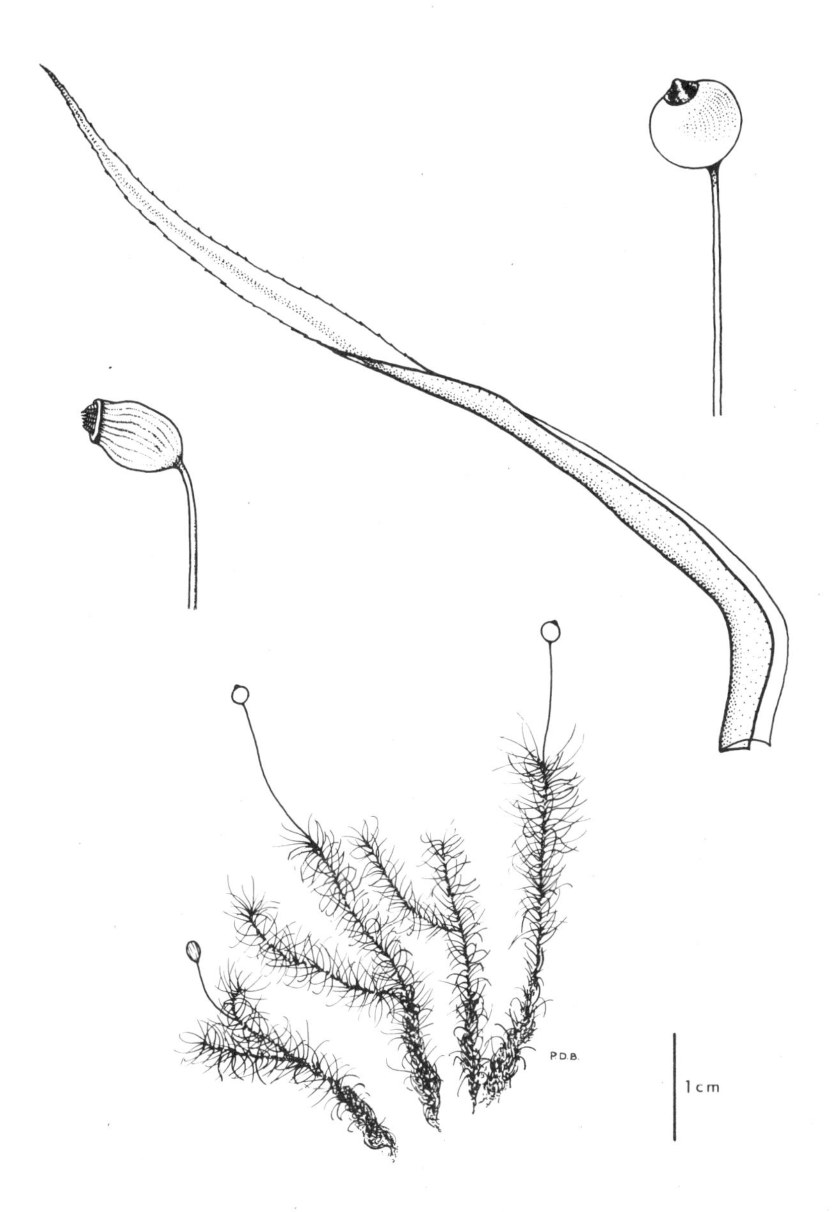

Bartramia pomiformis

Blindia acuta (Hedw.) B.S.G.

Name: Genus name in honour J.J. Blind, of a German pastor of the early 19th century. The species name derived from the acute leaf.
Habit: Forming tight turfs and tufts of dark green, brownish-green to golden-green plants. Leaves not changing in form from wet to dry condition. A clump of cells at the basal margin of leaves (alar cells) usually dark red-brown.
Habitat: Most frequent on damp to misted cliffs and boulder surfaces, especially near streams, lakes or ponds; occasionally on pebbles in seepage sites; on acidic or neutral rocks.
Reproduction: Sporophytes are often abundant in late spring to summer but some populations lack them. The seta is light coloured to dark brown, as is the sporangium. Sometimes the seta is curved but it is often straight, especially when mature.
World Distribution: Widely distributed in the arctic and boreal portions of the Northern Hemisphere.
B.C. Distribution: Map 11, page 324.
Distinguishing Features: The narrow, acute leaves, the short sporangia and, especially, the reddish or dark brown alar cells are useful features to separate this moss.
Similar Species: Although some species of *Campylopus* may resemble *B. acuta*, the midrib in genus *Campulopus* is usually ⅓ of the leaf's width while in *Blindia* the midrib is much narrower. *Campylopus* rarely has sporophytes in British Columbia. *Dicranum tauricum* rarely grows on rock. When it does it may resemble *Blindia* but its rock habitat is always dry and shaded, and its leaf tips break off easily; in *Blindia*, a plant of irregated rock, whole leaves often break off. Species of *Ditrichum* sometimes grow on damp rock but sporophytes are always long-cylindric and have red peristome teeth. Species of *Dicranella*, *Arctoa* and *Kiaeria* also have noticeably red-brown peristome teeth. *Arctoa fulvella* is a cliff-crevice species of subalpine and alpine elevations and the red-brown peristome teeth flare outward conspicuously when dry. *Blindia* has dark brown peristome teeth.

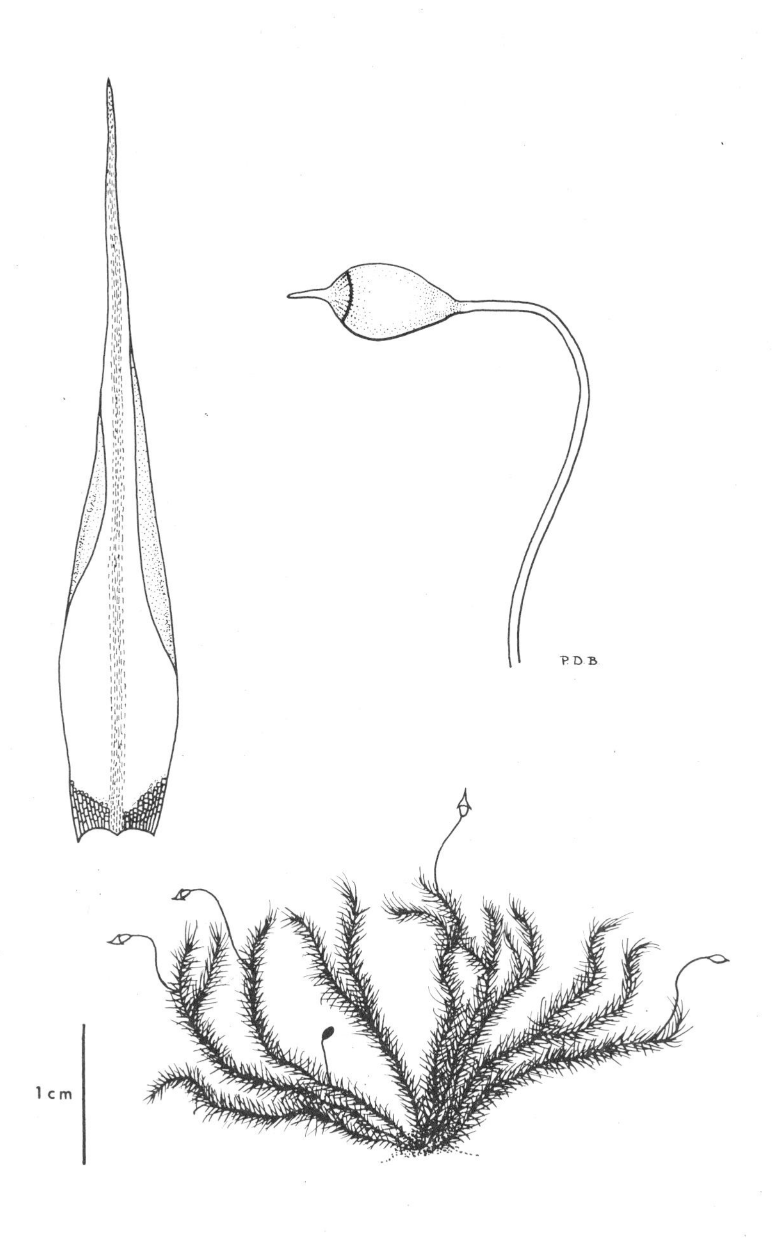

Blindia acuta

Brachythecium asperrimum (Mitt.) Sull.

Name: Genus name meaning short sporangium, a feature of many species. Species name meaning very rough, probably in reference to the rough seta.
Habit: Forming tangled mats of light green, somewhat glossy plants, sometimes with attenuate shoots and branches, especially when growing on branches of trees or shrubs.
Habitat: On branches and trunks of trees and shrubs of somewhat open forest, also on decaying logs and rock (see note on similar species).
Reproduction: Sporophytes common, red-brown, maturing in early spring.
World Distribution: Endemic to western North America from southern Alaska into California, mainly west of the western mountain chains. The closely related *B. frigidum* is more widely distributed in western North America and extends southward to the mountains of Mexico.
B.C. Distribution: Map 12, page 325.
Distinguishing Features: The genus *Brachythecium* is a very troublesome one. The shortened, inclined sporangium associated with a creeping leafy plant bearing glossy leaves with a single midrib will usually indicate this genus, although related genera can pose difficulties. The rough seta is also a useful feature in this species. This, combined with leaves that are not curved and the epiphytic habitat are usually enough to separate this species.
Similar Species: To name this species without sporophytes is hazardous, but sporophytes are frequent in spite of the separation of male and female plants. *B. curtum* usually has a rough seta but the leaves are strongly decurrent (weakly decurrent in *B. asperrimum*). *B. frigidum*, a terrestrial species, is extremely similar but is larger and the leaves are sometimes curved. Some workers consider *B. frigidum* to be identical to *B. asperrimum* but, besides the usual difference in habitat, *B. frigidum* often has a twisted leaf apex (not twisted in *B. asperrimum*) and the alar cells are often enlarged (not enlarged in *B. asperrimum*). The habit of *B. frigidum* shows leaves that are closely or slightly spread from the stem while in *B. asperrimum* leaves are wide-spreading and the shoots sometimes appear somewhat flattened.

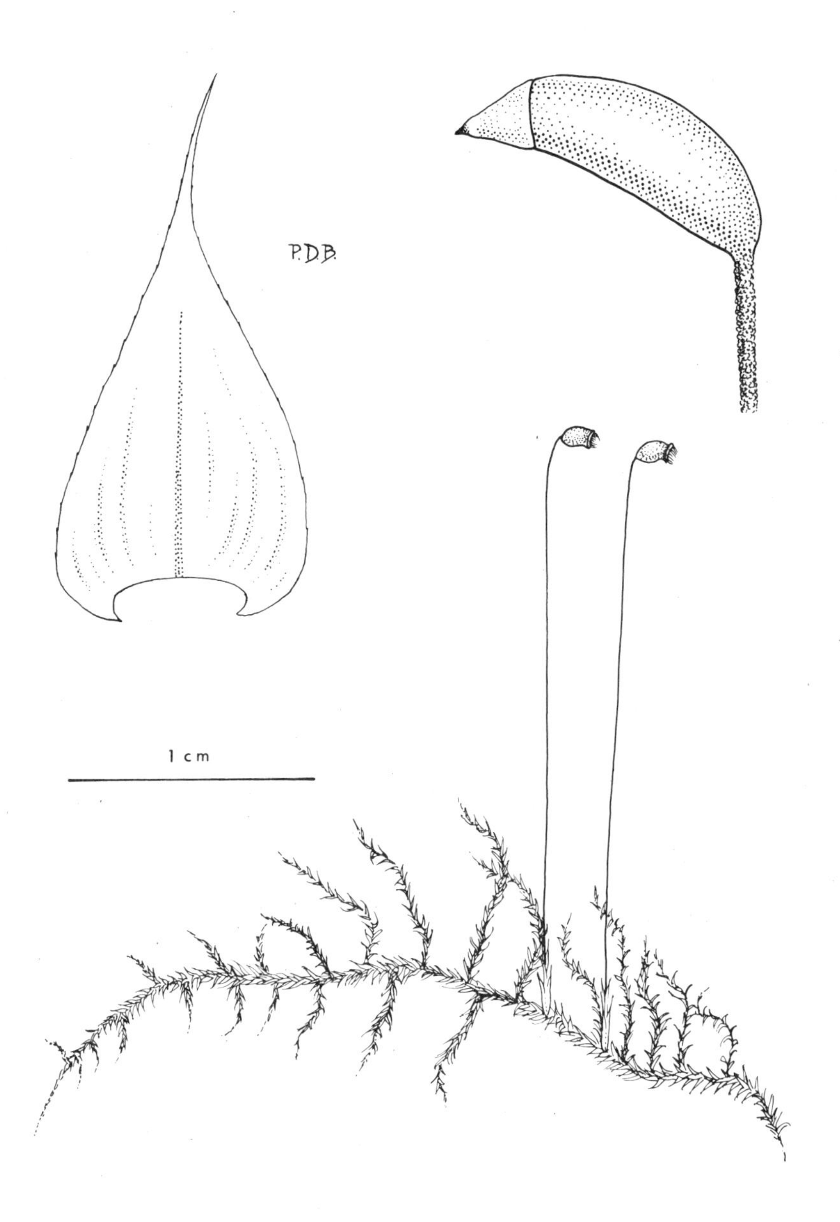

Brachythecium asperrimum

Bryum argenteum Hedw.

Name: Genus name derived from a Greek word meaning moss. Species name from the silvery appearance of the plants. Sometimes called "silver moss".
Habit: Forming silvery, whitish-green turfs, sometimes nearly white when dry; the erect shoots resembling erect worms.
Habitat: On mineral soil, concrete, stones and pavement, from near sea level to subalpine elevations; most frequent in urban areas on man-created substrata, but also in natural habitats; generally in sunny sites.
Reproduction: Sporophytes occasionally abundant in spring and brilliant scarlet when ripe: very striking against the silvery white leafy shoots. When without sporophytes plants sometimes produce an abundance of bud-like gemmae in leaf axils at the apex of the shoot.
World Distribution: Cosmopolitan from the Arctic to Antarctic, frequent in temperate and cold climates, less common in tropical climates. Throughout North America.
B.C. Distribution: Map 13, page 325.
Distinguishing Features: The silvery whitish-green, worm-like erect plants with nodding red sporangia are highly characteristic. Even without these sporangia, the glossy, worm-like, erect shoots are distinctive.
Similar Species: *Plagiobryum zierii*, an infrequent moss on calcareous cliffs and in tundra, resembles a very large *B. argenteum* but the silvery plants are often tinged with pink and the sporophytes have a very long tapered neck and are light brown when mature. *B. calobryoides* of subalpine elevations also resembles *B. argenteum* but is larger and the leaves tend to be less tapered and more widely spaced than in *B. argenteum*. *B. bicolor* is similar in size to *B. argenteum* and has similar sporangia but the plants are obviously yellow or dark green, with no hint of whitish-silver. *Myurella julacea*, which also forms dense tufts of silvery green, worm-like shoots, is usually confined to somewhat shaded, calcium-rich sites and the leafy shoots are about ½ the diameter of those of *B. argenteum*. In *Myurella* the shoots are generally branched and the leaves show no midrib.
Comments: Sometimes forms extensive colonies on the gravel of the flat roofs of high-rise buildings, attesting to the efficiency of its dispersibility as well as its tolerance of a seemingly inhospitable environment.

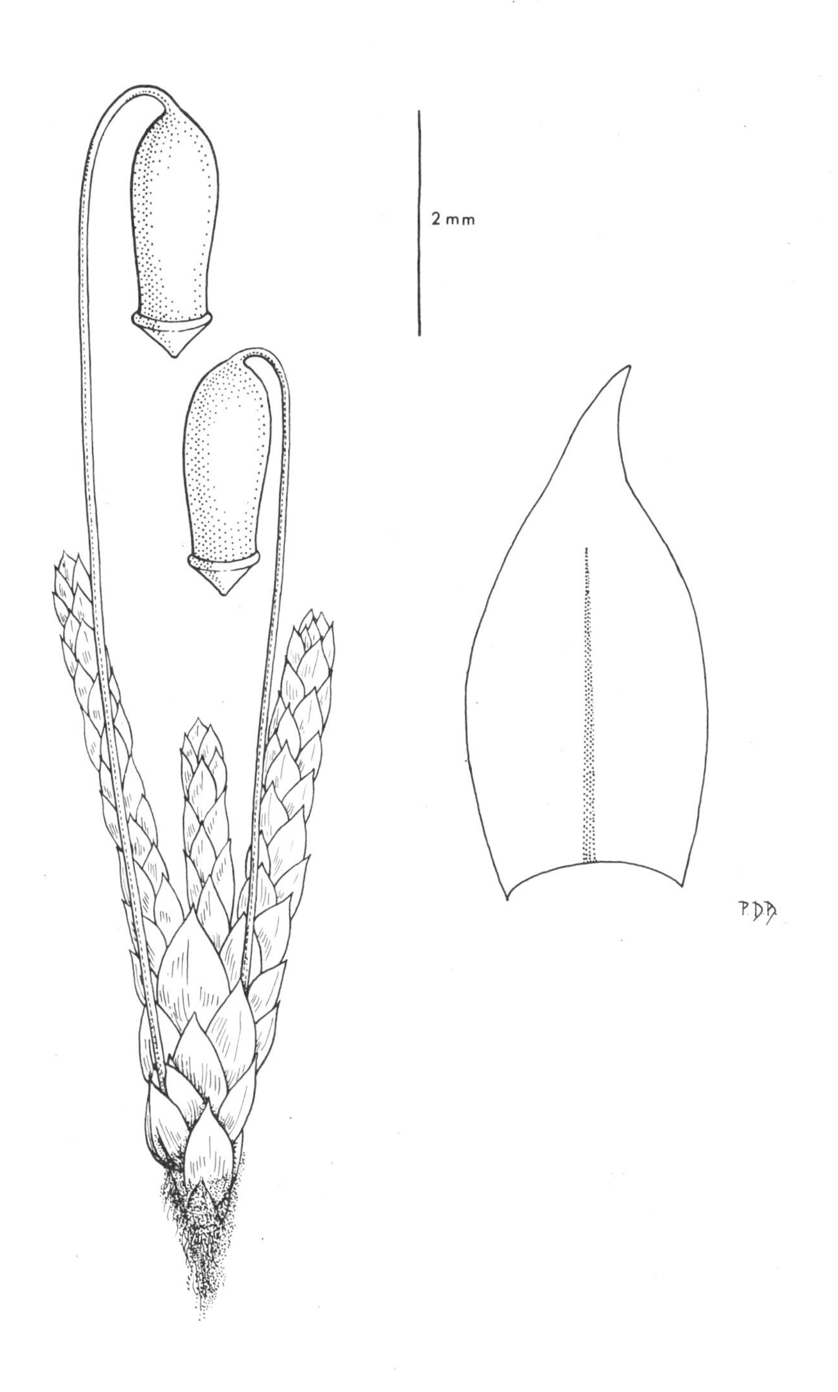

Bryum argenteum

Bryum capillare Hedw.

Name: The species name from the hair-like point of the upper leaves.
Habit: Forming loose to tight short turfs, usually medium to dark green to sometimes reddish-brown with shoots bearing red rhizoids on the lower part of the stem. Midrib often reddish, as is the differentiated leaf margin. Leaves often somewhat corkscrew twisted around the stem when dry.
Habitat: Open sites, usually on mineral soil or cliff crevices, also on concrete or stone walls and in gardens of urban areas; predominantly of lower elevations. Frequent near the coast; rare in the interior.
Reproduction: Sporophytes common, reddish-brown, maturing in spring. Reddish-brown, spherical rhizoidal gemmae are often present but usually are not obvious under a hand lens.
World Distribution: Cosmopolitan, but most frequent in temperate climates. Probably scattered throughout North America.
B.C. Distribution: Map 14, page 326.
Distinguishing Features: The most useful features are the hair-tipped leaves that are spirally twisted around the stem when dry; the reddish tint of midrib and differentiated leaf margin are also useful. Sporophytes are necessary in determining species of this genus. The flower-like rosette of upper leaves that flare outward when moist also serve as a useful feature for this species.
Similar Species: Of the many species of *Bryum* in the province, *B. pseudotriquetrum* somewhat resembles *B. capillare* but lacks hair tips and the leaves are not spirally twisted around the stem when dry. *Pohlia nutans* is similar but lacks differentiated leaf margins, has small marginal teeth, lacks hair points and the leaves are not spirally twisted around the stem when dry.

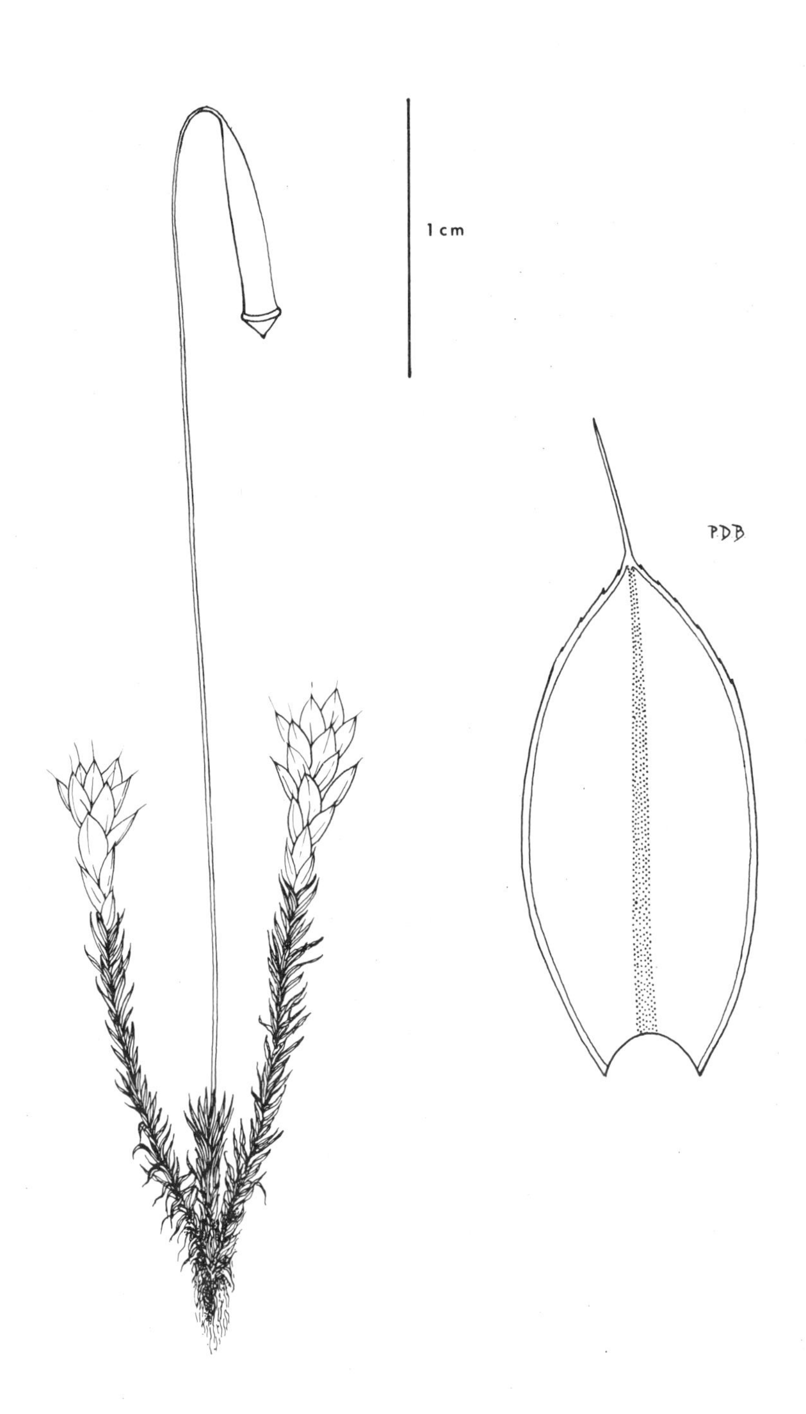

Bryum capillare

Bryum miniatum Lesq.

Name: Species name referring to the red colour of the leafy shoots.
Habit: Forming wine-red to dark green glossy turfs with the leaves closely overlapping when wet or dry.
Habitat: Frequent at low elevations near the coast on exposed rock surfaces that are periodically wet; extending less frequently to subalpine sites and into the interior.
Reproduction: Sporophytes common in spring, are very striking light green when immature in contrast to the wine-red leafy plants. When ripe the sporophytes are also bright red and, as they dry out, become red-brown.
World Distribution: Widely distributed in western North America from British Columbia southward into California and eastward to Montana and Nevada, reappearing infrequently in the Great Lakes, Ontario, Newfoundland, Greenland and the Faeroe Islands.
B.C. Distribution: Map 15, page 326.
Distinguishing Features: The glossy, wine-red plants, the blunt-tipped leaves, associated with the periodically wet exposed rock surface habitat are usually enough to identify this *Bryum*, even when sporophytes are absent. There are frequently swollen alar cells, not noted in other species of *Bryum*.
Similar Species: *Calliergon sarmentosum* is a similar colour to *B. miniatum* and the plants lack sporophytes, but the leaves are usually somewhat broader at the base than the leaves of the *Bryum*. The midrib in the *Bryum* is also more conspicuous and forms a ridge on the back of the leaf. The ridge is visible with a hand lens. In *Calliergen sarmentosum*, the midrib does not form a conspicuous ridge and the shoots usually have short lateral branches that are absent in the *Bryum*. *B. muehlenbeckii* occupies a similar habitat at alpine and subalpine elevations and is a similar colour but the leaf margins tend to be recurved up to half the length of the leaf, and swollen alar cells are lacking. In *B. miniatum*, the leaf margins are recurved only near the base and often not at all. Leaves of *B. muehlenbeckii* are sometimes pointed.

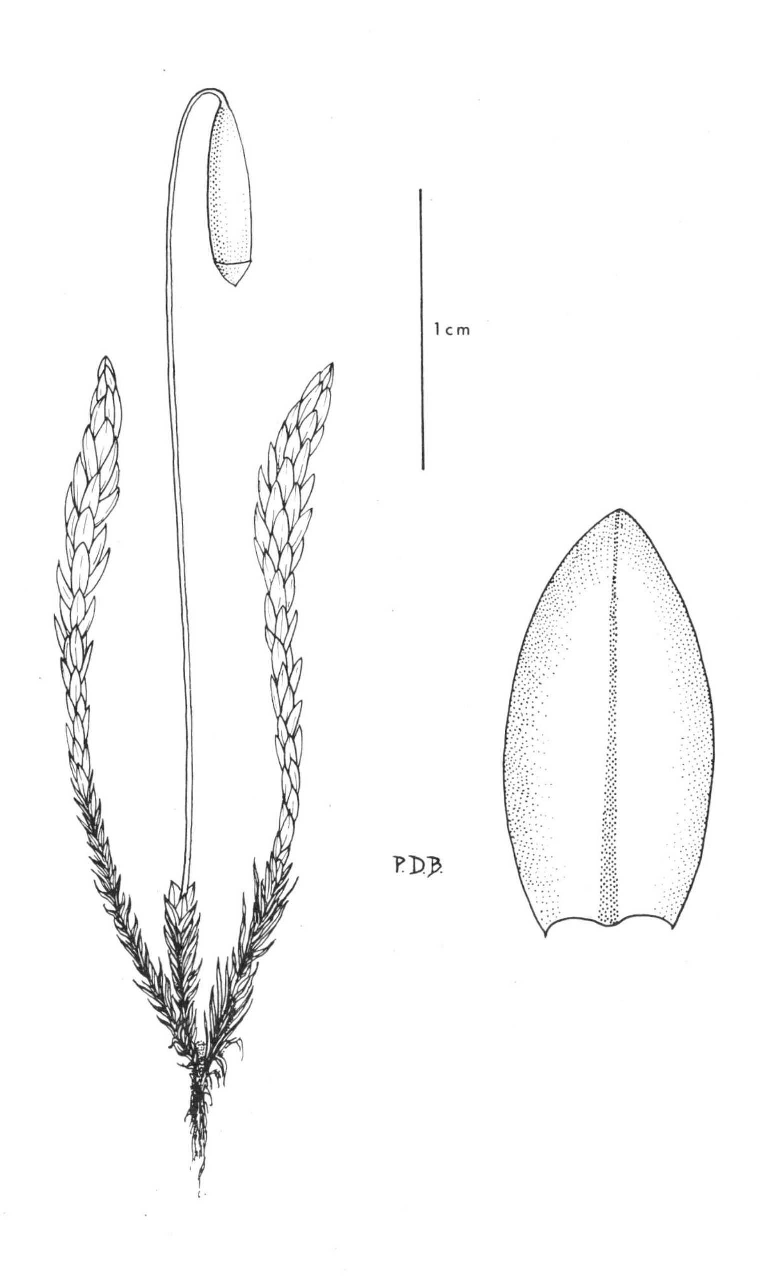

Bryum miniatum

Buxbaumia piperi Best

Name: Genus named in honour of the discoverer of the genus, J.C. Buxbaum, an 18th-century, German botanist. Species named in honour of C.V. Piper, an American botanist, who made major contributions to an understanding of the flora of Washington State.
Habit: Solitary, seemingly leafless plants noted only when sporophytes are present; usually widely scattered on substratum.
Habitat: Infrequent in coniferous forest, most often at subalpine altitudes, but occasionally near sea level; on stablized banks with some organic material or on rotten logs of forest floor, rarely on rock.
Reproduction: Sporophytes are light brownish-green when young, pale brown and not glossy when mature.
World Distribution: Endemic to western North America from southeastern Alaska and adjacent Yukon southward to northern California and eastward to Colorado in the Rocky Mountains.
B.C. Distribution: Map 16, page 327.
Distinguishing Features: The obliquely oriented, pale brown sporangium with a smooth curved surface on the oblique face and the very rough seta are strongly characteristic.
Similar Species: Two other species of *Buxbaumia* are found in British Columbia: *B. aphylla* and *B. piperi*. *B. aphylla* has glossy, chestnut-brown sporangia with the oblique face flattened, especially when dry; the outline of the sporangium is broadly ovate while in *B. piperi* it is very narrowly ovate, pale brown and not glossy when mature. *B. viridis* is similar in size and colour to *B. piperi* but the oblique face has the surface layer split and curling back, rather than smooth, as in *B. piperi*.

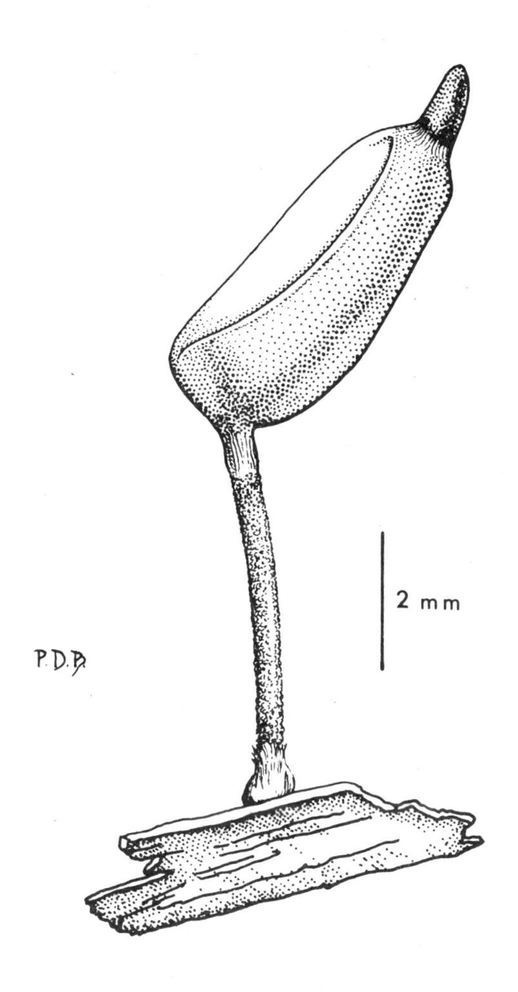

Buxbaumia piperi

Calliergon giganteum (Schimp.) Kindb.

Name: Genus name meaning "beautiful form" is reference to the frequently tidy branching and attractive appearance of the leafy plants. The species name denoting the very large plants that, in floating mats, can reach lengths of 350 mm or more.

Habit: Forming dense or loose mats of suberect to erect interwoven shoots that vary from light to dark green to often tinged with yellow or brown; plants often stiff and accumulating white coatings of calcium carbonate.

Habitat: Sometimes locally abundant, especially in northern and interior localities in wet depressions, lakeshores, pond margins and wetland in calcareous terrain, sometimes forming floating mats; rare near sea level, extending to subalpine elevations, usually in non-forested sites but sometimes margined by forest.

Reproduction: Sporophytes infrequent; seta red, smooth, twisted when dry; sporangium reddish-brown when mature, smooth but not glossy; peristome teeth yellow-brown.

World Distribution: Circumboreal in the Northern Hemisphere, extending northward in arctic latitudes; in North America extending southward, in the east to Pennsylvania and in the west to Colorado.

B.C. Distribution: Map 17, page 327.

Distinguishing Features: The regularly pinnate plants with blunt-tipped leaves bearing a midrib, associated with the wet habitat in calcareous substrata, are usually reliable features. Swollen alar cells are usually visible with a hand lens.

Similar Species: *C. giganteum* can be distinguished from *Calliergonella cuspidata* by its obscure midrib; the midrib in *Calliergon* is distinct. *C. cordifolium* is similar but usually little-branched and the alar cells are not obvious or swollen as is evident in *C. giganteum*. *C. sarmentosum* is usually wine-red, irregularly branched and mainly alpine or subalpine. *Pleurozium schreberi* has two midribs and is generally a terrestrial species of dry habitats; swollen alar cells are absent and the stems are red-brown (in *C. giganteum* they are green to brown).

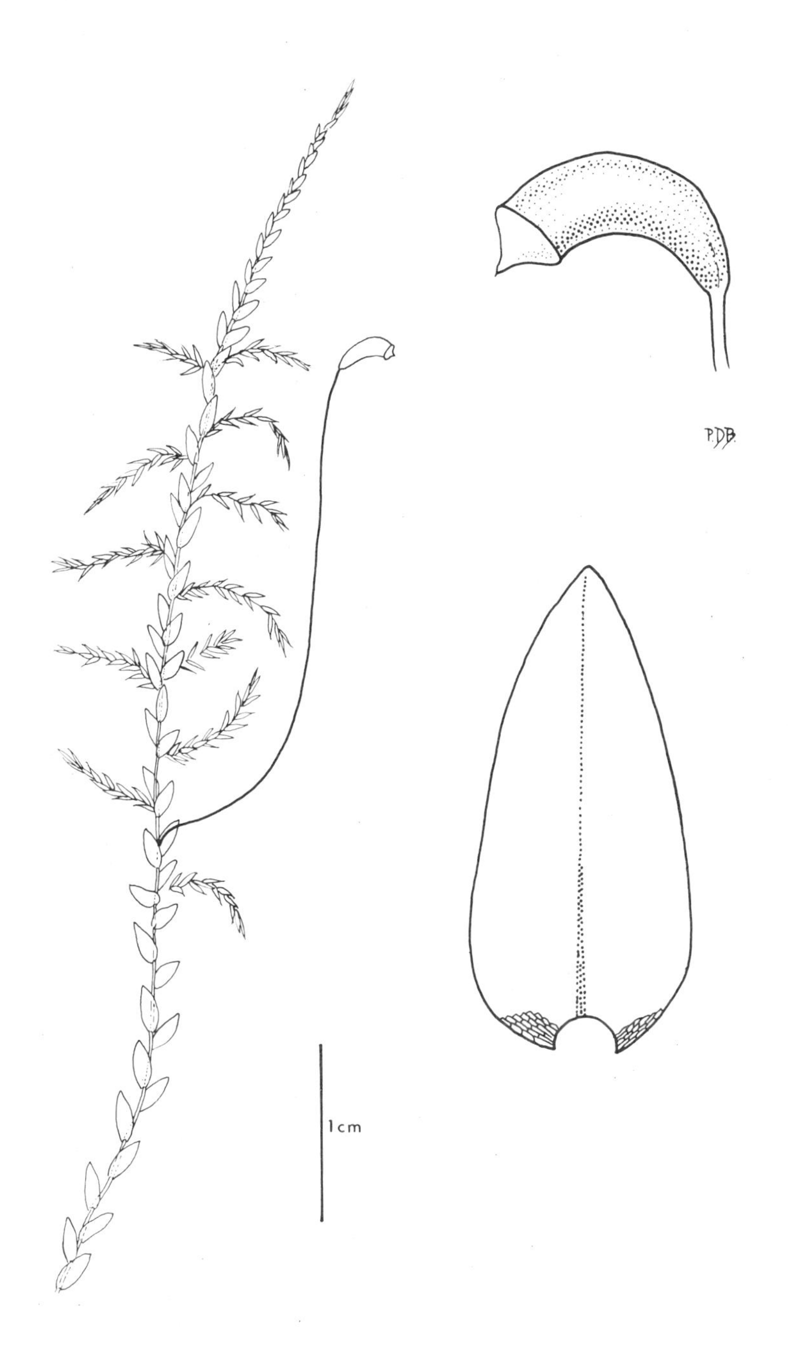

Calliergon giganteum

Calliergon stramineum (Brid.) Kindb.

Name: Species name meaning straw-coloured, based on the colour of the leafy plants.
Habit: Erect or reclining shoots sometimes forming loose or dense turfs, or shoots scattered among other mosses. Plants tending to be sparsely branched or unbranched, slender, glossy, soft and pale yellowish-green to golden yellow in sunny sites.
Habitat: In open, rich fens, in swamps, on marshy lake shores, seepy depressions and bog hummocks, from near sea level to alpine sites, usually in sunny sites.
Reproduction: Sporophytes infrequent, seta red-yellow to red, elongate; sporangia curved, cylindric to ovoid, brown when ripe. The plants are somewhat brittle when dry; the fragments may serve in vegetative reproduction.
World Distribution: Circumboreal (and reported from scattered localities at high elevations in the Southern Hemisphere), extending northward into the arctic. In North America extending, southward to the northern U.S.A.
B.C. Distribution: Map 18, page 328.
Distinguishing Features: The shiny, elongate, slender, pale yellow-green, glossy, unbranched or little-branched, leafy shoots that bear blunt leaves that have a midrib, combined with the wetland habitat, are distinctive.
Similar Species: From *Drepanocladus pseudostramineus*, which it resembles, *C. stramineum* differs by the blunt rather than acute leaf-tips and the more pronounced differentiated alar cells in *Drepanocladus*. From unbranched specimens of *Brachythecium* it can also be distinguished by the acute or acuminate leaves of *Brachythecium*.

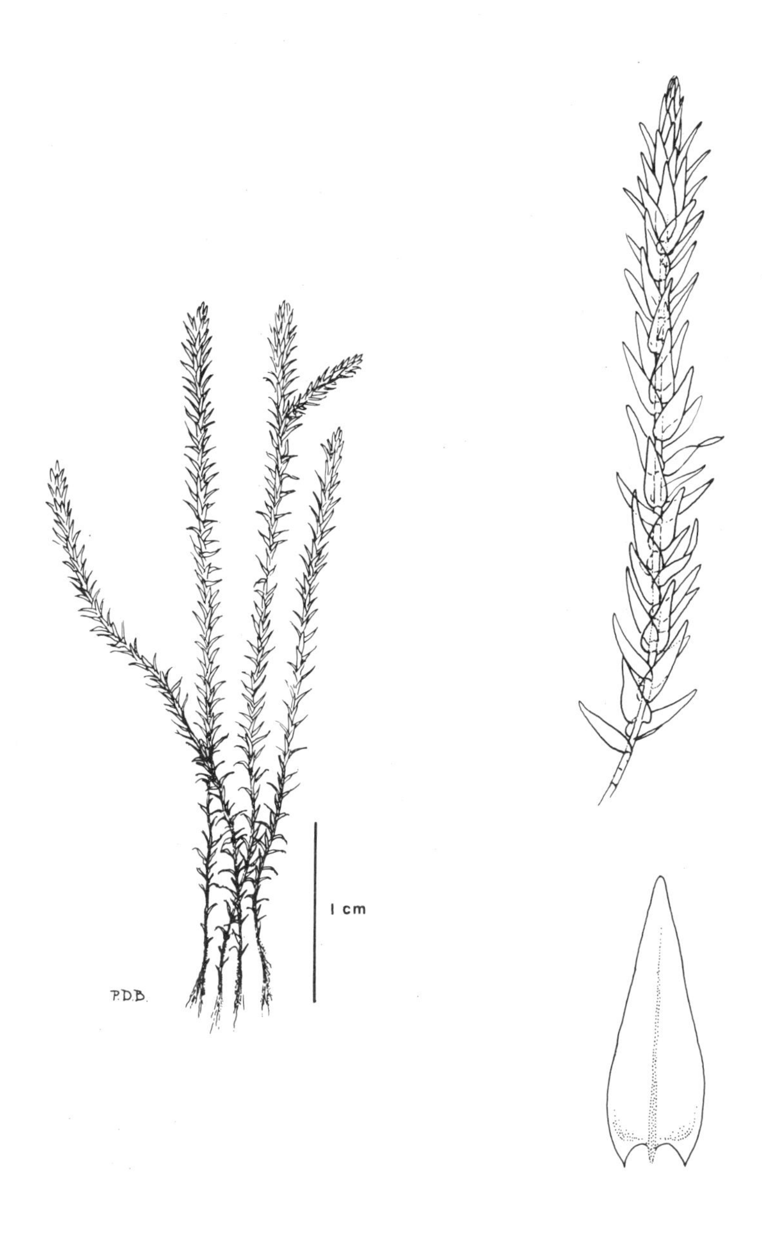

Calliergon stramineum

Campylium stellatum (Hedw.) J. Lange & C. Jens

Name: Genus name referring to curved, a possible reference to the curved sporangium; the species name describing the star-like appearance of the radiately spreading leaves of the stem and branch tips.
Habit: Forming golden green, brownish-green to dark green glossy mats of usually erect to suberect loosely interwoven shoots.
Habitat: Widespread in seepage or wet sites, especially on calcareous substrata, from sea level to subalpine elevations. In openings in subalpine forest it frequents springs while at lower elevations it is frequent on cliffs near waterfalls.
Reproduction: Sporophytes infrequent; sporangia curved-subcylindric, light to dark brown; seta red to reddish-yellow, elongate; possibly propagated mainly through fragmentation of the leafy shoots.
World Distribution: Widely distributed in the Northern Hemisphere, especially at arctic and boreal latitudes and in mountains, southward to mountains of Africa. In North America extending southward in mountains to Georgia in the east and to California in the west; also in Mexico and Guatemala.
B.C. Distribution: Map 19, page 328.
Distinguishing Features: The squarrose, narrowly triangular leaves and the relatively soft plants that grow in damp habitats usually make this plant readily recognizable.
Similar Species: Often *Rhytidiadelphus squarrosus* grows in the same habitat as *C. stellatum*; the leaves are also squarrose, but the stems are red-brown (in *Campylium* they are yellowish or green) and the shoots are more rigid than in *Campylium*. In the same habitat *Dicranella palustris* also has squarrose leaves but the leaves have a strong midrib and are lanceloate, the stems are never branched and the plants grow in tight turfs on mineral soil.

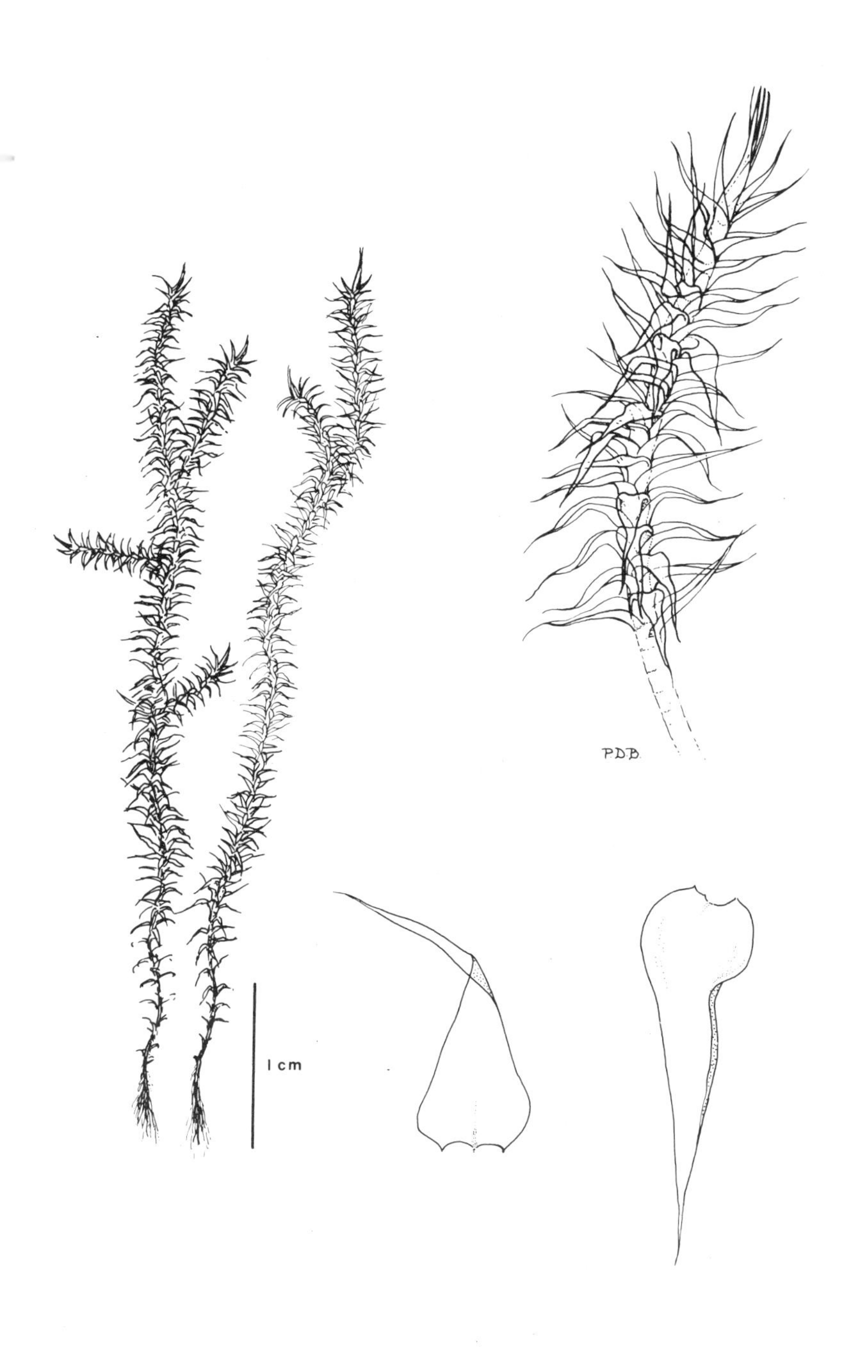

Campylium stellatum

Catoscopium nigritum (Hedw.) Brid.

Name: The genus name derived from the down-pointing sporangium mouth that "looks" down. The species name referring to the shiny black sporangium.
Habit: Forming densely packed, dark green turfs.
Habitat: In wet to damp calcareous sites, usually away from the coast, ascending to subalpine sites. Usually in marshes or fens, near water bodies or damp sites.
Reproduction: Producing abundant, glossy black sporophytes in summer.
World Distribution: Circumboreal; in North America extending southward to Iowa in the east and to Montana in the west.
B.C. Distribution: Map 20, page 329.
Distinguishing Features: With sporophytes, this plant is unmistakable: the tiny, black, subspherical sporangia that are inclined at right angles to the seta are unlike any other local moss. Without sporophytes, the usually regularly three-ranked, narrowly triangular leaves, and the dense, dark green tufts, found mainly in calcareous moist habitats, are enough to identify this moss.
Similar Species: In the seepage habitat *Philonotis* and *Pohlia*, if lacking sporophytes, might superficially suggest *Catoscopium* but their leaves are never in three distinct rows and plants are rarely dark green. *Meesia triquetra* has leaves in three rows but the plants are more than twice the size of *Catoscopium*, the leaves also diverge out strongly (rather than weakly, as in *Catoscopium*) and have toothed margins. *Meesia uliginosa* and *M. longiseta* generally have sporophytes that are pear-shaped, suberect and not glossy black.

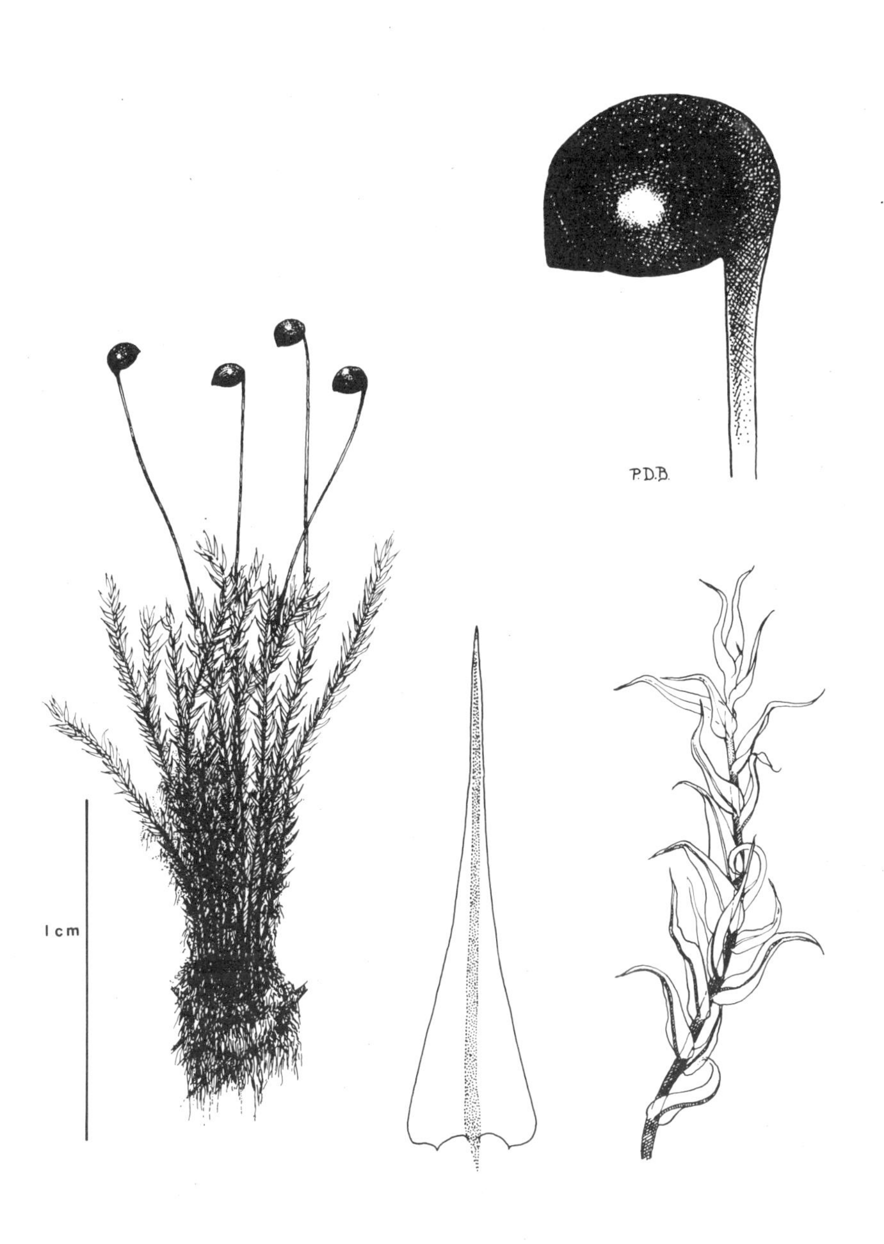

Catoscopium nigritum

Ceratodon purpureus (Hedw.) Brid.

Name: The genus name based on the forked peristome teeth that resemble the horns of an animal. The species name referring to the red-purple seta and the frequently reddish-purple plants.
Habit: Forming dense, short turfs varying from dark green to wine-red, with erect, usually unbranched shoots, stems red-brown; leaves somewhat contorted when dry.
Habitat: Sunny, usually sterile or disturbed soil, in areas that dry out rapidly. Common in urban areas, on stone or concrete walls, on road margins, even on rooftops. Sometimes forming extensive stands on roadside banks, most frequently at lower elevations but extending to alpine sites; infrequent in the wettest oceanic climates of the open coast.
Reproduction: Sporophytes common in spring when the deep red setae become conspicuous. Sporangia are glossy red-brown and 16-ribbed when mature, and oriented at right angles to the seta; the immature sporangium is smooth and erect or suberect.
World Distribution: Widespread at all latitudes, especially in temperate or colder climates, from the high arctic to antarctic. At tropical latitudes more frequent in urban areas or at higher elevations.
B.C. Distribution: Map 21, page 329.
Distinguishing Features: The sporophytes are usually present and form the most reliable feature for identification. The inclined, grooved, glossy red-brown sporangium and with a pronounced bulge at its base where it joins the seta, in combination with the dark green or reddish-brown leafy shoots, are characteristic.
Similar Species: Although *Aulacomnium* species possess grooved sporangia, these are light brown and not glossy, and taper gradually into the pale brown seta (dark red-brown in most specimens of *Ceratodon*). The leafy plants are pale yellow-green in *Aulacomnium*. *Aulaconium androgynum*, the species that might be mistaken for *Ceratodon*, usually grows on wood or humus and gemma-bearing shoots are often present. When without sporophytes *Ceratodon* can be confused with many genera, including *Didymodon*, *Barbula* and *Bryoerythrophyllum*. In these genera, sporangia are erect, cylindric, not grooved when ripe and the peristome teeth are often twisted. Sterile *Catoscopium*, when moist, show the leaves in three rows, especially near the stem apex, a feature never present in *Ceratodon*.

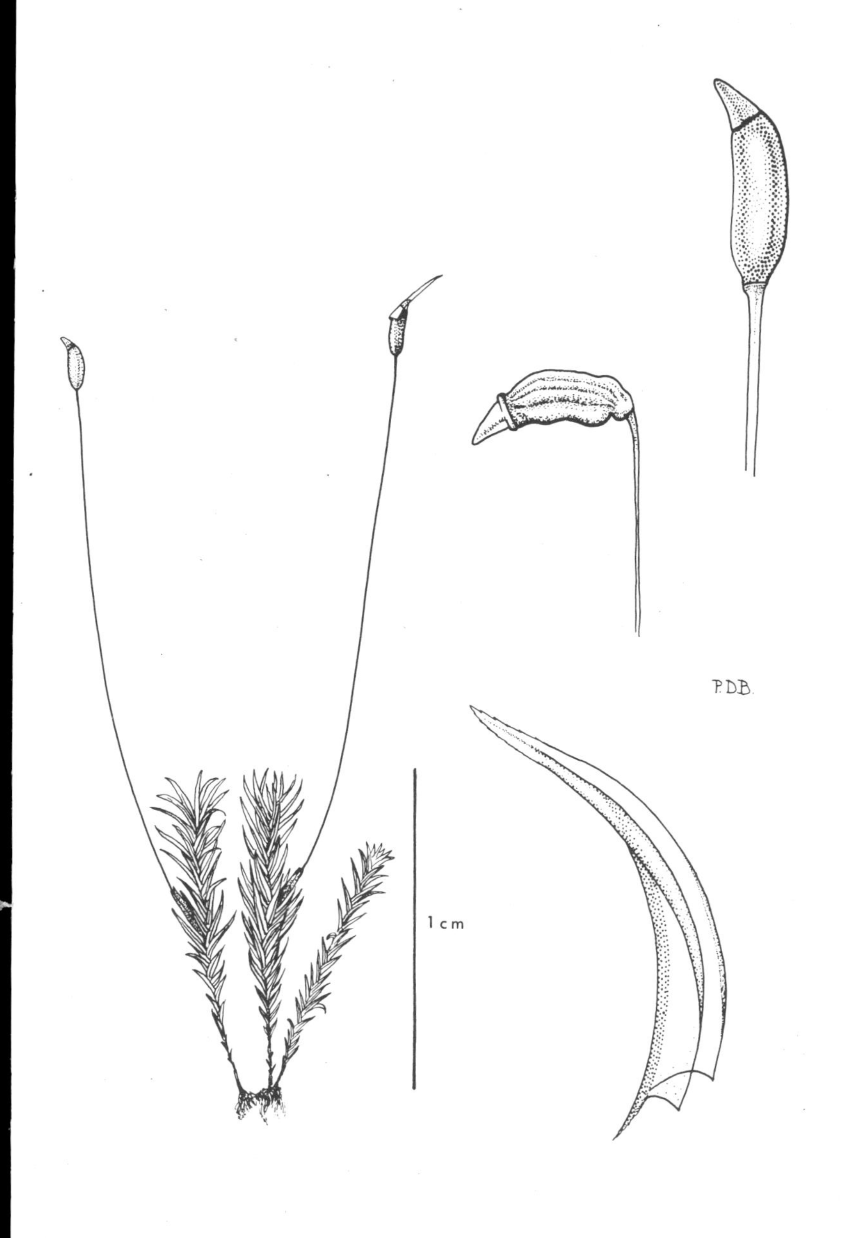

Ceratodon purpureus

laopodium crispifolium (Hook.) Ren. & Card.

Name: Genus name meaning to break off at the foot, the particular significance which is not clear. Species name referring to the leaves that become curled and twisted when dry.

Habit: Pale yellow-green to bright green mats, especially vivid when humid and during the spring, shoots often forming a dense weft over the substratum. The shoots usually bear many regularly arranged lateral branches. When dry the plants are dull yellow-green to dark green and the leaves are much contorted.

Habitat: Commonly epiphytic, especially frequent on broad-leafed maple tree trunks but also on conifers (yew, western red cedar, etc.), also on boulders and cliffs, infrequently terrestrial. Generally in partial shade, usually at elevations near sea level.

Reproduction: Sporophytes abundant in late winter and maturing in early spring; seta and sporangia red-brown and not shiny when mature and dry.

World Distribution: Confined in North America to near the Pacific coast; also in southeast Asia from Japan and the adjacent Asian mainland.

B.C. Distribution: Map 22, page 330.

Distinguishing Features: The regularly pinnate, bright yellow-green plants, the broadly triangular leaves, very contorted when dry, the rough seta and the long-snouted operculum are all distinctive.

Similar Species: It is impossible to distinguish some specimens of *C. bolanderi* from *C. crispifolium* without microscopic examination of the papillae on leaf cells: in *C. crispifolium* there is a single usually sharp papilla for each cell, while in *C. bolanderi* there are several papillae on each cell. *C. bolanderi* is the dominant species of this genus at subalpine elevations; it also occurs, infrequently, near sea level; east of the Coast Range, *C. bolanderi* is the likely species.

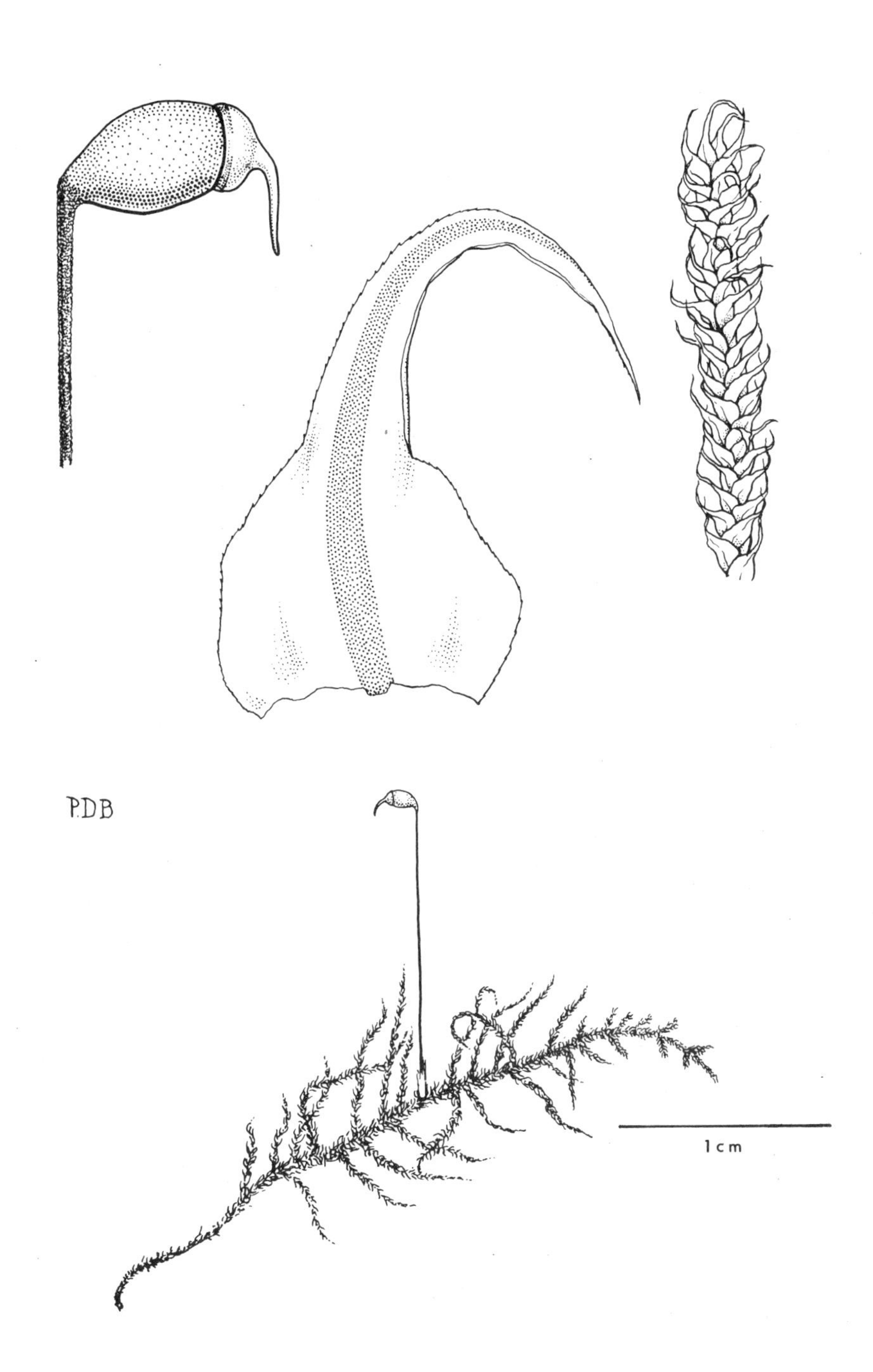

Claopodium crispifolium

Climacium dendroides (Hedw.) Web. & Mohr

Name: Genus name is derived from the broad perforations in the ladder-like, inner-peristome teeth. Species name describing the tree-like form of the plants.
Habit: Forming loose, tall turfs of miniature trees that arise from creeping shoots; dark green to light green with reddish-brown stems.
Habitat: Terrestrial or on logs (rarely on rock), in wet to marshy habitats, most luxuriant in areas subject to periodic flooding, from sea level to subalpine elevations; relatively widespread but not common.
Reproduction: Sporophytes infrequent except in floodplain areas where they are sometimes locally abundant in late winter and early spring; red-brown throughout.
World Distribution: Widely distributed in the Northern Hemisphere, extending in eastern North America southward to Wisconsin and Pennsylvannia and in western North America to California; also in New Zealand.
B.C. Distribution: Map 23, page 330.
Distinguishing Features: The miniature, tree-like plants that arise from a creeping stem are highly characteristic.
Similar Species: Although *Leucolepis acanthoneuron* has miniature tree-like plants, these do not arise from a creeping shoot. The leaves of the main stem of *L. acanthoneura* are whitish and narrowly triangular while those of *Climacium* are heart-shaped and green. Sporophytes of *Leucolepis* are frequent in spring and sporangia are nodding, compared to the erect sporangia of *Climacium*.

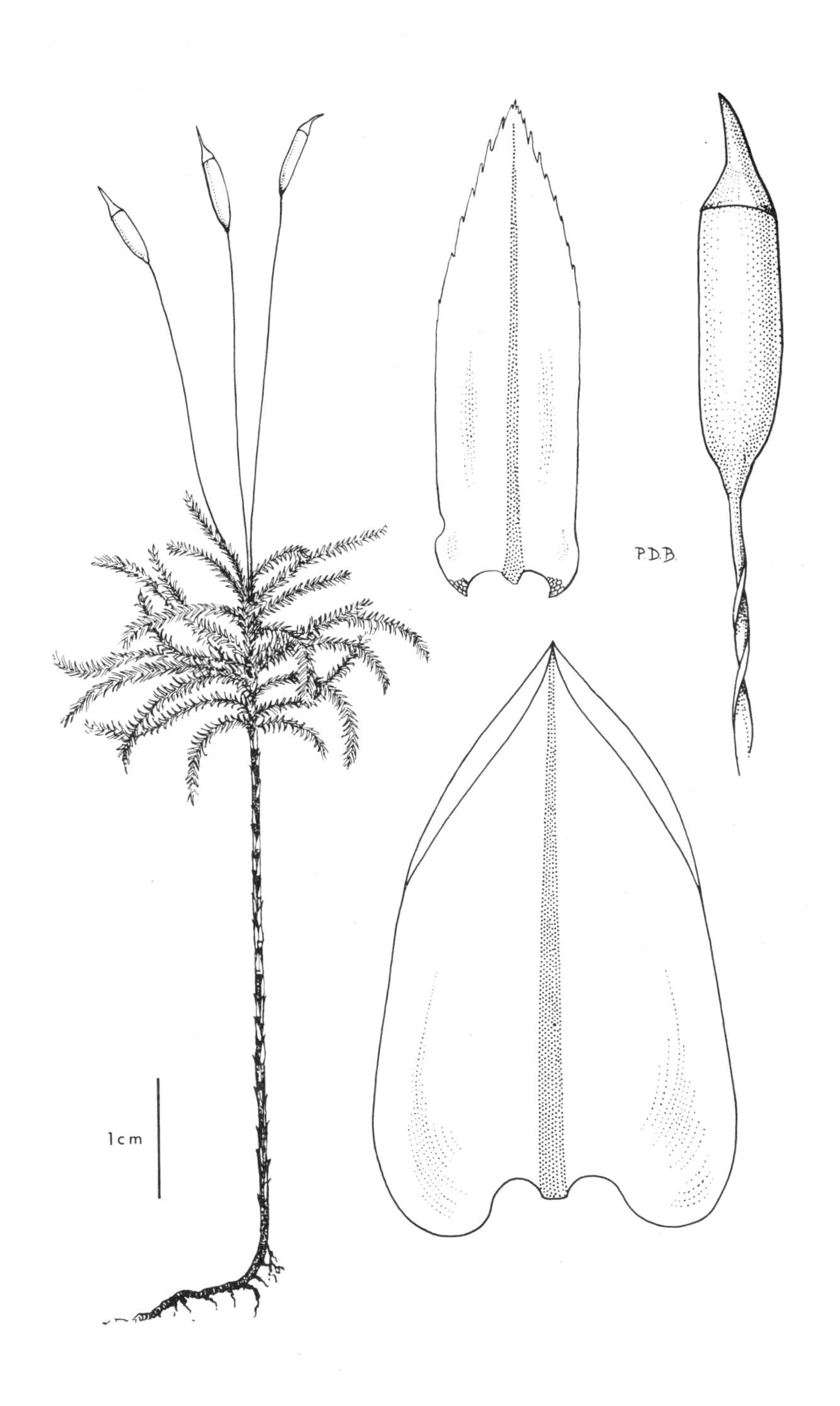

Climacium dendroides

Coscinodon calyptratus (Hook *in* Drumm.) C. Jens. *in* Kindb.

Name: Genus name referring to the irregular holes in the peristome teeth. Species name noting the large, hood-shaped calyptra.
Habit: Forming whitish to grayish, rounded tufts; the hair points on the leaves are conspicuous and create the grayish appearance of the tufts.
Habitat: On rock surfaces usually in open sites of semi-arid climates.
Reproduction: Sporophytes frequent in spring; the sporangia are smooth when ripe.
World Distribution: Confined to western North America, reaching its northern limit in the semi-arid interior of British Columbia and extending southward to southern California and Arizona; predominantly at lower elevations but extending occasionally to subalpine sites.
B.C. Distribution: Map 24, page 331.
Distinguishing Features: The rounded tufts with conspicuous hair points on the leaves, the semi-arid climate restriction, and the bell-shaped calyptrae will separate this species.
Similar Species: Several species of *Grimmia* resemble this moss, but in *Grimmia* the calyptrae are never bell-shaped and only partially enclose the sporangium. Technical microscopic features also distinguish *Coscinodon* from *Grimmia*. Unfortunately species of *Grimmia* of similar form are frequent in the semi-arid interior. *G. pulvinata*, which superficially resembles *C. calytratus*, has a curved seta unlike the straight seta in *Coscinodon*; the sporangia are grooved when ripe in *Grimmia pulvinata*, smooth in *Coscinodon*.

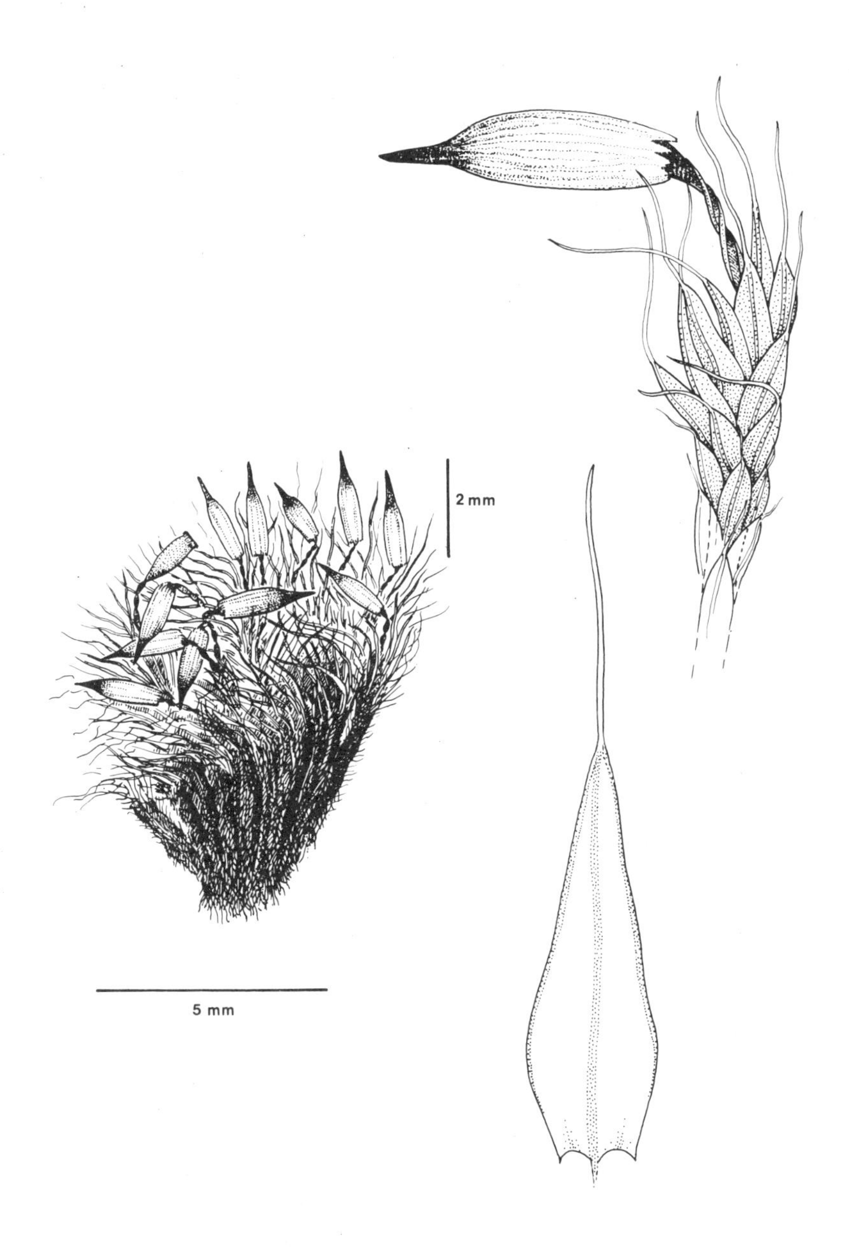

Coscinodon calyptratus

Cratoneuron commutatum (Hedw.)Roth

Name: Genus name referring to the strong midrib of the leaf. Species name derived from the tendency of the plants to accumulate lime and form tufa that encrusts them.

Habit: Forming turf-like mats of rich green to rusty, brownish-green to golden green; plants sometimes accumulating tufa; from sea level to subalpine elevations.

Habitat: Swampy to seepage areas, especially calcium-rich areas; most frequent near lakes and watercourses but also abundant around springs and on wet cliffs.

Reproduction: Sporophytes infrequent, sporangia curved-cylindric, seta elongate, appearing in summer; plants somewhat brittle, the fragments serving for vegetative reproduction.

World Distribution: Widely distributed in the Northern Hemisphere. In North America, extending in the east, southward to Illinois; in the west to Colorado and California.

B.C. Distribution: Map 25, page 331.

Distinguishing Features: This species, except on microscopic characters, closely resembles some species of *Drepanocladus* that grow in the same habitat. Microscopic examination, however, will expose the paraphyllia on the stem; these are absent in *Drepanocladus*. Paraphyllia are short, branched, hair-like structures on the stem, among the leaf bases. These can be seen on moist (not wet) material with a 10x hand lens. The leaves are markedly pleated, a feature shared by few mosses in this wet habitat.

Similar Species: Besides *Drepanocladus*, *C. commutatum* superficially resembles some species of *Hypnum*, but *Hypnum* has a double and obscure midrib, rather than a single obvious one, as in *Cratoneuron*. The pleated leaves and paraphyllia also separate *Cratoneuron* from *Drepanocladus* and *Hypnum*.

Comments: This species accumulates lime in hot-spring areas and also in some cold springs. Indeed, sometimes whole plants become encrusted and "petrified" with lime (tufa). Sometimes treated as *Palustriella commutata*.

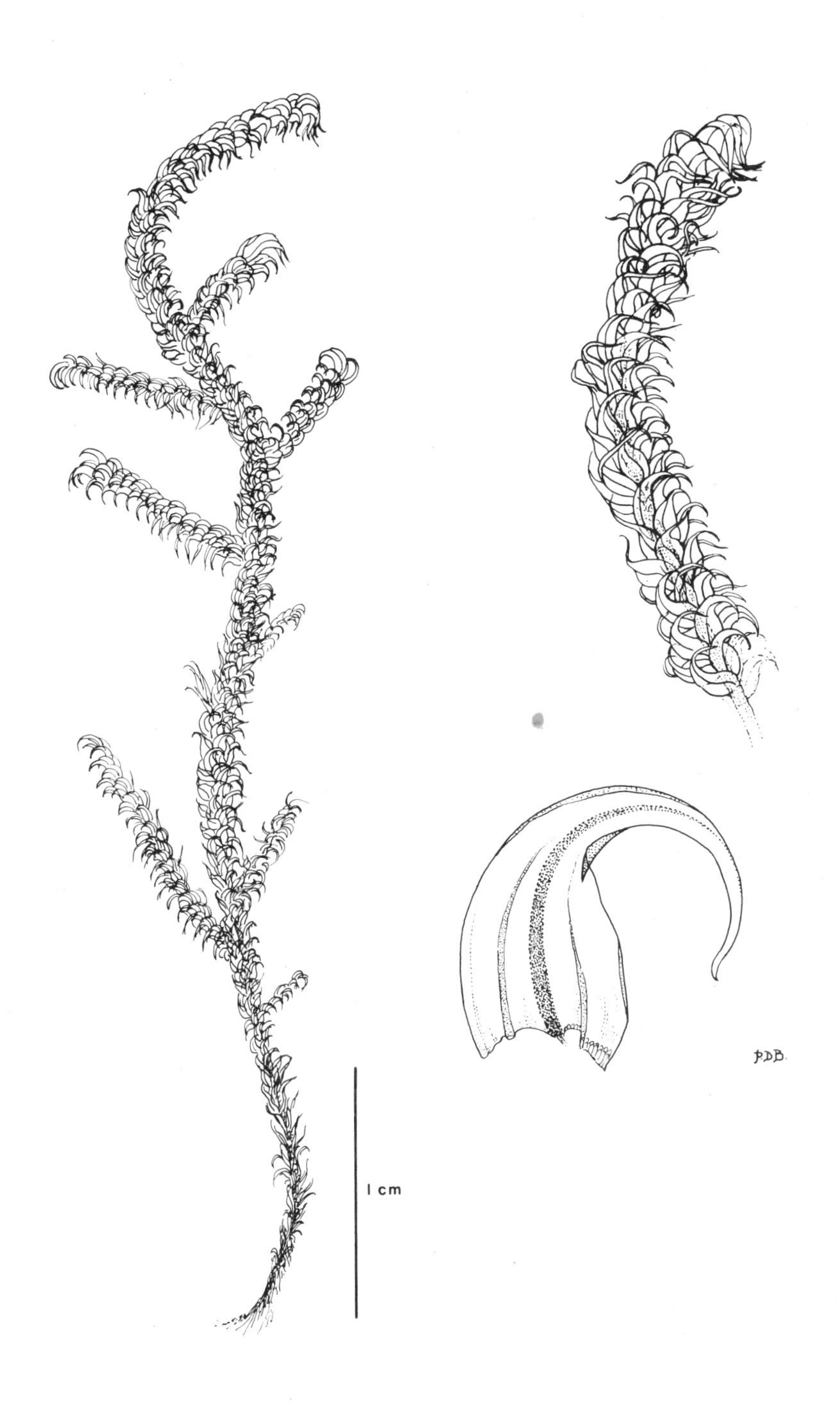

Cratoneuron commutatum

Cratoneuron filicinum (Hedw.) Spruce

Name: The species name refers to the fern-like appearance of the leafy plants.
Habit: Forming dense mats of erect to suberect shoots varying from pale green to brownish-green to golden green.
Habitat: In seepage areas, or near watercourses, also on wet cliffs; from near sea level to alpine sites, more frequent in calcareous areas but also on non-calcareous substrata; tolerant of greater periods of dryness than *C. commutatum*.
Reproduction: Sporophytes infrequent, maturing in summer; seta red, elongate, sporangium brown, curved, not glossy when dry. Vegetative reproduction by fragmentation is probably the main means of local dispersal.
World Distribution: Widely distributed in the Northern Hemisphere; in the Americas extending southward to Guatemala; also in New Zealand.
B.C. Distribution: Map 26, page 332.
Distinguishing Features: Usually plants have a fern-like form with many short, lateral branches that gradually become shorter toward the apex; thus the whole shoot resembles a fern frond. The broadly triangular leaves with distinct midrib and bulging alar cells are characteristic. These features, combined with the wet habitat, are usually sufficient to mark the species.
Similar Species: *Helodium blandowii* grows in similar habitats but the plants are usually yellow green, the pinnate branches are more than twice the length they are in the *Cratoneuron*, and they tend to emerge at right angles rather than acute angles; paraphyllia are obvious at 10X; differentated alar cells are absent. *Tomentypnum nitens* has conspicuously pleated leaves, and lacks differentiated alar cells. *Kindbergia praelonga* has more complex branching and lacks the conspicuous alar cells. *K. oregana*, besides the features of *K. praelonga*, grows in dry rather than wet habitats.

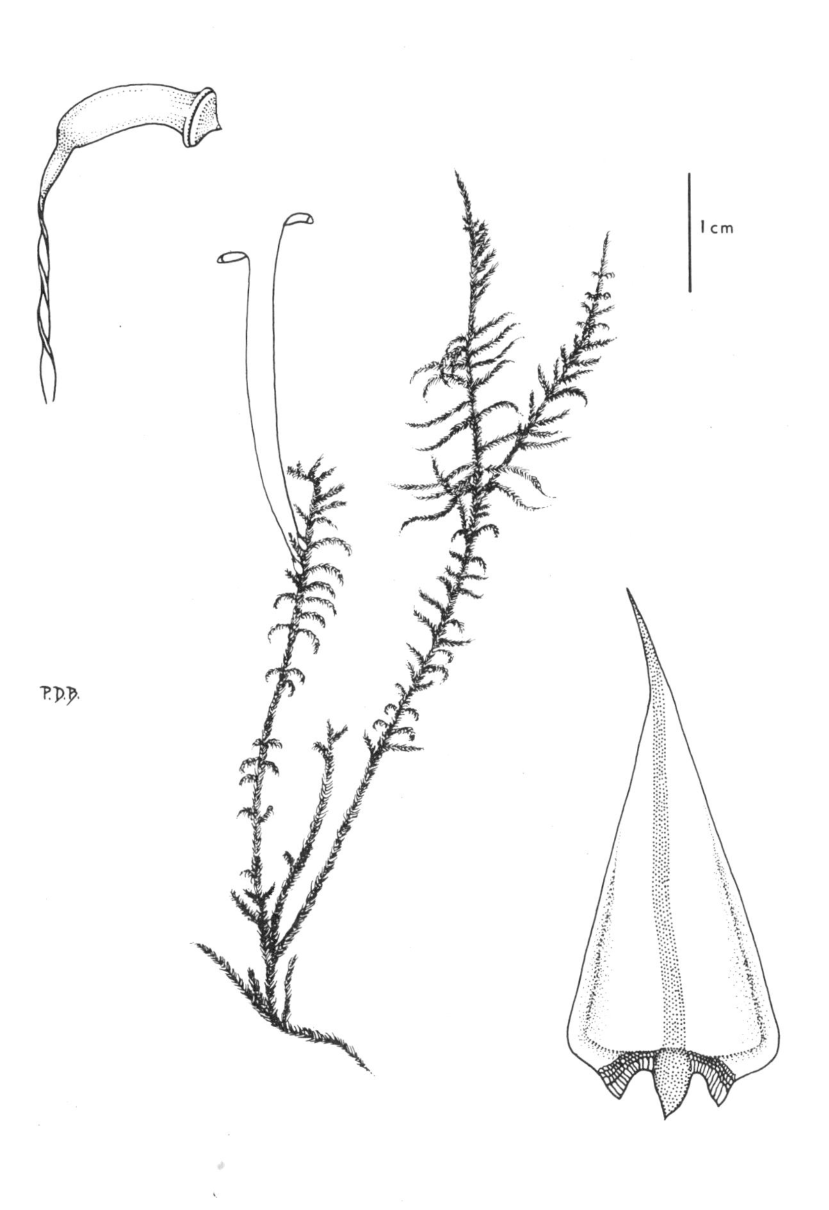

Cratoneuron filicinum

Dendroalsia abietina (Hook.) Britt.

Name: Genus name based on the tree-like outline of the branched shoots and the resemblance of the plants to the moss genus *Alsia*. Species name referring to the fancied resemblance of the plant outline to a fir tree (*Abies*).
Habit: Forming extensive downward-pointing, dark green, densely pinnate shoots in which the lower portion of the shoot lacks branches; branches mainly arising in a single plane. Plants coiling downward toward the substratum when dry. Male plants usually with numerous lateral, bulb-like, sexual branches that are light green to straw-coloured.
Habitat: Usually epiphytic, especially on tree trunks, most frequently on broad-leafed maple and Garry oak, but also on other trees, including poplar. Sometimes on somewhat shaded cliffs.
Reproduction: Sporophytes red-brown, frequent, maturing in late winter to early spring, arising on a short seta on the undersurface of the shoots. Male and female plants separate.
World Distribution: Restricted to western North America, reaching its northern limit in southern British Columbia, mainly near the coast, but found eastward to Idaho and southward to Baja California.
B.C. Distribution: Map 27, page 332.
Distinguishing Features: The distinctive dark green, epiphytic, miniature, feather-like plants are unlikely to be confused with any other moss in western North America. The dry, downward coiled plants are also characteristic.
Similar Species: Some specimens of *Isothecium stoloniferum*, when moist and without sporophytes, occasionally resemble *Dendroalsia*, but *I. stoloniferum* is pale rather than dark green and the sporophytes arise on elongate setae from the upper surface of the main shoot.

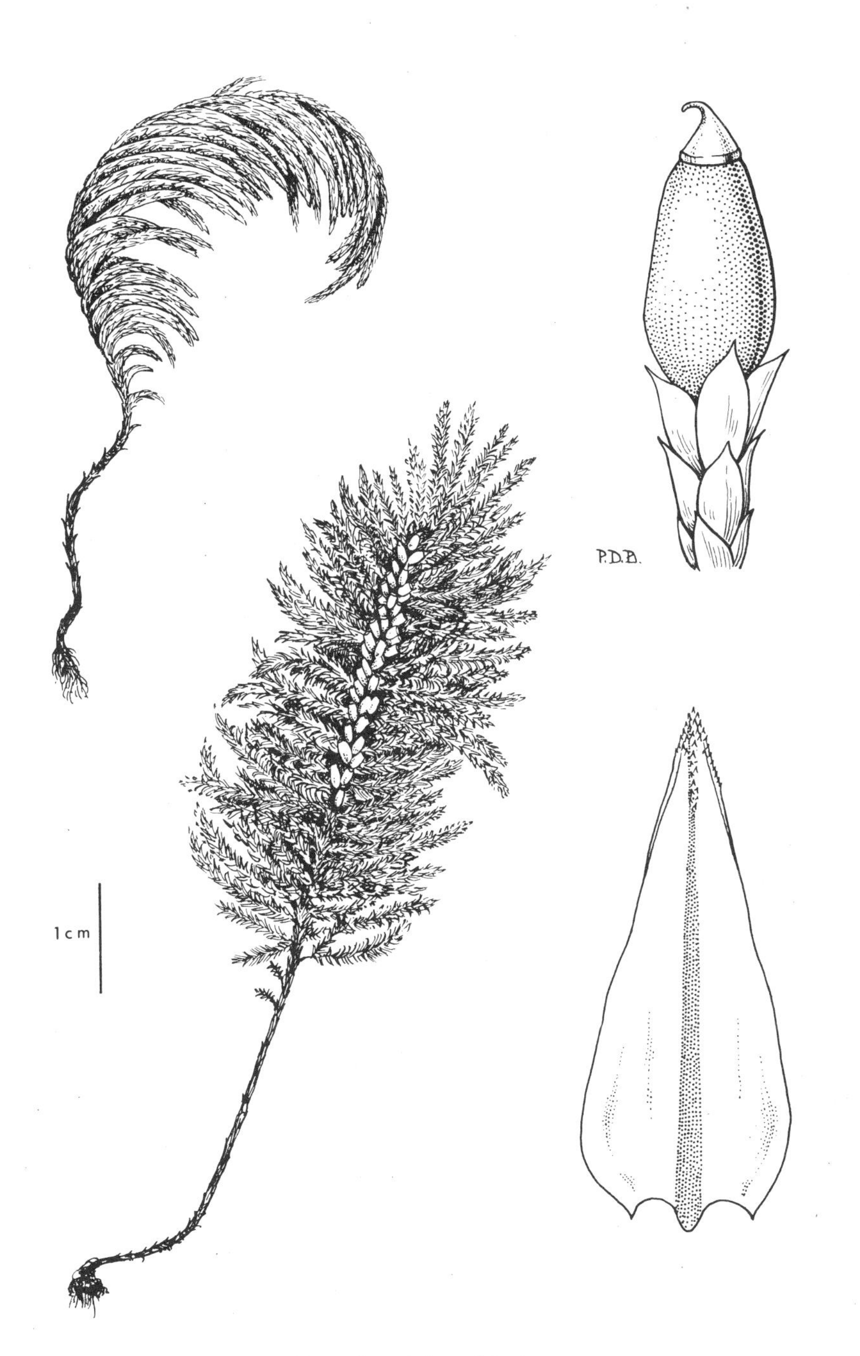

Dendroalsia abietina

Dichodontium pellucidum (Hedw.) Schimp.

Name: Genus name referring to the two-forked peristome teeth. Species name meaning transparent, presumably in reference to the leaves as viewed under the microscope.
Habit: Forming bright, yellow-green to dark green turfs.
Habitat: Frequent in wet, somewhat shaded, habitats, especially near streams, on rocks, soil, even on logs, but also on cliffs away from water bodies. Most frequent at lower elevations, but also in subalpine and alpine sites where it is found in sunny localities with mineral soil.
Reproduction: Commonly producing sporophytes in winter to early spring at lower elevations; at higher elevations sporophytes are rare but axillary gemmae are frequent.
World Distribution: Circumboreal; in North America extending southward to North Carolina and Tennessee in the east and to California in the west.
B.C. Distribution: Map 28, page 333.
Distinguishing Features: With sporophytes, this species is distinctive; the suberect sporangia associated with the pale green, toothed leaves are reliable characters.
Similar Species: *Dicranella palustris*, found in similar habitats, has sporangia that are inclined and curved and the leaves are markedly squarrose (subsheathing at base then abruptly diverging outward).
Comments: Western North American material is mainly *D. flavescens* (With.) Dixon, a species that shows a wide distribution independent of acidity of substratum. *D. pellucidum*, in the strict sense, is a calcicole. The plants tend to be bright yellow-green, compared to duller colours in *D. flavescens*. *D. pellucidum* is mainly subalpine, but may also be found at lower elevations, and occurs in calcium-rich sites. Many researchers do not consider *D. flavescens* to be an independent species but treat it as within the variability of *D. pellucidum*.

Dichodontium pellucidum

Dicranella heteromalla (Hedw.) Schimp.

Name: Genus name derived from the resemblance to a miniature *Dicranum* (p. 120), and bearing the two-forked teeth characterizing that genus. Species name referring to the leaves that spread in all directions although, in fact, they mainly curve in one direction.
Habit: Forming short turfs of silken plants that are dark to light green.
Habitat: On banks of mineral soil, in somewhat shaded sites, frequent on recently disturbed soil, more often on forest margins than within undisturbed forest, and extending from near sea level to subalpine elevations. Often on the soil of the roots of overturned trees.
Reproduction: Sporophytes abundant, maturing in early spring; the peristome teeth are bright red.
World Distribution: Circumboreal; in the Americas extending as far south as Colombia.
B.C. Distribution: Map 29, page 333.
Distinguishing Features: The asymmetric sporangium with an obliquely oriented mouth bearing bright red teeth and grooved surface when dry, are distinctive. These features plus the silken, usually curved leaves and short, turf-forming plants are usually enough to separate this species.
Similar Species: Other species of *Dicranella* resemble *D. heteromalla* but the sporophyte of the latter is distinctive (see above). *Arctoa fulvella* resembles *D. heteromalla* vegetatively, but the erect sporangium and the alpine or subalpine cliff crevice habitat will separate it. *Kiaeria* species also differ in habitat (usually on rocks) and sites (usually alpine or subalpine). *Ditrichum* species, that occur in similar habitats, have a straight erect sporangium that is never grooved.

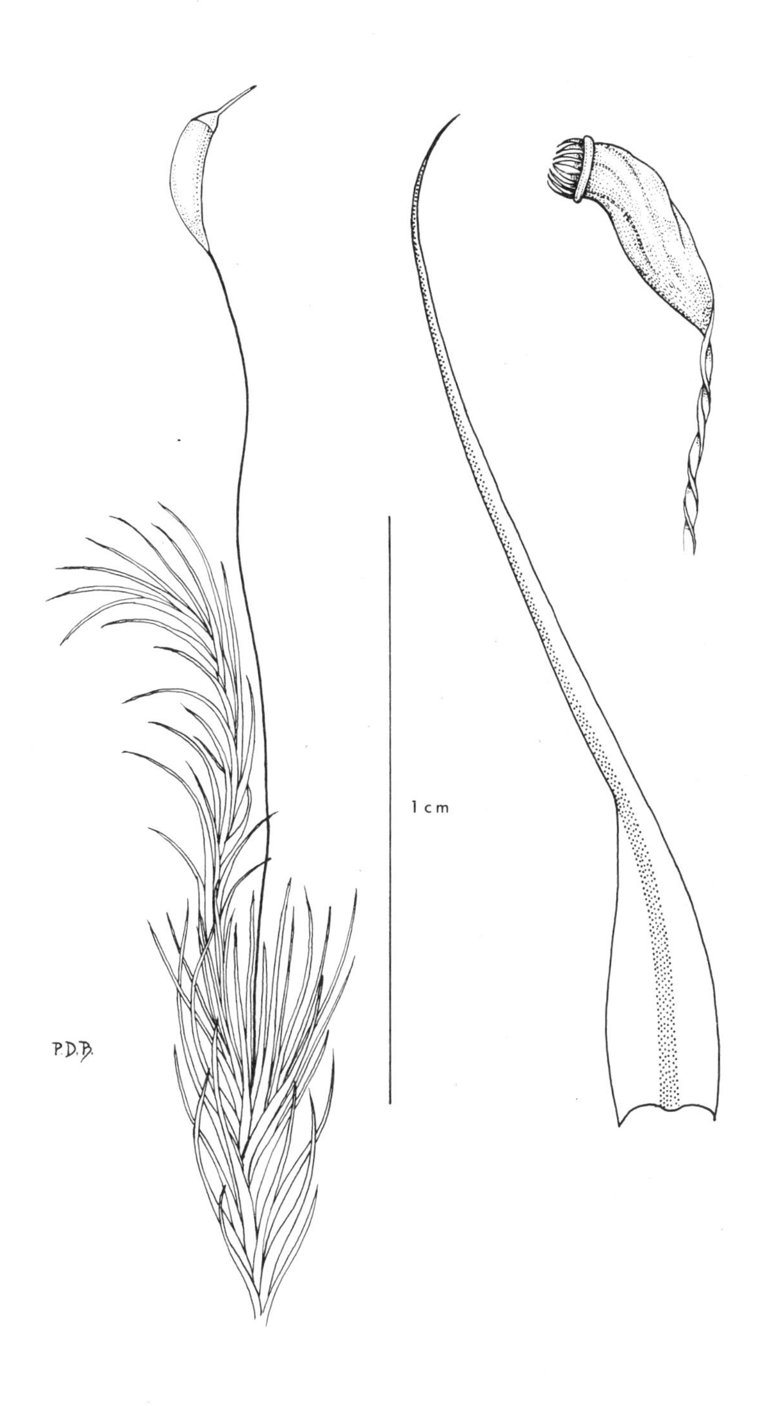

Dicranella heteromalla

Dicranella rufescens (With.) Schimp.

Name: Species name referring to the reddish colour of both sporophytes and leafy plants, especially when young.
Habit: Short turfs of widely scattered, reddish-brown to light green plants.
Habitat: Moist, usually somewhat shaded sandy banks, often characterized by some seepage water.
Reproduction: Sporophytes frequent in late winter to spring, bright red-brown.
World Distribution: Scattered in temperate areas of the Northern Hemisphere; in North America reaching its northern range in the east in Quebec and southward to Missouri, and in the west to southeastern Alaska and southward to California.
B.C. Distribution: Map 30, page 334.
Distinguishing Features: The plants, especially from autumn to spring, form red-brown turfs on sandy banks, sometimes sharply marking a horizontal seepage layer. The curious, stalked, bulbiform male shoots are also distinctive.
Similar Species: Unlikely to be confused with any other local species; the reddish-brown colour, the stalked bulbiform male plants, the usually erect sporangia, and the late autumn to winter appearance of the sporophytes are useful to separate this species from all other local mosses that resemble it (especially *Ditrichum* and other species of *Dicranella*).

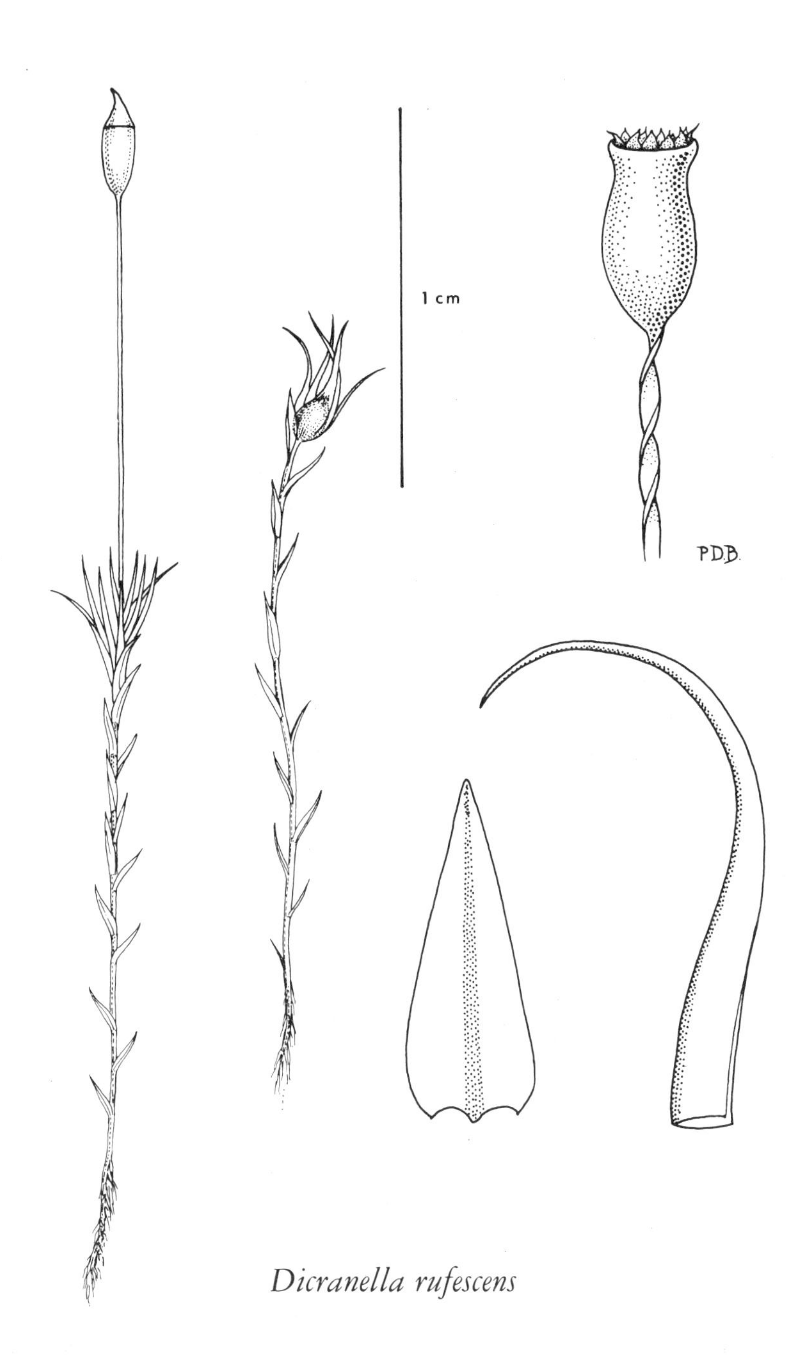

Dicranella rufescens

Dicranoweisia cirrata (Hedw.) Lindb. *ex* Milde

Name: Genus name refers to the relationship to *Dicranum* and the resemblance to another moss, *Weissia*. Species name referring to the sharp pointed leaves.
Habit: Forming turfs or tufts of dark green to light green plants.
Habitat: Usually on tree trunks or wood in early stages of decomposition, also on rooftops and stone or concrete walls and occasionally on rock, usually near sea level.
Reproduction: When without sporophytes it produces abundant dust-like gemmae in the axils of the leaves. These, plus the abundant sporophytes that mature in late winter to spring, disseminate the species widely in urban areas near the coast.
World Distribution: Scattered in temperate portions of the Northern Hemisphere. In North America, confined to the Pacific coast from southeastern Alaska to California.
B.C. Distribution: Map 31, page 334.
Distinguishing Features: The usual epiphytic habitat and turf-forming masses, added to the erect sporophytes with red peristomes, are distinctive.
Similar Species: *Zygodon viridissimus* is less common in the same habitat and also produces abundant gemmae. The tufts are not as condensed and the leaves coil helically around the shoot when dry; in *Dicranoweisia* the leaves are individually twisted in all directions. *Orthotrichum consimile*, occurring often on the same trees as *Dicranoweisia*, has short setae and grooved sporangia; peristome teeth in *Orthotrichum* are in two rows rather than one, as in *Dicranoweisia*.
Comments: This moss must be highly tolerant to exhaust fumes because it often abounds on trees along busy urban streets.

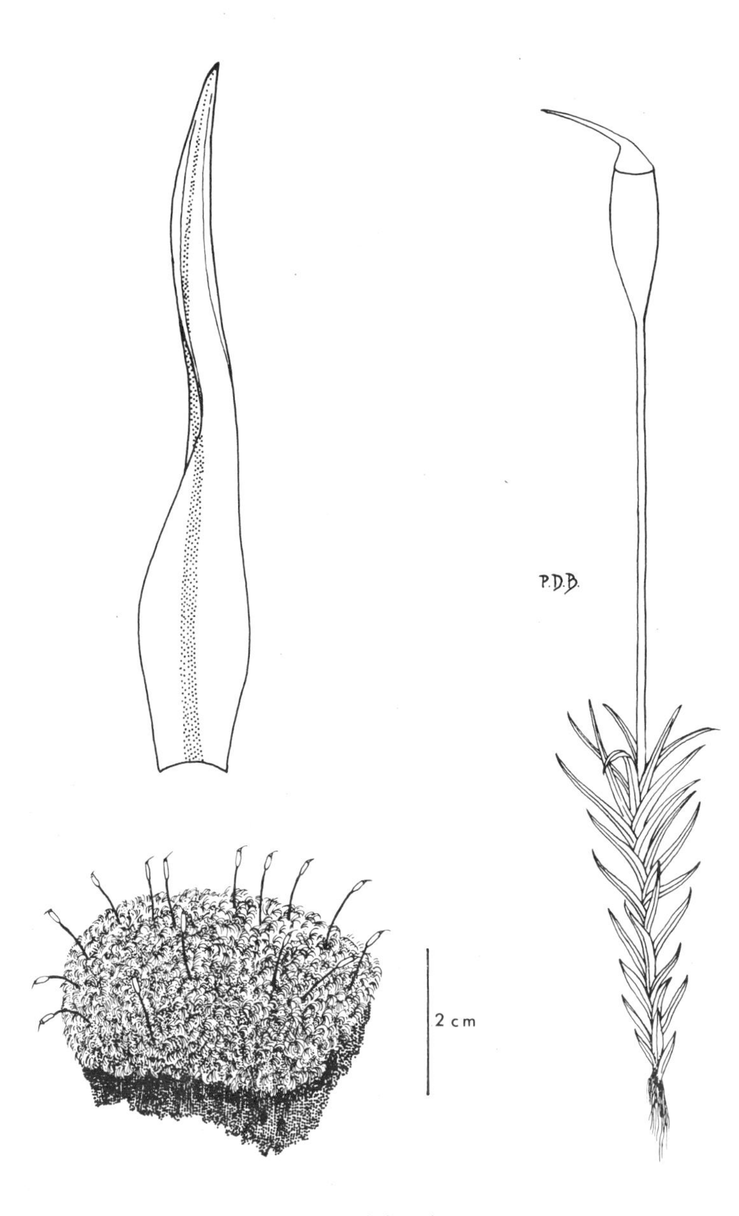

Dicranoweisia cirrata

Dicranoweisia crispula (Hedw.) Lindb. *ex* Milde

Name: Species name referring to the leaves that become much twisted when dry.
Habit: Forming dark green to yellow-green tufts.
Habitat: Confined to siliceous rock surfaces and crevices, generally at alpine to subalpine elevations, usually in sunny sites.
Reproduction: Sporophytes frequent in summer, yellowish to pale brown.
World Distribution: Widespread in arctic and boreal regions, particularly in mountains. In North America extending southward in the east to New Hampshire and in the west to California.
B.C. Distribution: Map 32, page 335.
Distinguishing Features: With sporophytes, and based on its habitat, the very contorted leaves and suberect to erect sporangium are diagnostic.
Similar Species: Some species of *Kiaeria*, found in similar habitats and at the same elevation, can be confused with *D. crispula*. In *Kiaeria*, however, the leaves are usually curved on a single side of the shoot and the sporangia are not erect. *Amphidium*, which also occurs in rock crevices, is easily distinguished when with sporophytes by the short seta and the sporangium that is grooved when dry. *Amphidium* also lacks peristome teeth. Sterile tufts of *Amphidium* are dense and are confined to crevices.

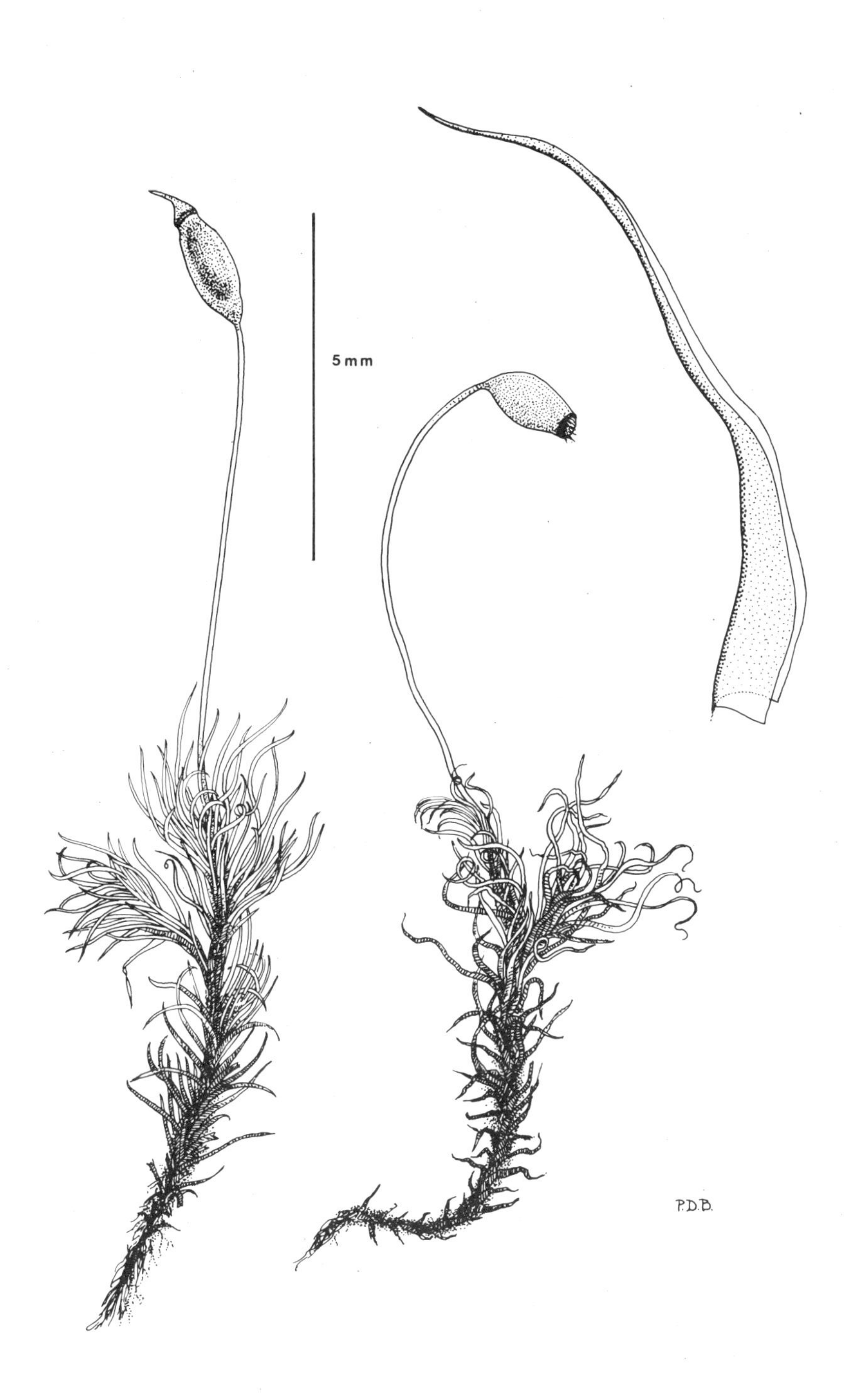

Dicranoweisia crispula

Dicranum fuscescens Sm.

Name: The genus name refers to the forked peristome teeth; the species name to becoming brown with age.
Habit: Forming tufts or turfs of dark green plants.
Habitat: Commonly on wood and humus, occasionally epiphytic on living trees, especially Douglas-fir and alder. Usually somewhat shaded, and in forests, from sea level to subalpine elevations.
Reproduction: Sporophytes frequent, maturing in spring, dark brown when mature.
World Distribution: Circumboreal, extending into arctic regions. In eastern North America extending southward into the mountains of Tennessee; in the west to California.
B.C. Distribution: Map 33, page 335.
Distinguishing Features: The dull green plants with leaves that are twisted when dry, added to the brown, somewhat grooved, mature, dry sporangium are useful features to separate *D. fuscescens* from the many variable species of *Dicranum* found in the province.
Similar Species: Of the many species of *Dicranum*, *D. pallidisetum* is most similar and, in vegetative characteristics, is not easy to separate from *D. fuscescens*. In *D. pallidisetum*, however, the consistently yellow seta and paler brown sporangium, added to the leaves that are not so strongly twisted, when dry, as in *D. fuscescens*, are usually reliable features, especially in combination with the usual restriction of *D. pallidisetum* to subalpine forest and heath slopes. *D. pallidisetum* is commonly terrestrial. *Paraleucobryum* and *Dicranodontium* both are somewhat glossy, although they are similar in form to *D. fuscescens*. The same is true for *Ditrichum crispatissimum*, a species confined mainly to calcareous cliffs.

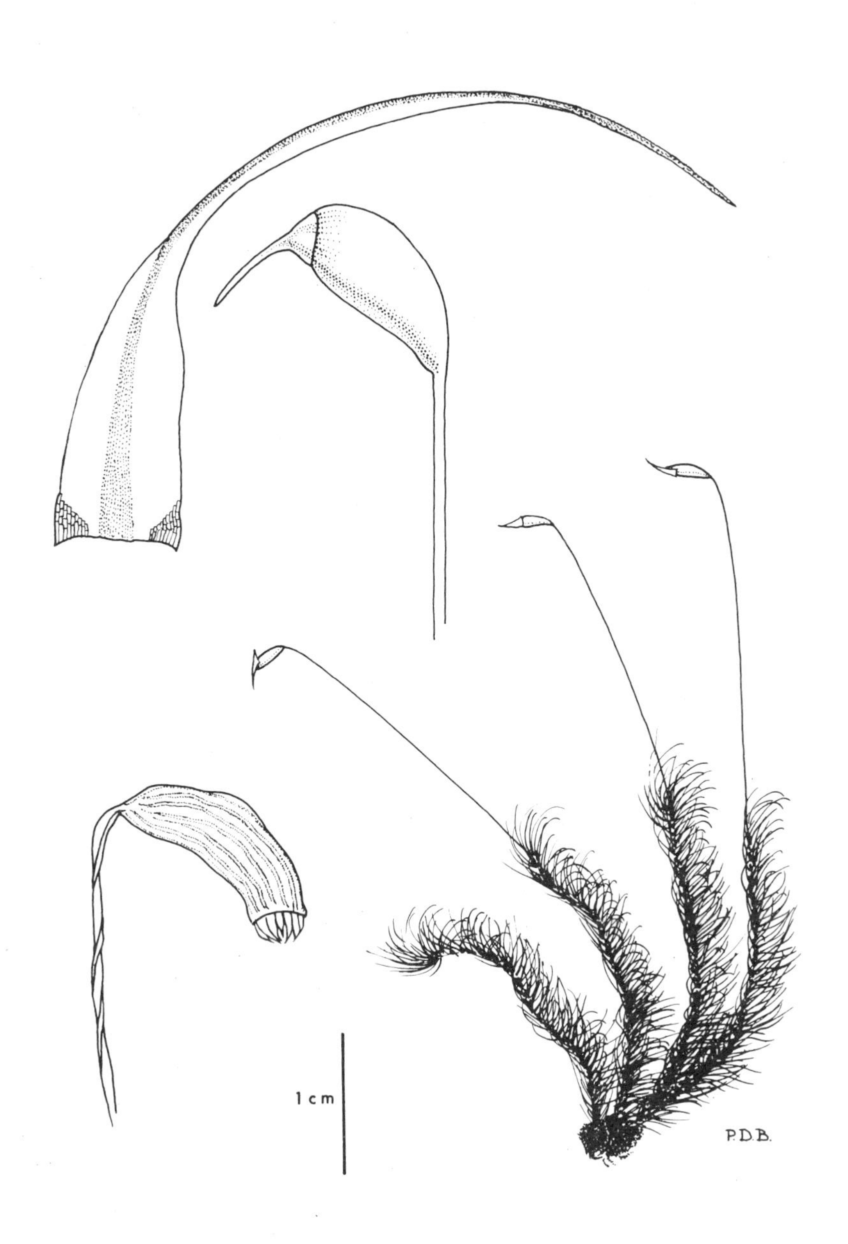

Dicranum fuscescens

Dicranum polysetum Sw.

Name: Species name referring to the many setae supporting the sporangia that emerge from the tips of each sporophyte-bearing plant and support the sporangia.
Habit: Forming tall, loose turfs of light green, somewhat glossy, erect plants densely clothed in pale rhizoids on the stems at the leaf bases.
Habitat: Usually on humus in relatively open forest dominated by conifers. Occasionally on mineral soil or over rock. Most frequent in dry woodland in well-drained sites away from the coast from middle to subalpine elevations.
Reproduction: Sporophytes common in spring and summer, the several sporophytes arising on each shoot being characteristic. Setae yellowish and sporangia light brown when mature.
World Distribution: Circumboreal, following the range of the boreal coniferous forest. In North America extending southward in the east to North Carolina and Missouri; in the west to Wyoming.
B.C. Distribution: Map 34, page 336.
Distinguishing Features: The markedly undulate leaves, the dense matting of rhizoids, the multiple sporophytes and the general restriction to well-drained sites in drier coniferous forest, are usually reliable features.
Similar Species: *D. majus* also has multiple sporophytes but, unlike *D. polysetum*, the leaves are not strongly undulate or are the stems densely matted with rhizoids. *D. majus* is confined mainly to extremely humid shaded forests near the coast. *D. affine* has undulate leaves but the plants form dense tufts in peatland and sporophytes are commonly solitary on each shoot.

Dicranum polysetum

cranum scoparium Hedw.

Name: Species name refers to a broom, from the swept appearance of the leaves on the stem.
Habit: Forming loose to dense, tall turfs of glossy, pale to dark green, generally unbranched or irregularly branched plants that show considerable variability in leaf orientation.
Habitat: On rotten logs, exposed cliff edges, sometimes on forest floor and tree trunks, from sea level to subalpine and alpine elevations; predominantly in forest but also in open sunny sites.
Reproduction: Sporophytes common, maturing in spring. In exposed areas the plants often produce brittle shoots that are undoubtedly important in propagation.
World Distribution: Circumpolar in the Northern Hemisphere; also in New Zealand. In North America southward in the east to Florida and Arkansas and in the west to California and Arizona.
B.C. Distribution: Map 35, page 336.
Distinguishing Features: This is an extremely variable species with considerable habitat diversity; therefore, it is readily confused with several other species of *Dicranum*. The most useful features include the non-undulate, usually curved leaves (except in drier open nutrient-poor sites), the usual lack of abundant matting of rhizoids, and the occurrence of single sporophytes on each shoot.
Similar Species: From *D. majus*, *D. scoparium* is readily distinguished on the basis of the multiple sporophytes in the former. When vegetative, these species are difficult to separate, even on microscopic characters. From species of *Campylopus* and *Paraleucobryum* the very broad midrib in these two genera will separate *D. scoparium* with its narrower midrib. In these genera the midrib occupies ⅓ or more of the leaf width. *Comments*: Sometimes called "stork's bill moss" in reference to the strongly snouted operculum, or "broom moss" based on the frequently curved swept-to-one-side appearance of the leaves.

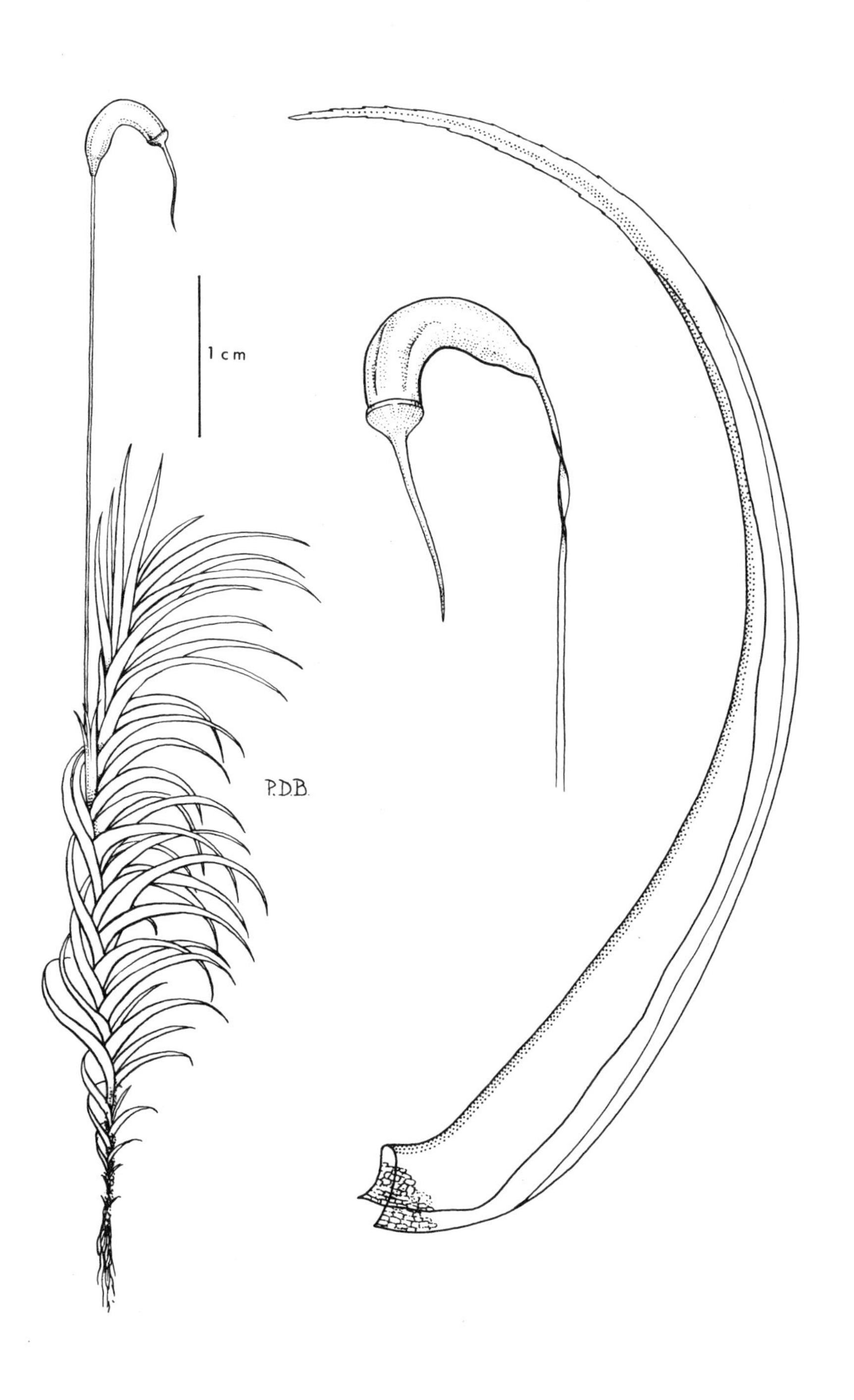

Dicranum scoparium

Dicranum tauricum Sap.

Name: Species name from the Taurus Mountains, Turkey, where the specimen was collected.
Habit: Forming turfs or tufts of light to dark green, rather rigid plants with brittle leaf tips.
Habitat: Most commonly on rotten logs in coniferous forest but also on living trees, tree stumps, and rarely on cliff crevices, from sea level to subalpine forests.
Reproduction: Sporophytes infrequent near the coast but rather common in interior humid forests. The broken-off leaf tips probably serve in propagation.
World Distribution: Widespread, but never frequent in Europe; widespread in western North America from southeastern Alaska to middle California along the Pacific coast and extending eastward to northern Colorado and South Dakota.
B.C. Distribution: Map 36, page 337.
Distinguishing Features: The rigid, straight leaves with very brittle tips that are often broken off, combined with the usual forested sites where this moss occurs are very diagnostic. The erect sporophytes are also useful when present.
Similar Species: *D. fragilifolium* also has brittle leaf tips but the sporangia are curved and grooved when mature rather than erect and smooth, as in *D. tauricum*. Material lacking sporophytes cannot be distinguished from *D. tauricum* on hand lens characters. *Tortella fragilis* also has brittle leaf tips but has yellow-green, dull leaves and is found on calcareous rock rather than wood.

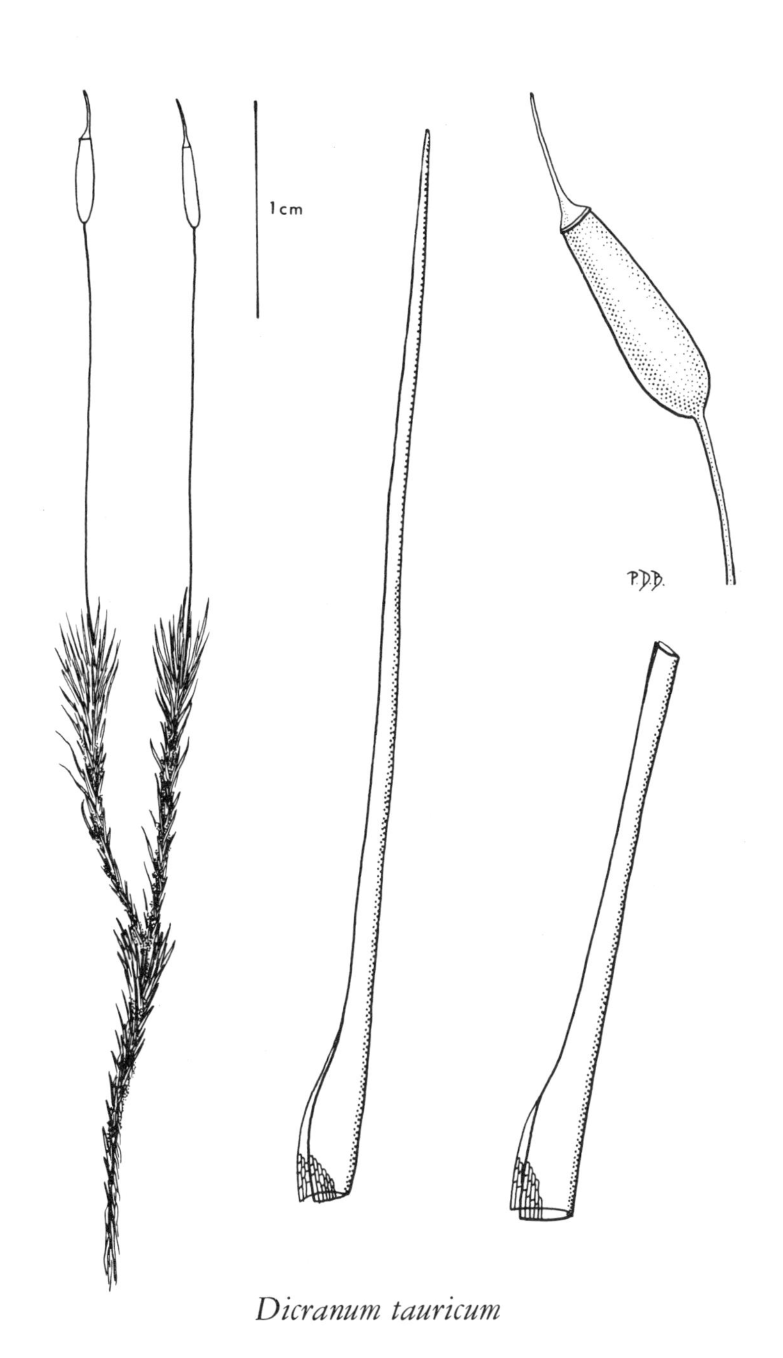

Dicranum tauricum

Didymodon insulanus (DeNot.) M. Hill

Name: Genus name derived from the twin division of the peristome teeth. The species name pertaining to islands, a possible reference to the source of the type specimen.
Habit: Forming dark green to brownish to dark, red-brown turfs. Leaves often much twisted and contorted when dry.
Habitat: On mineral soil, rock and occasionally up tree bases. Frequent near the coast. In urban areas in gardens, on sidewalk margins and stone or concrete walls. Usually in sunny sites, mainly at lower elevations.
Reproduction: Sporophytes frequent in spring with both sporangium and seta red-brown and the twisted peristome teeth pinkish-brown to red-brown.
World Distribution: On the western coasts of the Northern Hemisphere southward to areas of Mediterranean climate in North Africa and western Asia. In North America, frequent along the Pacific coast from southwestern Alaska southward to Baja California and inland to the Rocky Mountains.
B.C. Distribution: Map 37, page 337.
Distinguishing Features: When without sporophytes, this is a troublesome species, resembling several mosses of very similar form. The very twisted leaves and associated dark colour are helpful, but not reliable. The sporophytes, when associated with a gametophyte of this colour and form, are usually enough to distinguish *D. insulanus*.
Similar Species: *D. occidentalis* is very similar but the sporangia lack peristome teeth; without sporophytes these two species are difficult to distinguish on hand-lens features, but specimens of *D. occidentalis*, when dry, have the leaves tightly appressed to the stem and not markedly contorted; those of *D. insulanus* are strongly divergent and markedly contorted. *Ceratodon purpureus*, when without sporophytes, is superficially similar but *Ceratodon* usually has sporangia and these are grooved, usually glossy, and inclined when mature, rather than smooth, not glossy, and erect as in the *Didymodon*; peristome teeth of *Ceratodon* are not twisted. *Bryoerythrophyllum recurvirostrum* is usually smaller and the peristome teeth are not twisted.
Comments: *D. insulanus* is sometimes treated under the name *Barbula cylindrica* (Tayl.) Schimp.

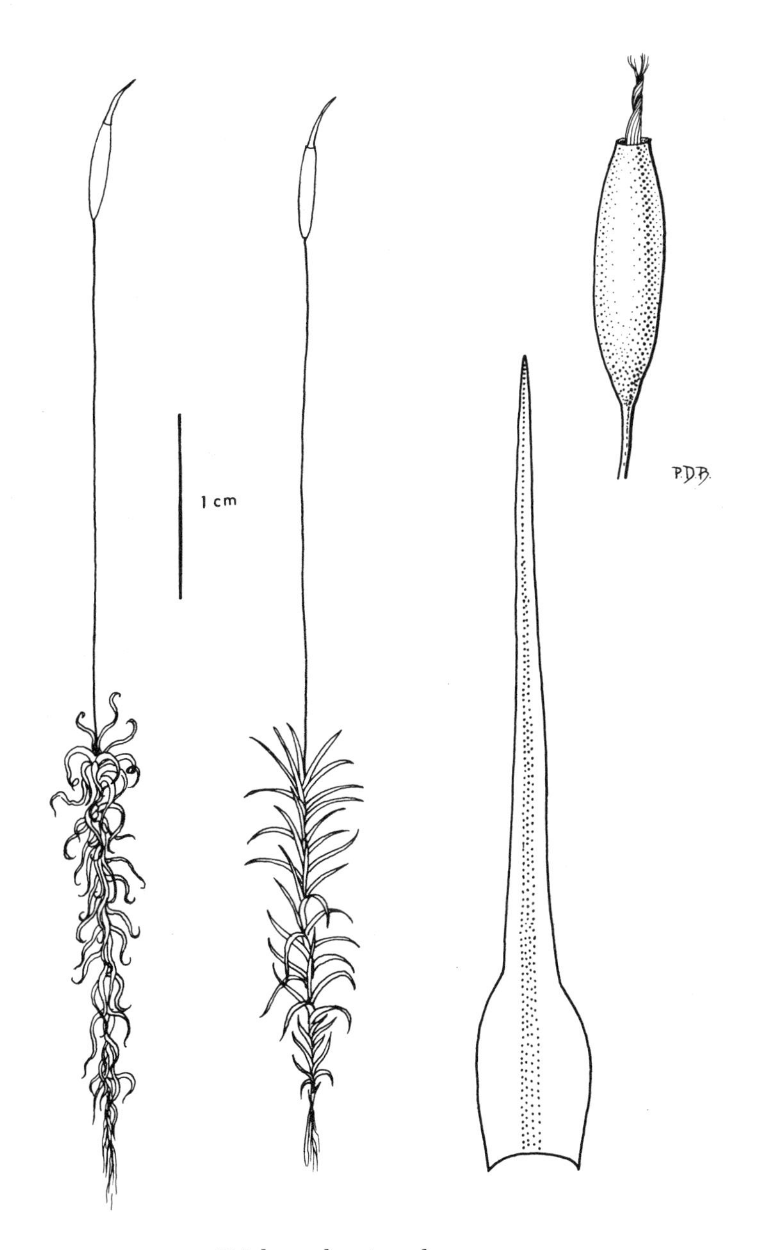

Didymodon insulanus

Distichium capillaceum (Hedw.) B.S.G.

Name: Genus name meaning two rows, in reference to the leaf arrangement. Species name denoting hair-like, in reference to the slender leaves.
Habit: Forming dense turfs of glossy, dark green to light green, unbranched, somewhat flattened, erect shoots.
Habitat: Usually in cliff crevices, particularly on calcium-rich substrata, but also terrestrial, from sea level to alpine elevations.
Reproduction: Sporophytes frequent and maturing in summer, light brown.
World Distribution: Circumpolar in the Northern Hemisphere, also extending southward, especially in the mountains, in the Southern Hemisphere; in North America extending southward to New York State in the east and to Arizona in the west.
B.C. Distribution: Map 38, page 338.
Distinguishing Features: The erect, strongly flattened, leafy plants with glossy, overlapping, sheathing leaf-bases give each shoot the distinctive superficial appearance of a miniature ear of barley.
Similar Species: *D. inclinatum* differs in the usually shorter plants that have a short, stout inclined sporangium (compared to the erect, narrowly cylindric sporangium of *D. capillaceum*). Some specimens of *Ditrichum* resemble this species but the leaves lack the flattened, sheathing bases.

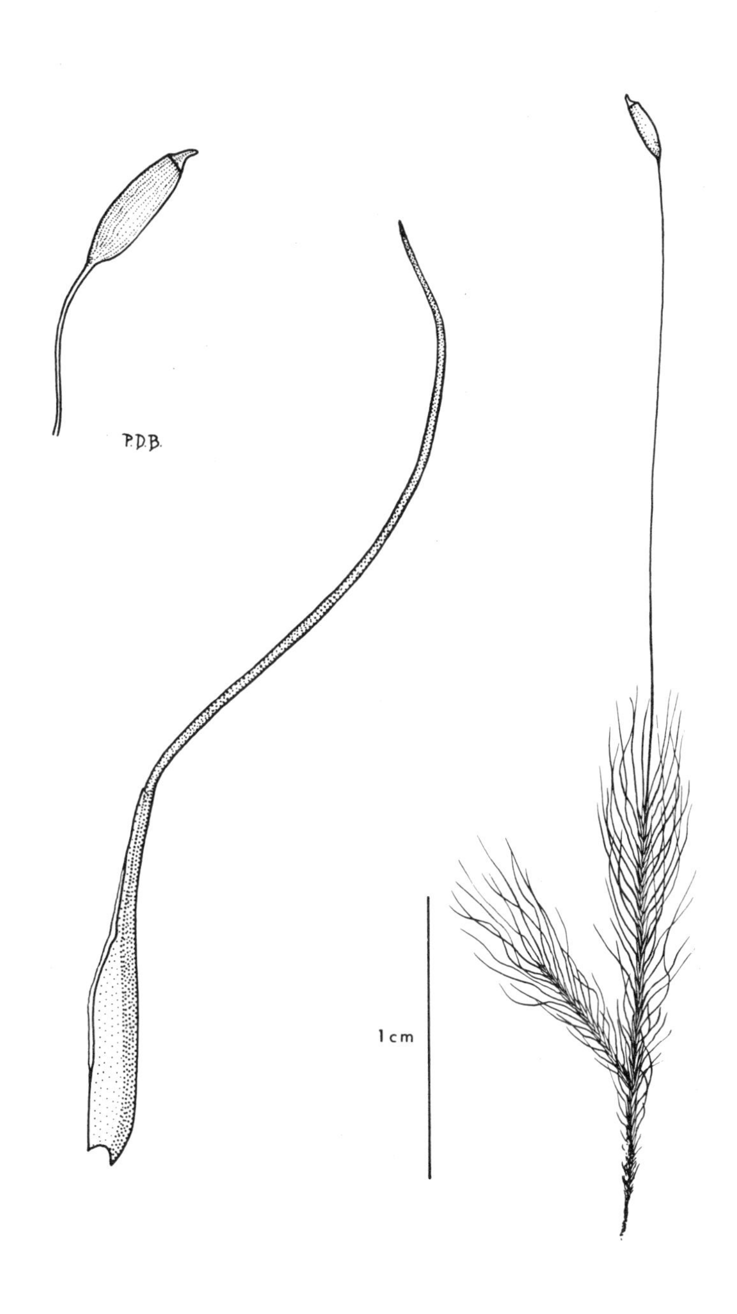

Distichium capillaceum

Ditrichum flexicaule (Schwaegr.) Hampe

Name: Genus name refers to the peristome teeth that are split into two hair-like divisions. Species name refers to the long flexuose stems noted in the specimen upon which the name was based.
Habit: Forming short turfs of erect, dark green to light green, unbranched shoots that often bear brittle branches near the apex of vegetative plants.
Habitat: On mineral or humic soil of open, often disturbed sites, sometimes in cliff crevices or tundra.
Reproduction: Sporophytes frequent in summer but the vegetative plants produce brittle apical branches that probably serve in propagation.
World Distribution: Circumpolar in the Northern Hemisphere. In North America extending southward in the east to Michigan; in the west to Colorado.
B.C. Distribution: Map 39, page 338.
Distinguishing Features: *D. flexicaule* is not readily separable from other species of *Ditrichum* based on non-microscopic features. Plants that bear brittle, stiffly erect apical branches with reduced leaves give the plants an appearance unique to this species. These forms are more frequent in open or disturbed sites. In tundra, the plants form dense tufts that become blackened within the tuft.
Similar Species: Most species of *Ditrichum*, at least in some variants, can be confused with each other, even with microscopic features. See also notes under *D. capillaceum*.
Comments: A closely related species, *D. crispatissimum*, is consistently without sporophytes, forms tall, silky turfs (to 100 mm tall, and with leaves to 10 mm long) especially on humid, calcareous cliffs and is often considered synonymous. *D. crispatissimum* is frequent, especially in southern B.C.

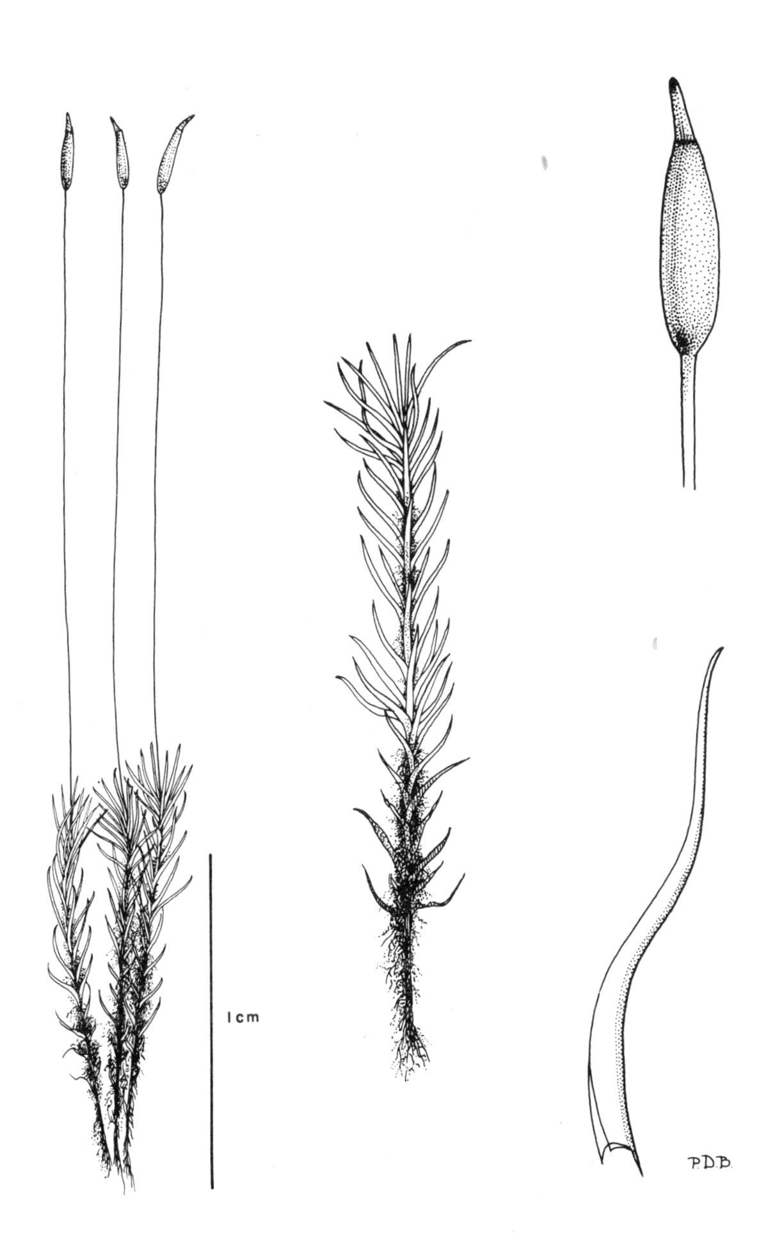

Ditrichum flexicaule

Ditrichum heteromallum (Hedw.) Britt.

Name: Species name referring to the leaf arrangement: pointing in many directions.
Habit: Forming dense, dark green turfs of erect plants with erect or sometimes secund leaves.
Habitat: Frequent on disturbed mineral soil of open sites, predominantly at lower elevations.
Reproduction: Sporophytes with reddish setae and dark brown sporangia, produced abundantly and maturing in late winter to early spring.
World Distribution: Western Europe and western North America. In North America from southeastern Alaska southward to California.
B.C. Distribution: Map 40, page 339.
Distinguishing Features: On disturbed sandy banks, this is the most likely species of the genus to be collected in southern British Columbia and southward. The winter-maturing sporophytes is also useful.
Similar Species: *D. ambiguum*, with the sporangia nearly twice as long as than those in *D. heteromallum*, is a less common species near the coast. From *D. schimperi*, which has yellowish setae, *D. heteromallum* is usually easily distinguished. *D. zonatum* is always sterile and restricted to subalpine cliff crevices or windswept outcrops near sea level in oceanic areas; these features are usually sufficient to separate *D. zonatum* and *D. heteromallum*. Material without sporophytes can be confused with several species (see notes under *Dicranella heteromalla*).

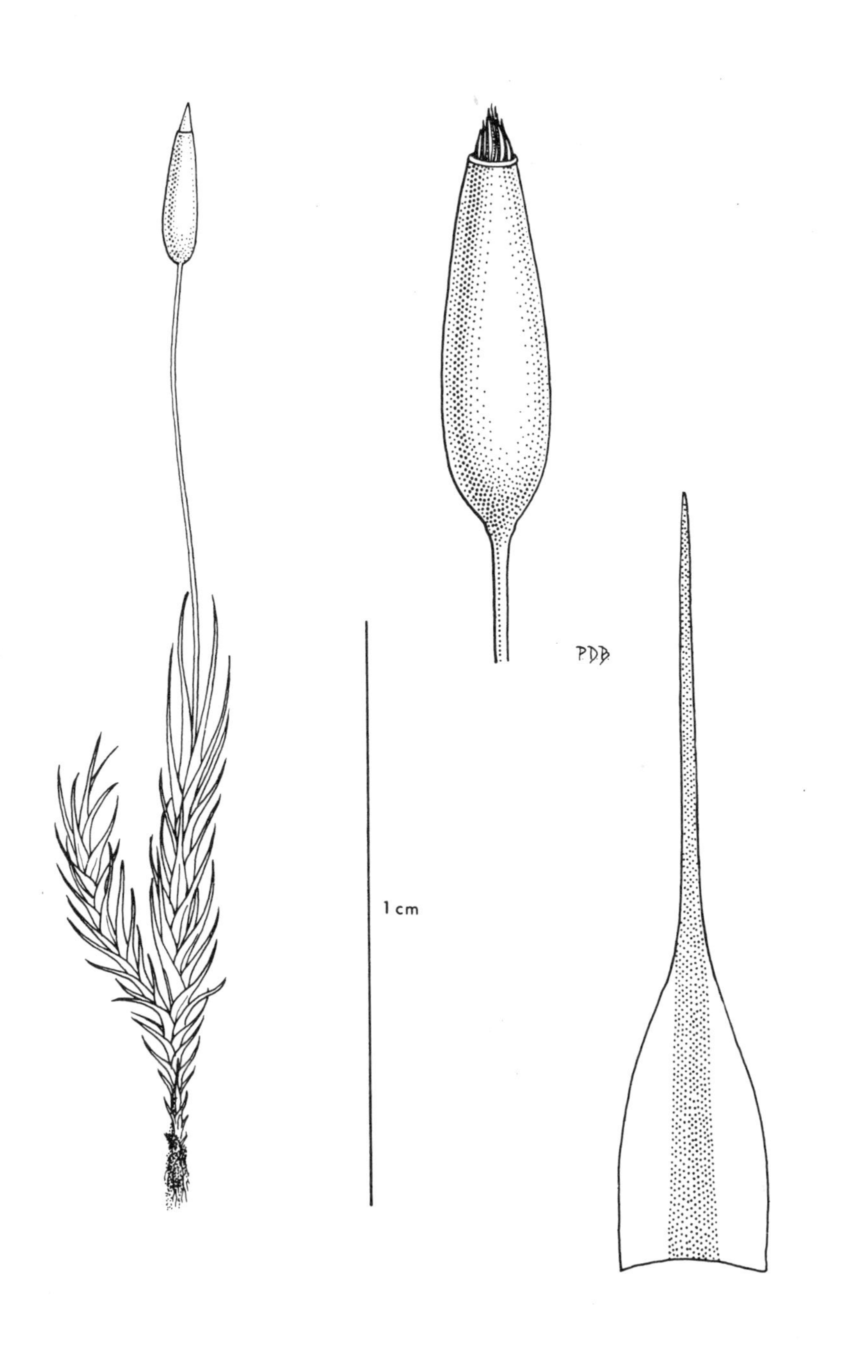

Ditrichum heteromallum

Drepanocladus exannulatus (B.S.G.) Warnst.

Name: Genus name referring to the curved branches, exaggerated by the curved leaves on the stems. Species name based on the absence of a ring of specialized cells (annulus) around the mouth of the sporangium.
Habit: Loosely to densely interwoven, reclining, dark green to purplish-brown mats of glossy plants.
Habitat: In swamps, peatland, seepage sites and submerged to floating in the quiet waters of pools, lake margins and backwaters of water-courses.
Reproduction: Sporophytes occasional, with red-brown setae and dark brown to light brown sporangia. Plants also probably readily fragmented.
World Distribution: Circumpolar in the Northern Hemisphere and scattered in the Southern Hemisphere. In North America extending southward in the east to New York State and Wisconsin; in the west to Arizona.
B.C. Distribution: Map 41, page 339.
Distinguishing Features: The wet habitat, combined with the well-defined group of alar cells and the curved leaves with a midrib are distinctive. It is the most common aquatic and wetland species of the genus in the Province.
Similar Species: *D. fluitans* grows in similar habitats, but is less common. It lacks the distinctive alar cells and the leaf apices are generally markedly toothed. *Cratoneuron filicinum*, which shows obvious differentiated alar cells, does not exhibit the strongly curved leaves; the plants also tends to be more obviously pinnately branched. *D. crassicostatus* also lacks the distinctive alar cells and tends to be confined to calcium-rich sites; also tends to be more rigid than the rather soft plants of *D. exannulatus*.

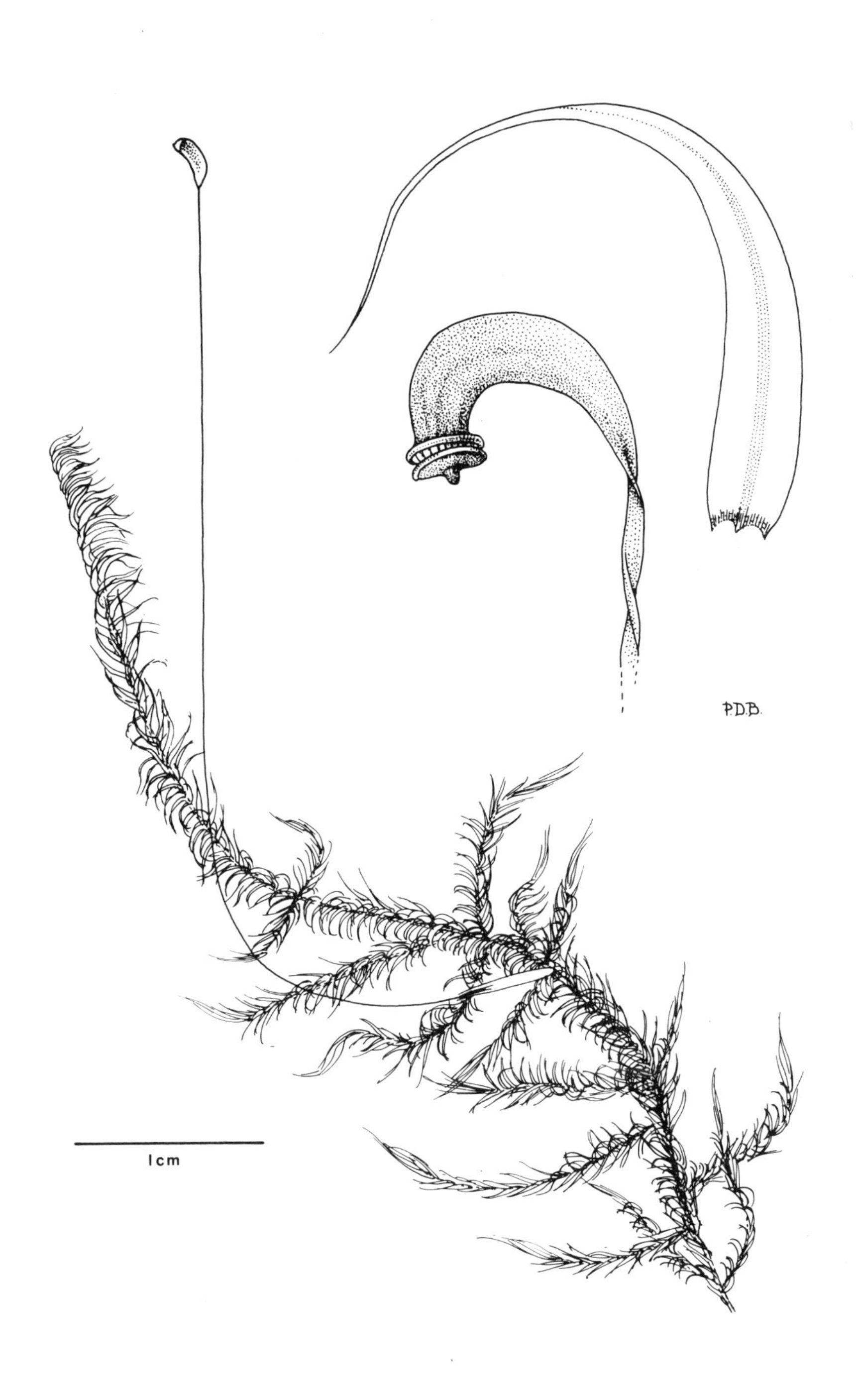

Drepanocladus exannulatus

Drepanocladus uncinatus (Hedw.) Warnst.

Name: Species name emphasizing the hooked or curved leaves.
Habit: Soft, light green mats, either loosely, or firmly, affixed to the substratum; varying from regularly pinnate to extremely irregular branching, from erect or suberect to reclining shoots.
Habitat: Extremely variable, but predominantly in well-drained sites; epiphytic, especially up tree bases, on rock, on humus, rotten logs and even sand. Most frequent in coniferous forests but also extending into tundra and to sunny sites near sea level.
Reproduction: Sporophytes frequent, maturing in spring to summer. Sporangia are erect in some (usually epiphytic) populations but are generally inclined and curved.
World Distribution: Extremely widespread in cooler climates at all latitudes. In eastern North America extending southward to New York State but, in the west, to South America, especially at higher elevations.
B.C. Distribution: Map 42, page 340.
Distinguishing Features: The well-drained habitat, combined with the single midrib in pleated, curved leaves, are usually enough to separate this species. Unfortunately some forms lack the pleats and can be puzzling. The pleats, even when present, can be confused as multiple midribs.
Similar Species: Species of *Hypnum* are superficially similar, but lack the single midrib. *Tomentypnum falcifolium* closely resembles this species but it is confined to calcareous wetland; the main stem is also usually densely clothed with red rhizoids. *Ptilium crista-castrensis* is more regularly and abundantly branched and the leaves lack the single midrib.
Comments: Sometimes treated as *Sanionia uncinata*.

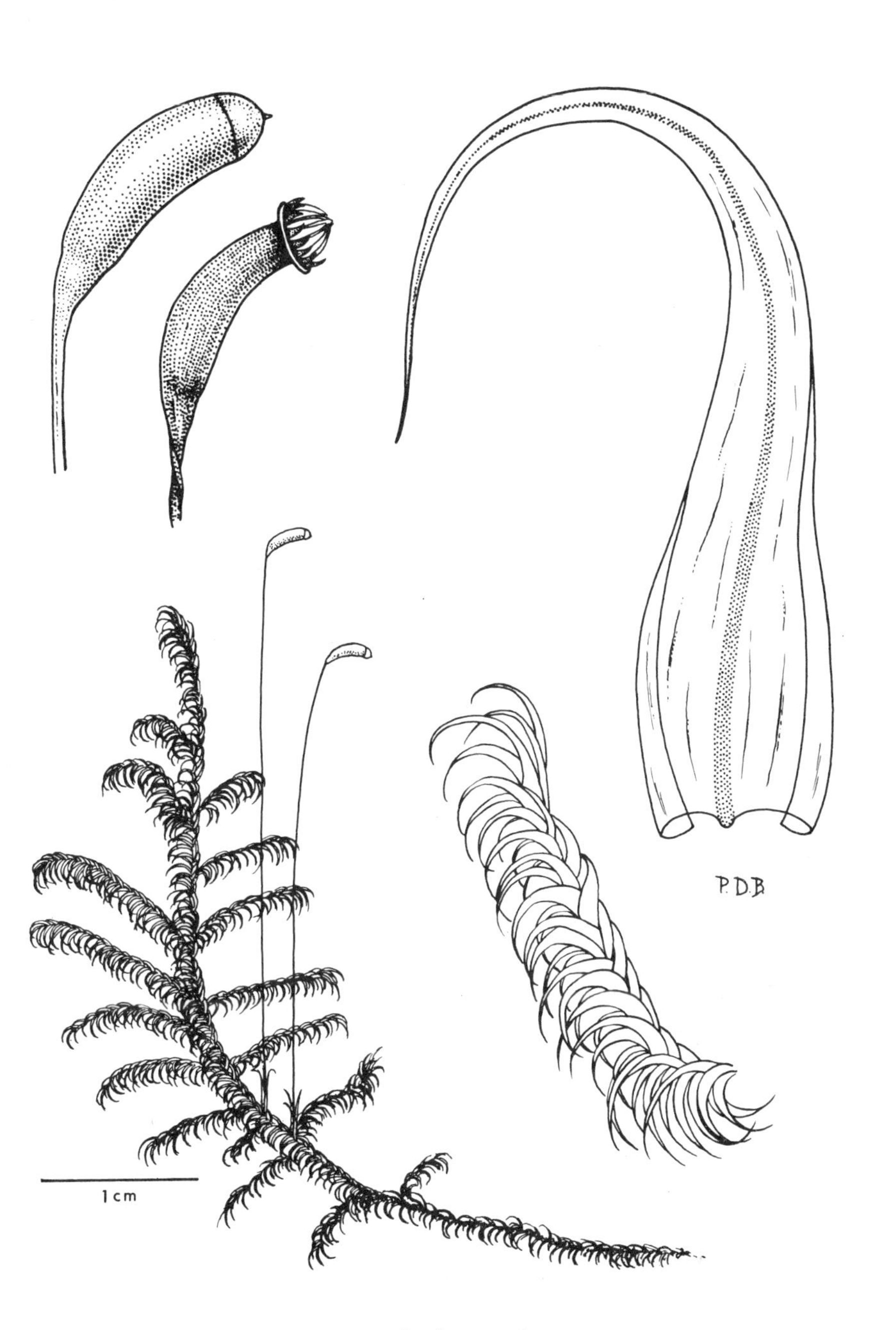

Drepanocladus uncinatus

Encalypta ciliata Hedw.

Name: Genus name meaning covered with a veil, in reference to the large cone-shaped calyptra. Species name based on the slender tip of the calyptra.
Habit: Forming loose turfs of bluish-green to yellowish-green, non-glossy, erect shoots.
Habitat: Usually on humus, often over rock, in somewhat shaded sites particularly on calcium-rich substrata but not confined to these, predominantly in forested habitats, but occasionally in tundra.
Reproduction: Sporophytes frequent, brown when mature, maturing in spring to summer.
World Distribution: Widespread in cooler climates of the Northern Hemisphere; also occurring in Africa and South America. In North America extending southward in the east to New York State and southern Wisconsin; in the west, almost continuously to Central and South America.
B.C. Distribution: Map 43, page 340.
Distinguishing Features: This is the only common species that has a fringed calyptra, obvious red peristome teeth and a smooth-walled sporangium. The usually shaded habitat is also diagnostic.
Similar Species: Several species of *Encalypta* have a fringed calyptra but, of these, *E. brevipes* lacks peristome teeth and has hair-pointed leaves, *E. brevicolla* has whitish, rather than reddish peristome teeth and hair-pointed leaves, *E. affinis* also has hair-pointed leaves and the peristome teeth are in two rows, and *E. alpina* lacks peristome teeth. *E. procera*, which commonly lacks sporophytes, has leaves that have a blunt apex. Also, sporophytes of *E. procera* have spirally oriented grooves.
Comments: Sometimes called "extinguisher moss" because of the resemblance of the calyptra to a candle snuffer. This is not a common moss and usually generates excitement when it is first seen because, when the calyptra is present, it is distinctive and highly attractive.

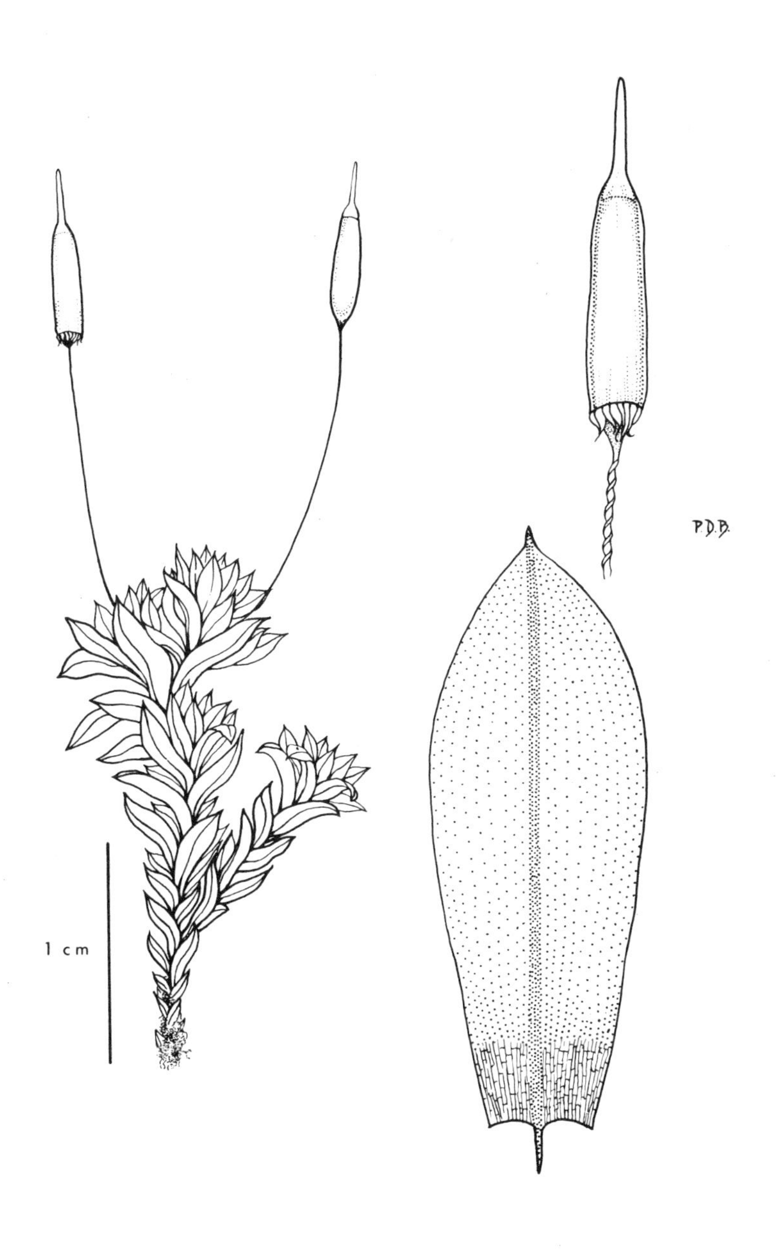

Encalypta ciliata

Encalypta rhaptocarpa Schwaegr.

Name: Species name denoting seamed, presumably in reference to the grooved sporangium.
Habit: Short, loose turfs of erect, yellow-green to bluish-green shoots with hair-pointed leaves.
Habitat: Usually open calcareous soil or cliff shelves and crevices from semi-arid grassland to alpine elevations.
Reproduction: Sporophytes common in spring to summer.
World Distribution: Widespread in arctic and temperate climates of the Northern Hemisphere. In North America extending southward to the northern Great Lakes area in the east and to northern Colorado in the west.
B.C. Distribution: Map 44, page 341.
Distinguishing Features: The conspicuous red-brown ribs on the sporangium wall, the presence of reddish peristome teeth, the unfringed, golden, opaque calyptra and the hair-pointed leaves are useful features.
Similar Species: Of species of *Encalypta* in which the sporangium is ribbed, *E. spathulata* differs in its absence of peristome teeth and in the pale, nearly transparent calyptra, *E. intermedia* also lacks peristome teeth and the leaves are generally without hair points, *E. procera* has obvious peristome teeth, the grooves on the sporangium are oblique, rather than vertical, and the leaves are blunt tipped. *E. vulgaris*, a relatively widespread species, has ungrooved sporangia and lacks peristome teeth.

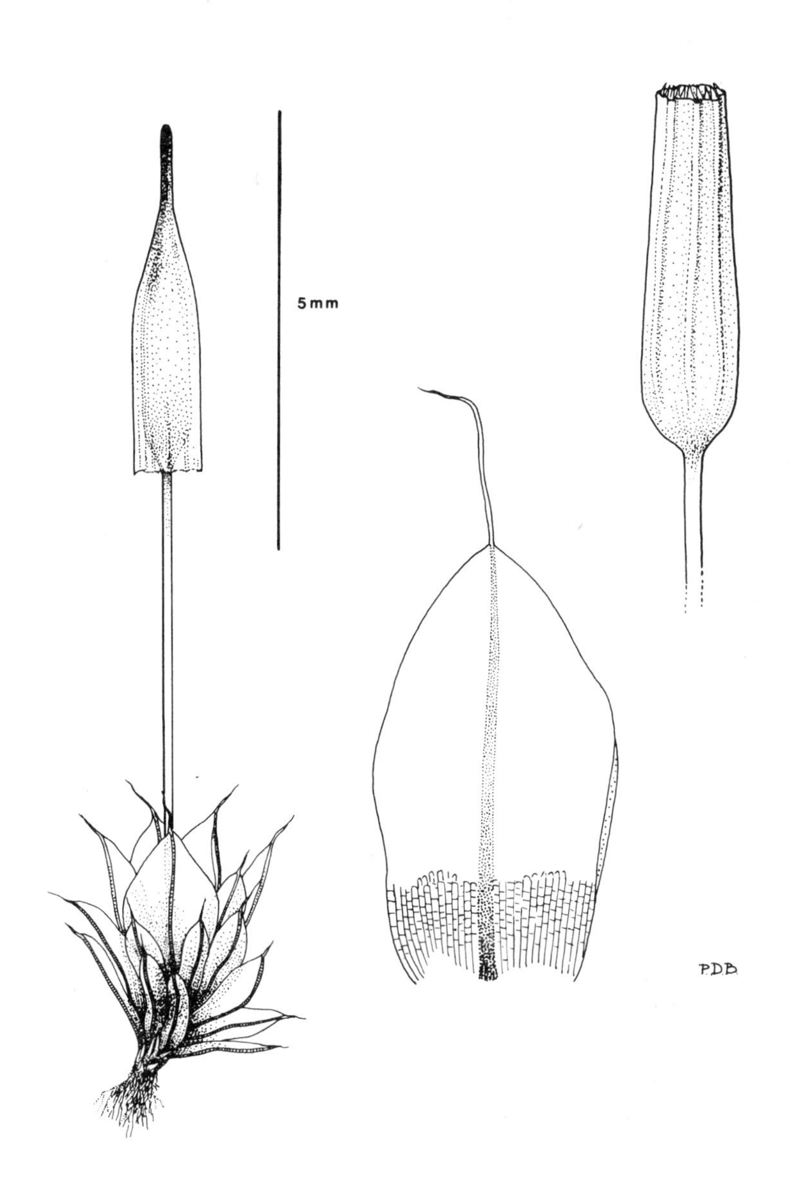

Encalypta rhaptocarpa

Eurhynchium pulchellum (Hedw.) Jenn.

Name: Genus name based on the conspicuous snout of the operculum. Species name indicating pretty and small, perhaps referring to the neat leaf arrangement on the shoots.
Habit: Forming loose to dense mats of reclining to suberect, light to dark green plants that vary considerably in density of branches and in size of the plants.
Habitat: Usually shaded calcareous mineral soil and rock surfaces but also on rotten wood and ascending tree bases; from near sea level to alpine elevations.
Reproduction: Sporophytes relatively frequent, maturing in spring to summer, dark to red-brown.
World Distribution: Widely distributed in the Northern Hemisphere but also extending southward to northern South America. In North America widely distributed throughout.
B.C. Distribution: Map 45, page 341.
Distinguishing Features: Of the creeping, branching mosses, the blunt-tipped leaves with toothed margins and a single midrib are characteristic.
Similar Species: The species of *Kindbergia* have sharply pointed leaves that, on the main stem, have the basal part somewhat incurved to the stem and the upper portion curving outward, while in *Eurhynchium* the whole leaf diverges. The blunt leaves distinguish *E. pulchellum* from *Brachythecium*, although some specimens of *B. fendleri* are troublesome to distinguish from small specimens of *E. pulchellum*, but the leaves of the *Brachythecium* tend to be sharp pointed and margins are not toothed.

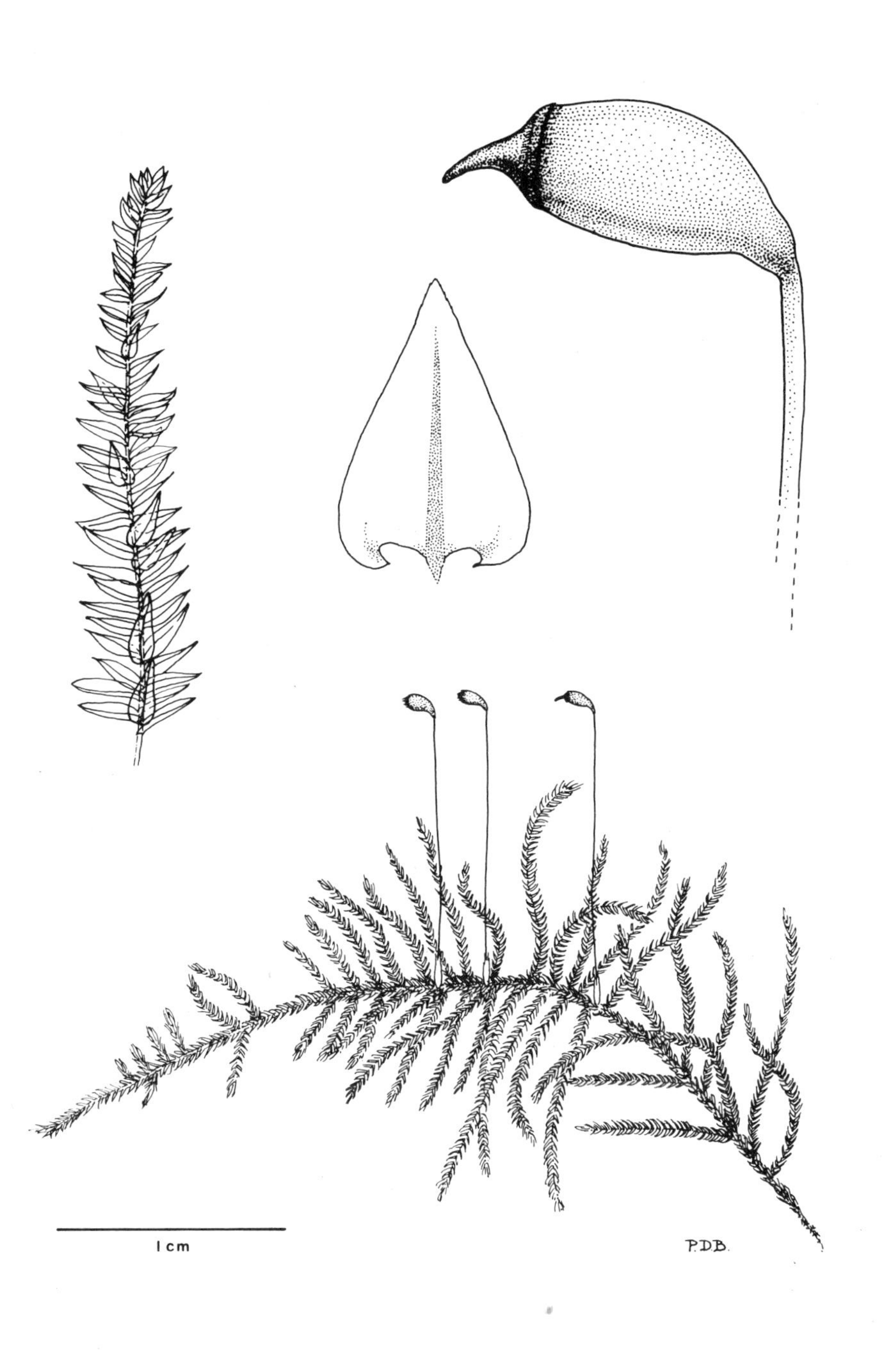

Eurhynchium pulchellum

Fissidens adianthoides Hedw.

Name: Genus name meaning split tooth, referring to the forked peristome teeth. Species name based on the fancied resemblance of the moss to the fern *Adiantum*.
Habit: Forming short to tall turfs of erect to suberect conspicuously flattened, dark green to light green plants.
Habitat: Usually somewhat shaded sites: damp cliffs, earth of lakeshores and streams, swampy areas, sometimes on logs in wooded floodplains. Predominantly in forested regions at lower elevations.
Reproduction: Sporophytes dark brown when ripe, peristome bright red-brown, occasional but, when present, abundant in that population, maturing in spring.
World Distribution: Widely distributed in the forested portion of the Northern Hemisphere but extending to arctic, alpine and prairie regions, usually in more sheltered sites. Widely distributed in North America.
B.C. Distribution: Map 46, page 342.
Distinguishing Features: The curious leaves with the unique flap immediately indicate this genus. The species shows pronounced teeth on the leaf margins and the plants tend to be soft with leaf points that curl downward when dry.
Similar Species: *F. osmundioides* is similar, but is half the size and usually forms very dense turfs; *F. grandifrons* is of similar size but the plants are aquatic and essentially opaque and very dark green while *F. adianthoides* leaves are translucent and the plants are not aquatic.

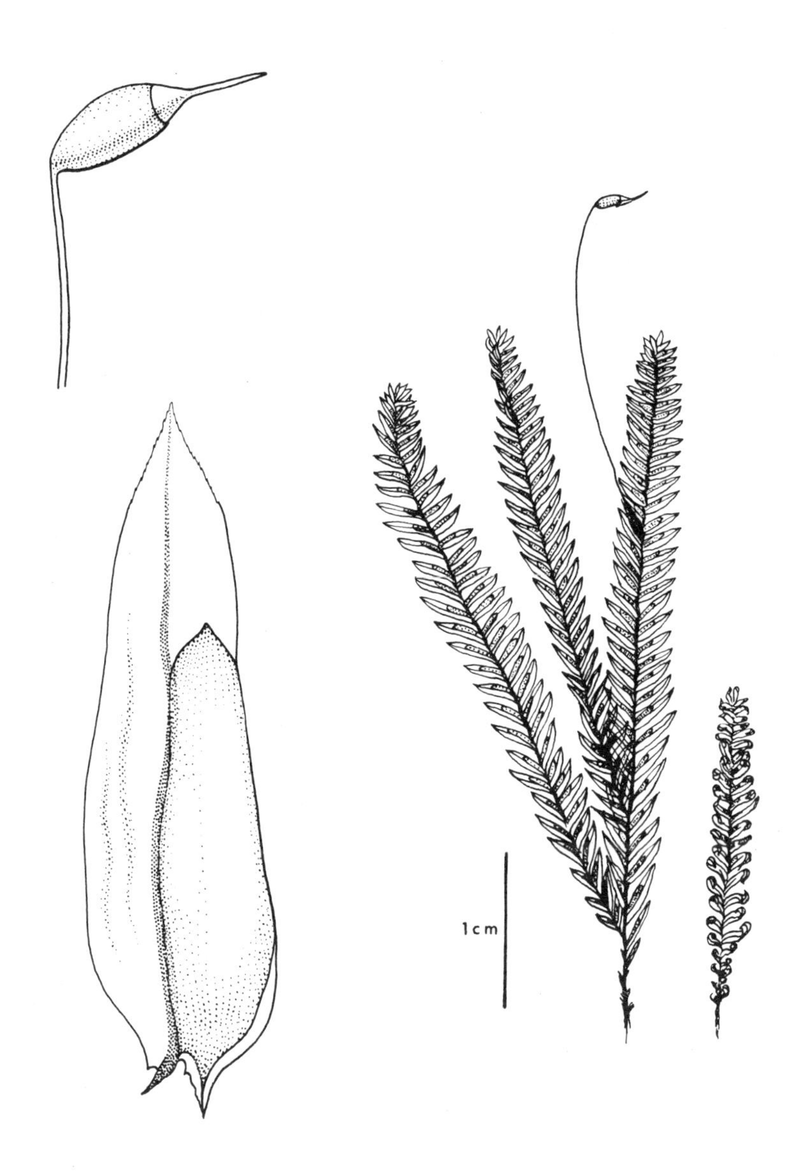

Fissidens adianthoides

Fissidens limbatus Sull.

Name: Species name meaning bordered, based on a microscopic difference in the border, compared to the widespread *Fissidens bryoides*.
Habit: Loose turfs or scattered individuals of unbranched, light green plants.
Habitat: Disturbed banks of usually shaded mineral soil in forests, or soil on shaded ledges of outcrops.
Reproduction: Sporophytes frequent in spring; the bright red peristome teeth contrast elegantly with the pale brown sporangium.
World Distribution: Apparently confined to the western part of North America including northwestern British Columbia to Mexico; also noted from northern South America. In North America it extends eastward to western Alberta, Montana and Utah.
B.C. Distribution: Map 47, page 342.
Distinguishing Features: The maturation of inclined sporophytes in spring, coupled with the tiny, flattened, leafy shoots are distinctive; the clear leaf margin that lacks teeth is also useful.
Similar Species: *F. bryoides* has erect sporophytes that mature in fall to winter, rather than spring. *F. ventricosus* is confined to water-splashed rocks and the sporangium is erect from a short seta.

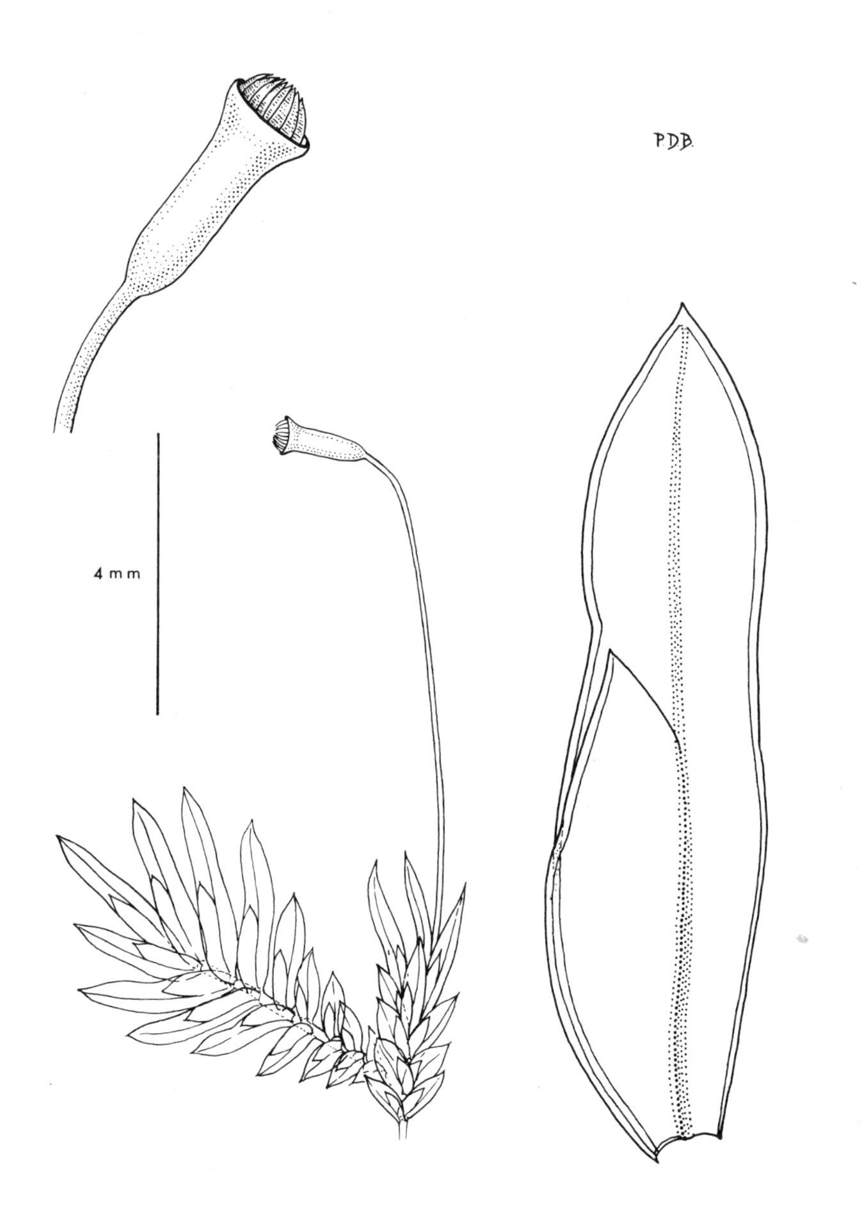

Fissidens limbatus

tinalis antipyretica Hedw.

Name: Genus name denoting fountains; the habitat of all species is aquatic. Species name meaning against fire, reflecting the former use of the moss as a caulking between the chimney and the walls around the chimney.

Habit: Very long (up to 1 m) reclining dark green plants in which the leaves tend to be strongly three-ranked. Predominantly of low elevations but extending to the subalpine.

Habitat: A submerged aquatic, either attached to rocks or logs in moving water or floating loose in the stagnant water of pools and backwaters, commonly in somewhat shaded sites.

Reproduction: Sporophytes occasional, immersed, maturing in spring to autumn. Also disseminated through fragmentation of the leafy plants.

World Distribution: Circumboreal but extending to the Southern Hemisphere in Africa. In North America extending southward to Pennsylvania in the east and to Arizona in the west.

B.C. Distribution: Map 48, page 343.

Distinguishing Features: The aquatic habitat, the distinctly 3-ranked, keeled, straight leaves and the large size are usually enough to separate this species.

Similar Species: *F. howellii* is extremely similar to *F. antipyretica* but the leaves tend to be very narrow, tapering gradually to the apex. In *F. antipyretica*, the leaves are broadly boat-shaped and curve (on the keel) to the insertion. *F. neomexicana* has slender shoots with rigid overlapping leaves. Aquatic forms of *Calliergonella cuspidata* do not have keeled leaves and alar cells are clearly defined.

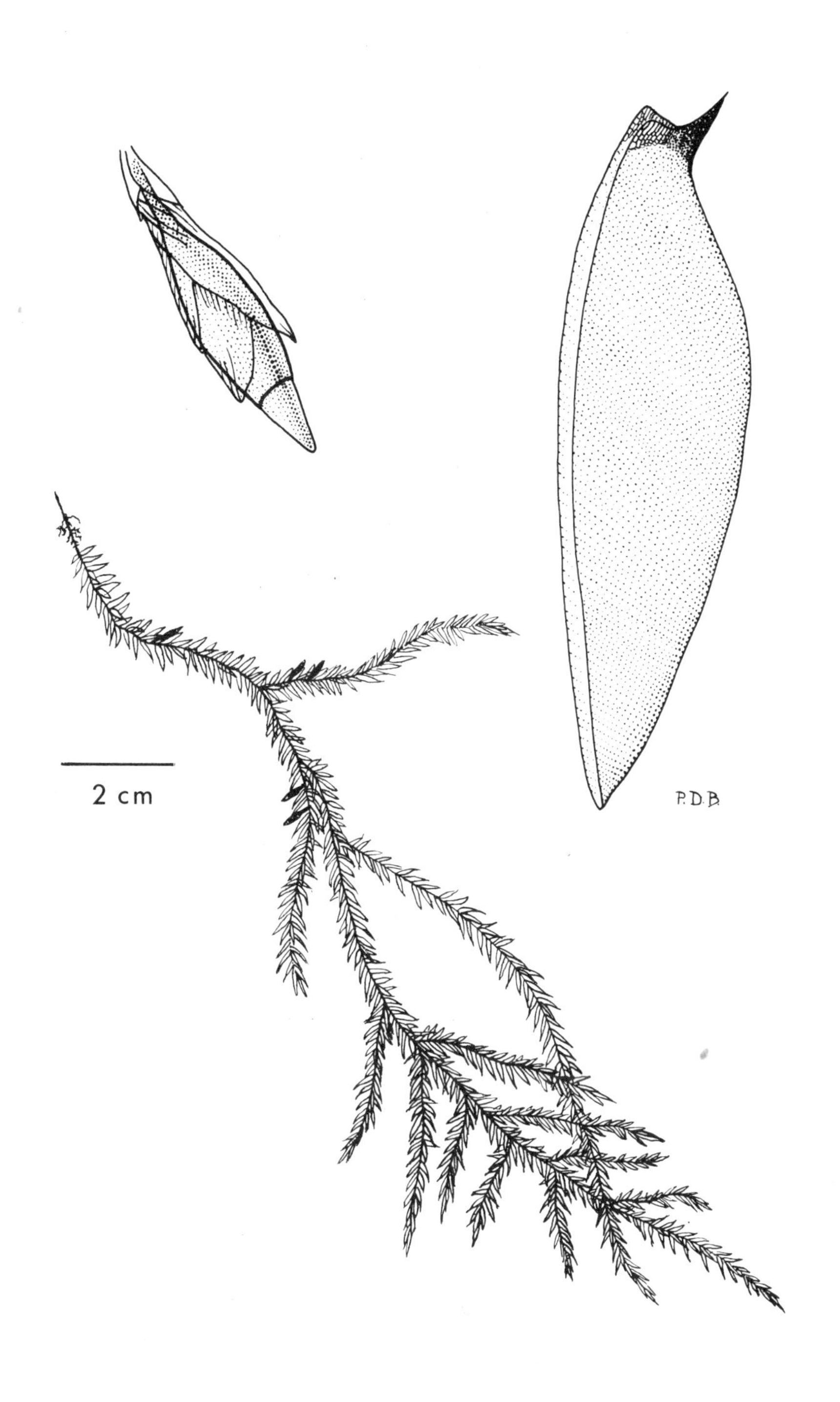

Fontinalis antipyretica

Funaria hygrometrica Hedw.

Name: Genus name referring to the cord-like twisting of the drying seta (hence "cord moss" as a popular name). Species name also denoting the twisting and curling of the seta of this species in response to changing moisture in the air.
Habit: Short loose turfs of yellow-green, glossy, often bulb-like plants.
Habitat: On bare disturbed soil, especially frequent on burned-over sites but also in gardens and neglected agricultural land. A common greenhouse weed.
Reproduction: Sporophytes common in spring to summer.
World Distribution: Worldwide, probably most frequent in areas of human habitation but also in natural vegetation, mainly at lower elevations.
B.C. Distribution: Map 49, page 343.
Distinguishing Features: The pale yellow, extremely curved seta, when young and the very inflated lower part of the long-snouted calyptra, added to the sheathing leaves at the seta base, provide characters that quickly identify this moss. In mature sporangia the grooves are distinctive.
Similar Species: *F. muhlenbergii* lacks the curved seta and grows on somewhat shaded, usually fine-textured soil.
Comments: The consistent appearance of this moss on the sites of abandoned bonfires and after forest fires is intriguing.

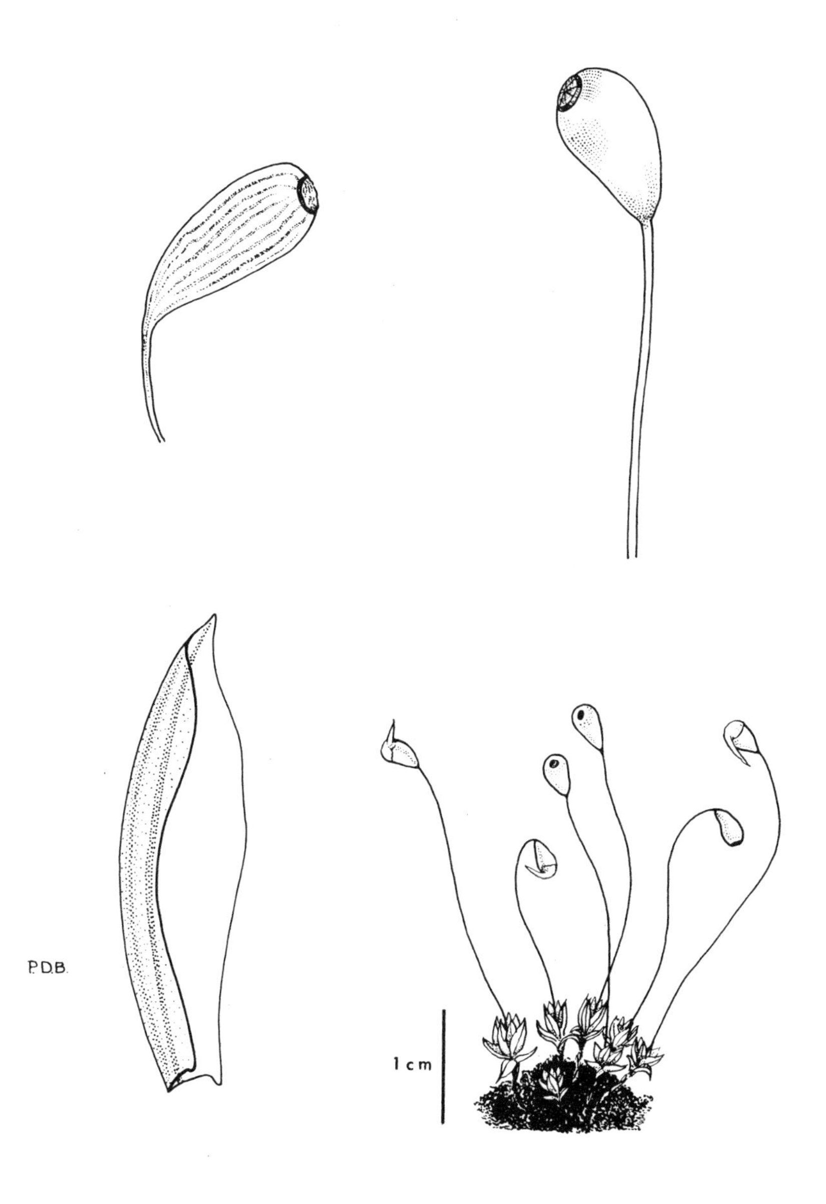

Funaria hygrometrica

mia pulvinata (Hedw.) Sm. *ex* Sm. & -rby

Name: Genus named in honour of F.J.C. Grimm, a German physician and botanist. Species name describing the cushion-like habit.
Habit: Forms rounded, grey-green cushions in which both the dried leaves and white hair points contribute to the grey-green effect.
Habitat: Commonly on rocks and concrete surfaces in urban areas; also frequent in drier climates, thus its frequency in the southwestern portion of the province and its presence in the semi-arid interior where it occurs on rock.
Reproduction: Sporophytes common in late winter to early spring; when immature the seta arches and the sporangium is buried, mouth downward, among the leaves. Sporangia are grooved when mature and dry.
World Distribution: Widespread, but scattered, in the temperate portions of the Northern Hemisphere; also in temperate Australasia. In North America, rare in the east from Ontario south to Missouri; in the west frequent from southern British Columbia to California and New Mexico.
B.C. Distribution: Map 50, page 344.
Distinguishing Features: The rounded, greyish cushions and the curved seta that arches so that sporangia are often buried, mouth downward in the cushion, plus the grooved sporangia, are usually enough to distinguish this species.
Similar Species: See *Coscinodon calyptratus*. From similar species of *Grimmia*, *G. pulvinata* is not easy to distinguish. From *G. trichophylla*, the tufted habit, when present, is usually distinctive; unfortunately this feature is not consistent. In *G. trichophylla*, the hair point tapers gradually to the body of the leaf and forms less than ¼ of the leaf length; in *G. pulvinata* the hair point tapers abruptly and is ⅓–½ the length of the leaf. Several high-elevation species of *Grimmia* are similar but their setae are not curved nor are the mature sporangia grooved. *Tortula muralis* grows in a similar habitat but plants are usually not tufted, sporangia are erect, long-cylindric and peristome teeth are spirally twisted.

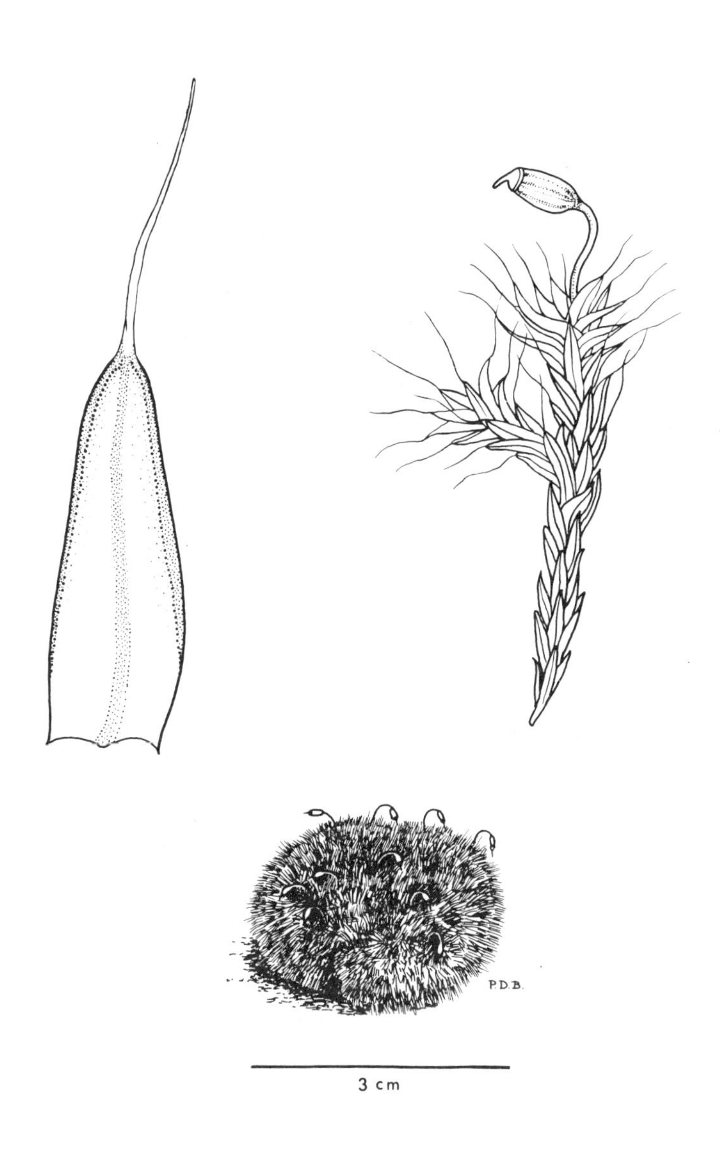

Grimmia pulvinata

Hedwigia ciliata (Hedw.) P. Beauv.

Name: Genus named in honour of J. Hedwig whose original work formed the foundation of the understanding of mosses. Species name referring to the remarkable hairs that fringe the perichaetial leaves that surround the sporangium.
Habit: Forming mats or tufts, firmly to loosely attached and irregularly branched.
Habitat: Usually sunny, siliceous rock surfaces in well-drained sites.
Reproduction: Sporophytes abundant, golden brown, immersed, usually apparent when plants are moist, appearing in spring to summer.
World Distribution: Almost cosmopolitan, found on all continents except Antarctica but rare in frigid climates.
B.C. Distribution: Map 51, page 344.
Distinguishing Features: The much-branched plants that are yellowish-green when humid, nearly white when dry, the white leaf tips, the lack of midrib in the leaves, the ciliate leaves surrounding the immersed sporangium, and the lack of peristome teeth make this moss distinctive.
Similar Species: Some species of *Racomitrium* are similar in colour and habitat but all have a midrib in the leaf, sporophytes with a long seta and conspicuous peristome teeth. From *Schistidium*, the *Hedwigia* differs in colour; *Schistidium* also usually has peristome teeth.

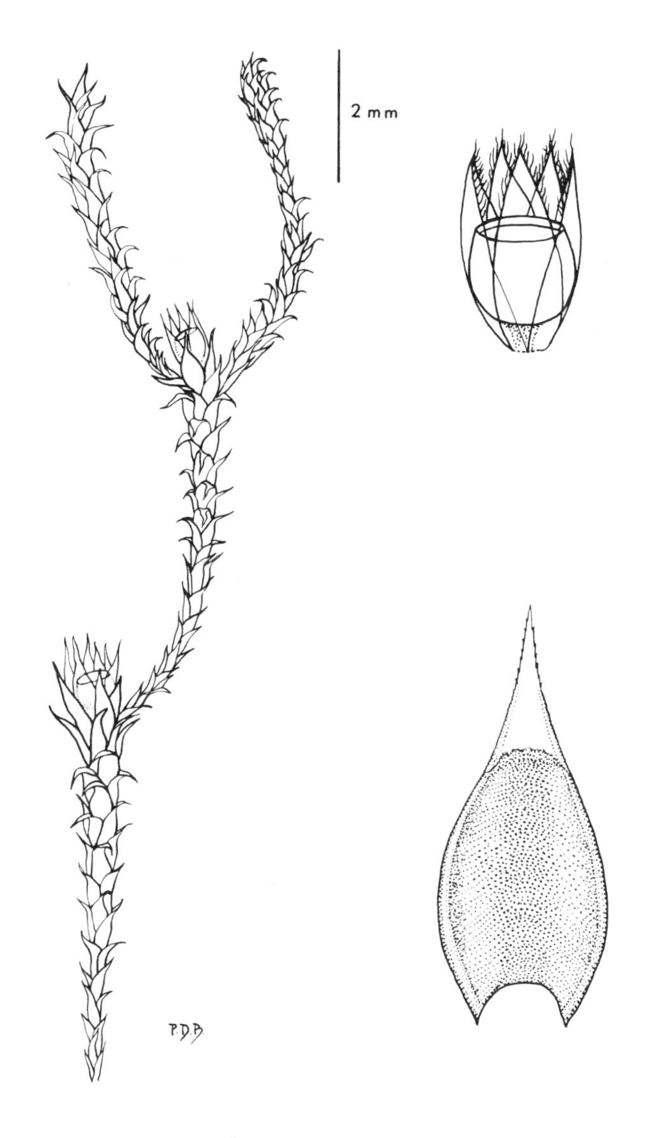

Hedwigia ciliata

Heterocladium macounii Best.

Name: Genus name refers to the differentiated branch and indicates the difference in form of the leaves in the main stem and the branches. Species named in honour of J. Macoun, a Canadian botanist of the late 19th century.
Habit: Forming tight to loose, dark green to light green mats that vary considerably in abundance of branches; those of deeply shaded and well-drained surfaces tending to have fewer and more elongate shoots and branches than those on horizontal damp surfaces.
Habitat: Usually on rock in shaded areas, especially in woodland and near watercourses but also on tree trunks in moist climatic areas.
Reproduction: Sporophytes occasional, maturing in late winter and spring.
World Distribution: Restricted mainly to western North America from southeastern Alaska to California and eastward to Idaho; reappearing, but rarely, in the southern Appalachian Mountains of eastern North America.
B.C. Distribution: Map 52, page 345.
Distinguishing Features: The slender, usually regularly branched shoots with leaves bearing a single midrib and lacking a hair point, the non-glossy appearance and the somewhat inclined, short, sporangium are useful features.
Similar Species: From *H. dimorphum*, *H. macomii* is not easily distinguished on hand lens characters, although *H. dimorphum* tends to be more regularly branched and is not firmly affixed to its substratum (frequently terrestrial while *H. macounii* is usually on rock). From *Claopodium* it differs in the very contorted dried leaves of that genus. *Leskea polycarpa* usually has erect-cylindric sporangia while in *Heterocladium* they are short and inclined. Species of *Leskeella*, although similar, tend to be smaller and are confined to calcareous rock while *Heterocladium* is frequent on non-calcareous rock.
Comments: *H. heteropteroides* is synonymous with *H. macounii*.

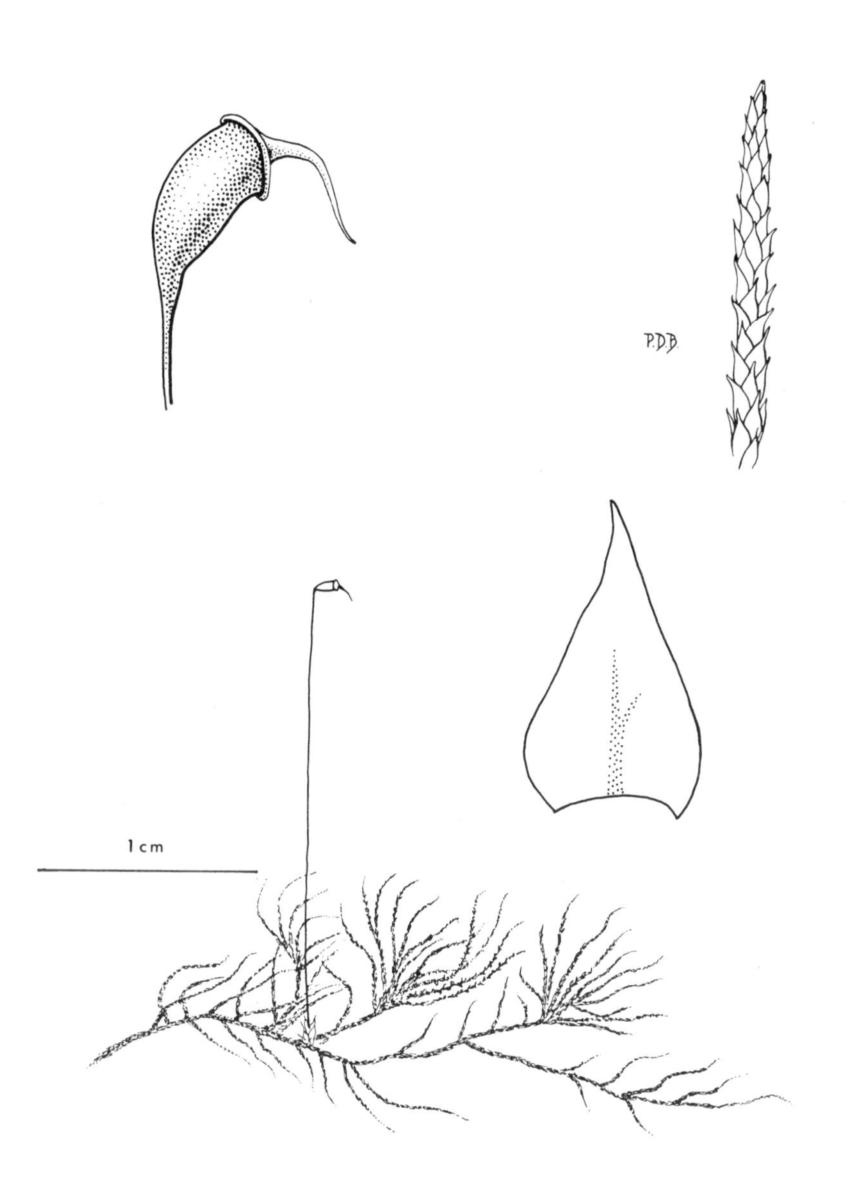

Heterocladium macounii

Homalothecium aeneum (Mitt.) Lawt.

Name: Genus name referring to straight sporangia that characterize some species. Species name meaning bronze, in reference to the colour that sometimes characterizes the plants.
Habit: Forming mats of interwoven, glossy, yellowish to golden-green, irregularly to regularly branched shoots, usually loosely affixed to substratum. Plants curling upward from substratum when dry.
Habitat: In sunny to somewhat shaded sites on cliffs, boulders and soil, most frequent in the semi-arid regions but also in the humid interior forests.
Reproduction: Sporophytes occasional, maturing in spring to early summer, dark red-brown.
World Distribution: Restricted to western North America, mainly west of the Rocky Mountains, from southeastern Alaska southward to California and eastward to Wyoming.
B.C. Distribution: Map 53, page 345.
Distinguishing Features: The yellow-green, glossy plants, irregular branching, very pleated leaves and the curling upward of dried shoots from the substratum are useful characteristics. It is the most frequent species of the genus in the interior of British Columbia.
Similar Species: *H. aenum* is difficult to distinguish from *H. arenarium* and *H. nevadense* on non-microscopic features. When bearing sporophytes, the erect to suberect sporangia of *H. nevadense* is distinctive; *H. nevadense* is also separated by habitat being, predominantly, a species of perpendicular rock surfaces. *H. arenarium* is restricted to the coast and is infrequent. *H. fulgescens* tends to be larger and is predominantly epiphytic (especially on broad-leafed maple). The sporangia of *H. fulgescens* are suberect and long-cylindric, thus differing from the short-cylindric inclined sporangia of *H. aeneum*. *Brachythecium albicans* superficially resembles *H. aeneum*, but the dried shoots never curl upward and are generally pale yellow-green; it is also generally terrestrial.

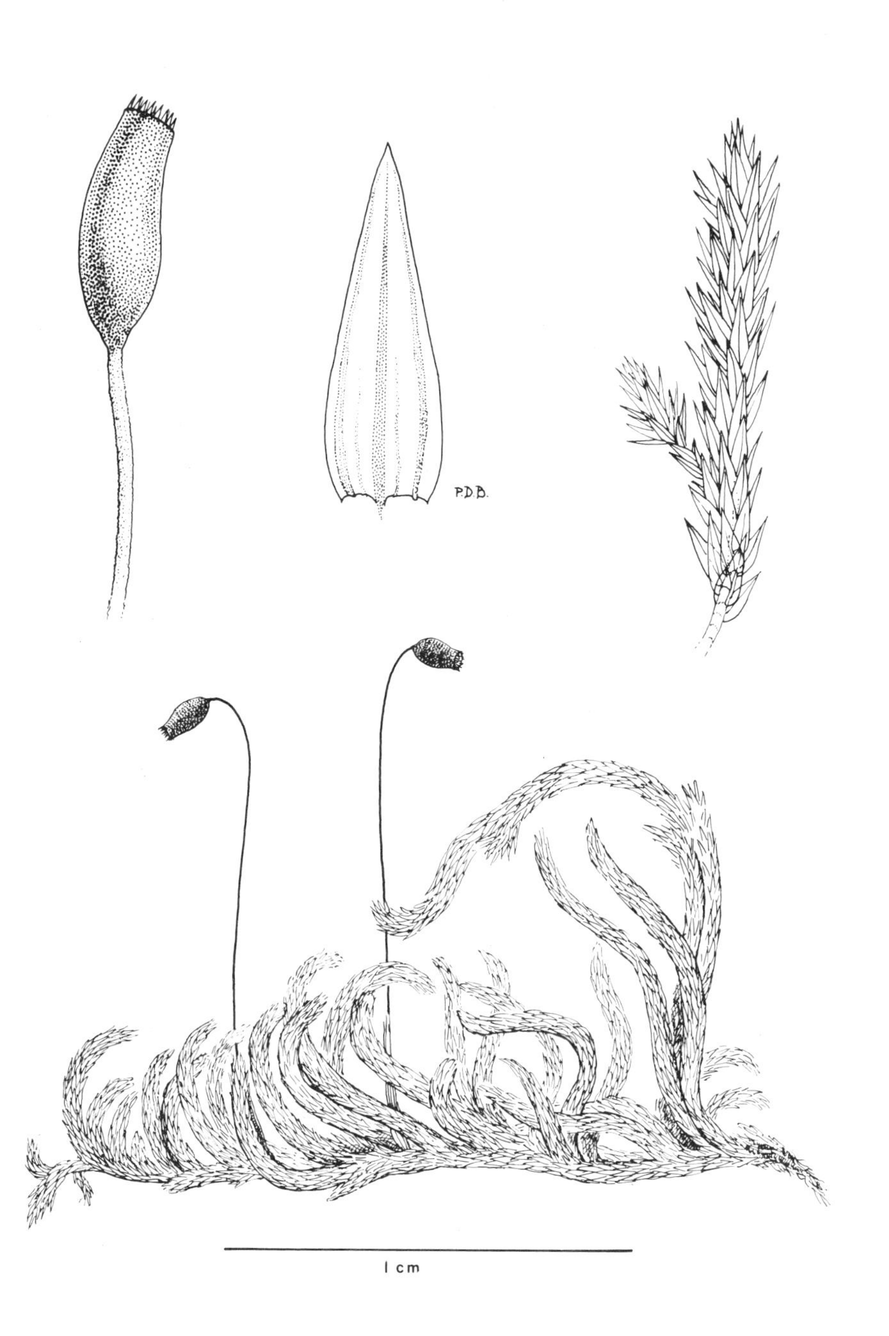

Homalothecium aeneum

Homalothecium fulgescens (Mitt. *ex* C. Muell.) Lawt.

Name: Species name describing the shining, brightly coloured plants.
Habit: Forming interwoven mats of yellow-green to dark green, glossy, creeping, much-branched plants 50–170 mm long; usually attached to substratum by red rhizoids. Leaves strongly pleated with the midrib resembling a central pleat. Plants not curling upward markedly when dry.
Habitat: Predominantly on living tree trunks and branches (especially broad-leafed maple), also on rock and rarely on stabilized sand.
Reproduction: Sporophytes frequent, red-brown, with long-cylindric, suberect sporangia, maturing in late winter to spring.
World Distribution: Confined to western North America from southeastern Alaska to California and eastward to western Montana.
B.C. Distribution: Map 54, page 346.
Distinguishing Features: The usually irregularly branched, coarse plants that produce long-cylindric, somewhat curved sporangia, the pleated leaves and the usual habitat on tree trunks are useful features to separate this species.
Similar Species: The irregular branching and the usual coarseness of the plants are enough to separate *H. fulgescens* from *H. nuttallii*. On populations on rock, the sporophyte shape distinguishes *H. fulgescens* from *H. aeneum* in which sporangia are short-cylindric and bent at a sharp angle at the seta apex. *H. nevadense* of the interior has erect sporangia, which mature in summer rather than spring, but the plants are on shaded cliffs and usually are formed of densely compacted branches that form a turf-like mass over the substratum.
Comments: *H. lutescens*, a European species, was once considered to be identical to *H. fulgescens*.

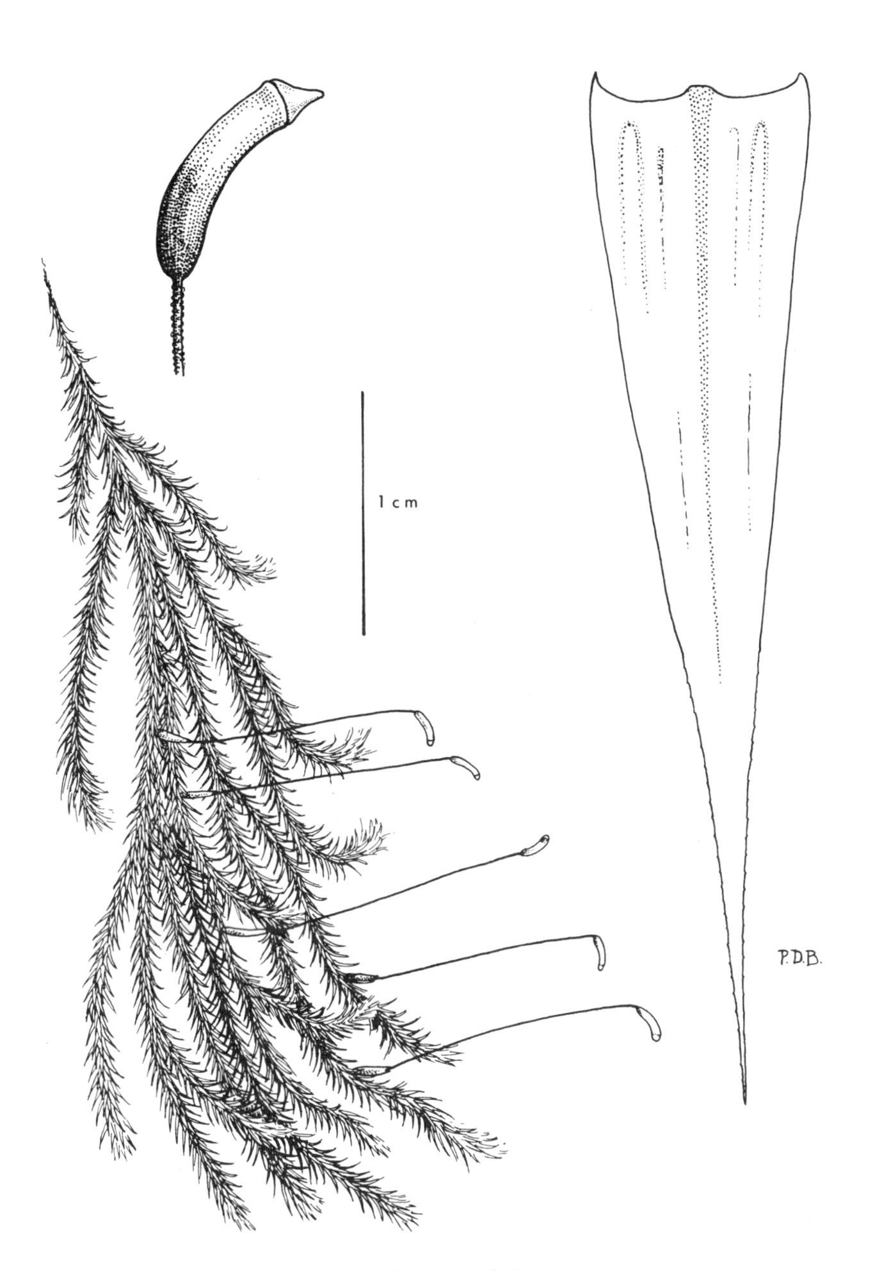

Homalothecium fulgescens

Homalothecium nuttallii (Wils.) Jaeg. & Sauerb.

Name: Species named in honour of T. Nuttall, an important American botanist of the early 19th century.
Habit: Forming bright glossy, yellow-green strands or mats closely or loosely affixed to substratum, usually densely and regularly short branched and the branches coiling upward from the substratum when dry. Sometimes pendent and weakly branched.
Habitat: Usually epiphytic, especially on broad-leafed trees (especially poplar, maple and oak) but also on western red cedar and yew, also on cliffs; most frequently on perpendicular surfaces.
Reproduction: Sporophytes frequent, maturing in spring, suberect, long-cylindric, and red-brown when mature.
World Distribution: Confined to western North America, especially near the Pacific coast, from southeastern Alaska to southern California.
B.C. Distribution: Map 55, page 346.
Distinguishing Features: The usually slender shoots and branches, the glossy, pale yellow-green colour, the frequent branching, the suberect cylindric sporangia and the usually epiphytic habitat distinguish this species.
Similar Species: See notes under *Homalothecium aeneum*. *H. pinnatifidum*, a species of rock and among grasses is usually very regularly pinnate and is not firmly attached to the substratum; sometimes the apical portion of the plant, including its branches, coils upward when dry. In *H. nuttallii* the main shoot is affixed to the substratum, thus the branches curl. *H. nevadense*, a species of the interior, is confined to rock surfaces, and is larger and forms dense turf-like masses of congested branches; *H. nuttallii* is unknown from the interior.

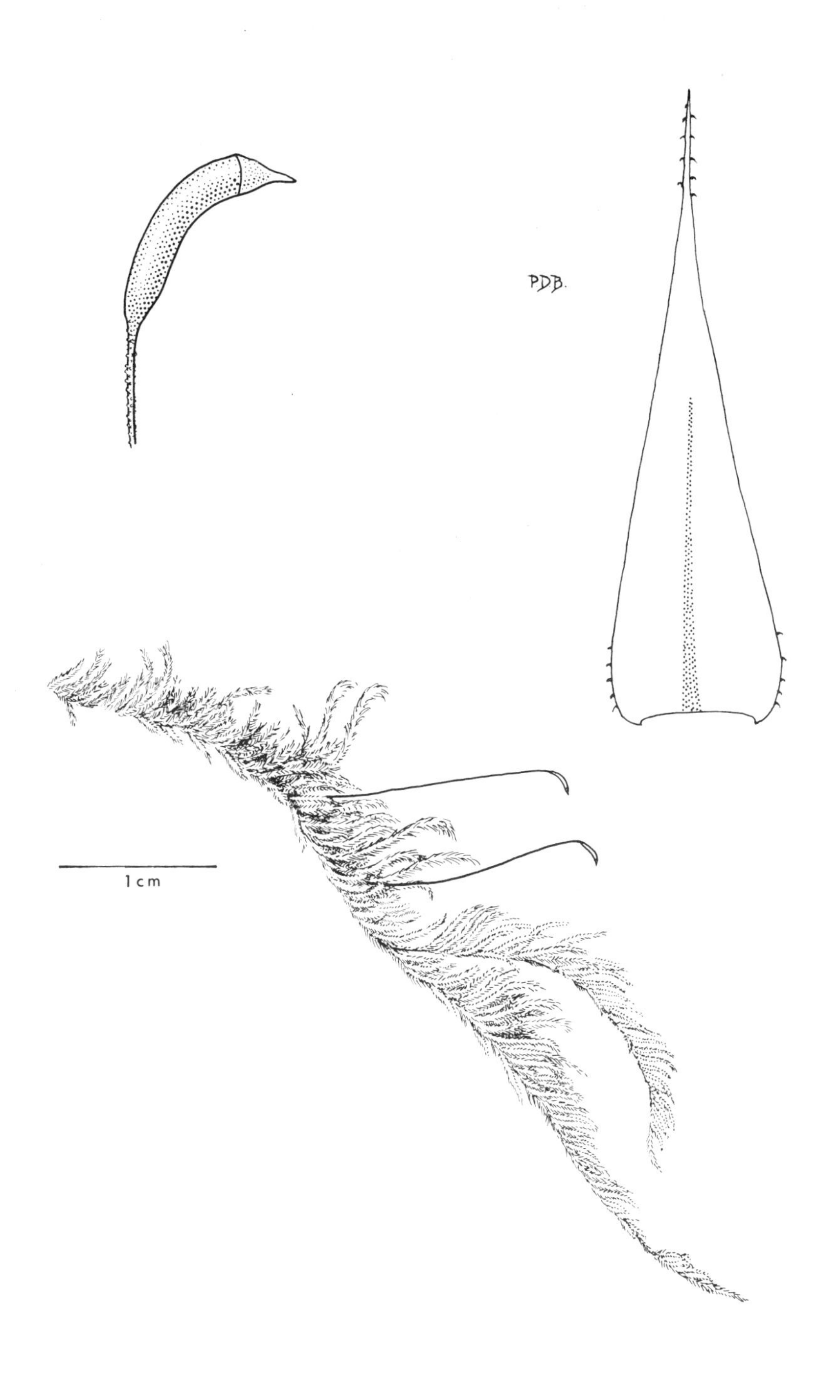

Homalothecium nuttallii

Hookeria lucens (Hedw.) Sm.

Name: Genus named in honour of W.J. Hooker, a 19th-century British botanist who made major contributions to the understanding of mosses and hepatics. Species name meaning polished, indicating the opalescent sheen of the dried plants.
Habit: Loose mats of pale green to dark green, translucent to glossy plants, loosely to firmly attached to substratum.
Habitat: Predominantly in wet sites, especially in humid coniferous forests, occasionally submerged in pools in depressions, on damp soil or rotten wood.
Reproduction: Sporophytes infrequent, red-brown, maturing in early spring to summer. The brittle plants, especially in water, also serve in propagation and the leaf apices sometimes bear filamentous gemmae, especially in plants subject to submergence sometime during the year. The calyptra bears a much-dissected, flaring frill.
World Distribution: Western and central Europe, westernmost Asia, to Madeira and northwestern Africa; in North America from southeastern Alaska to California, predominantly near the coast.
B.C. Distribution: Map 56, page 347.
Distinguishing Features: The flattened leafy plants in which the leaves show the cell network conspicuously (at 10X magnification) and the somewhat blunt leaf apices, combined with the damp, often swampy forest habitat are characteristic features. The plants resemble leafy liverworts superficially but the leaves are in more than three rows and sporophytes are clearly of a moss.
Similar Species: Can be confused with *H. acutifolia*, a rather uncommon species but one which has sharply acute leaves and usually grows on clayey soil patches in shaded sites.

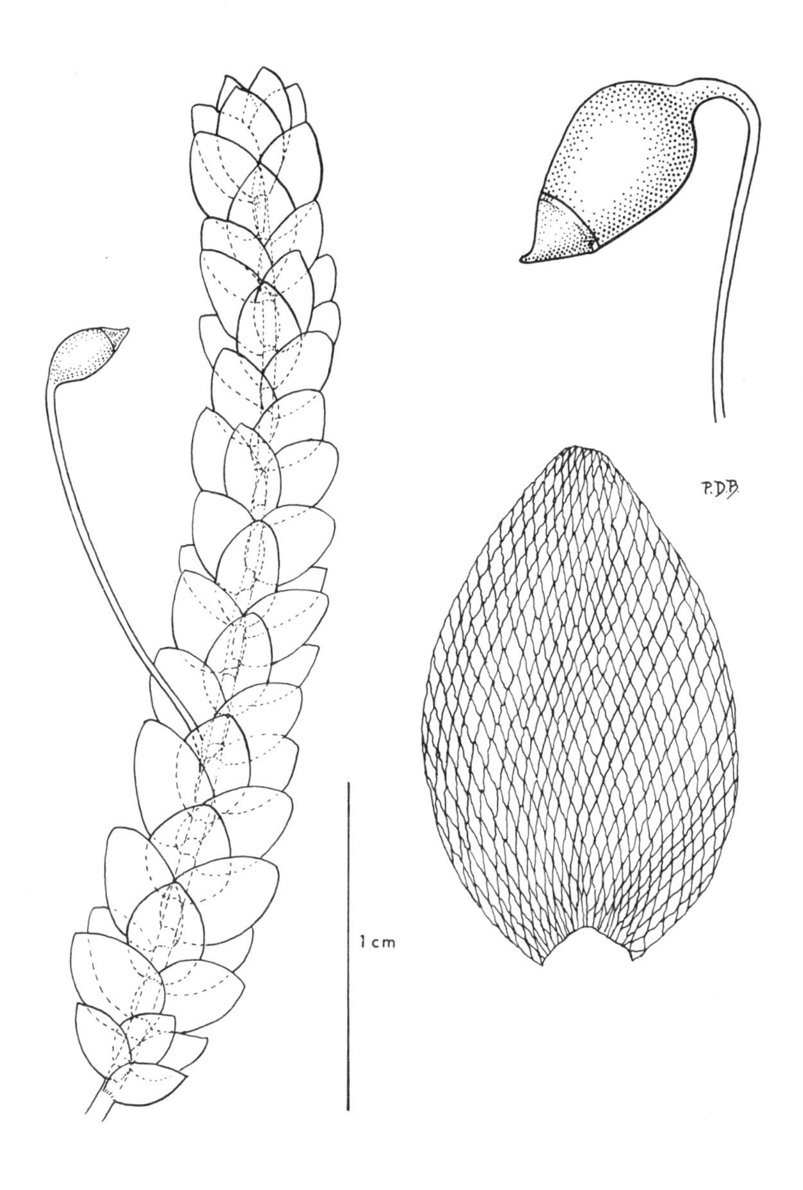

Hookeria lucens

Hygrohypnum ochraceum (Turn. *ex* Wils.) Loeske

Name: Genus name meaning aquatic *Hypnum*, describing the habitat. Species name meaning yellow-brown, describing plants of sunny sites.
Habit: Forming soft mats of golden green, yellow-green or, rarely, dark green, irregularly branched shoots usually loosely affixed to the substratum.
Habitat: Aquatic or in seepage sites; in wetlands of subalpine to alpine areas; on rocks and logs in streams, on wet cliffs, especially near water-courses and on lake margins; from sea level to alpine sites.
Reproduction: Sporophytes infrequent, maturing in spring to summer; brown when ripe.
World Distribution: Widely distributed in the cooler climates of the Northern Hemisphere. In the Western Hemisphere from southern Greenland southward to the southern Appalachian Mountains of North America and westward around the Great Lakes; in western North America from northern Alaska southward to Arizona, eastward along the Rocky Mountains.
B.C. Distribution: Map 57, page 347.
Distinguishing Features: Specimens with yellow-green, falcate-secund leaves, loosely or not attached to the substratum and with stout inclined sporangia are diagnostic, but see below.
Similar Species: This is a highly variable species and can be confused with some species of *Hypnum* that are irregularly branched; the characters that set them apart, except for the wet habitat, are largely microscopic. From *Drepanocladus* species that occur in the same habitat, the pronounced midrib of that genus is usually sufficient to separate them. It is often difficult to distinguish from most other species of *Hygrohypnum*, except that in cross-section, the stem of *H. ochraceum* shows enlarged, thin-walled cells in the epidermal layer; most other species of *Hygrohypnum* lack these enlarged cells.

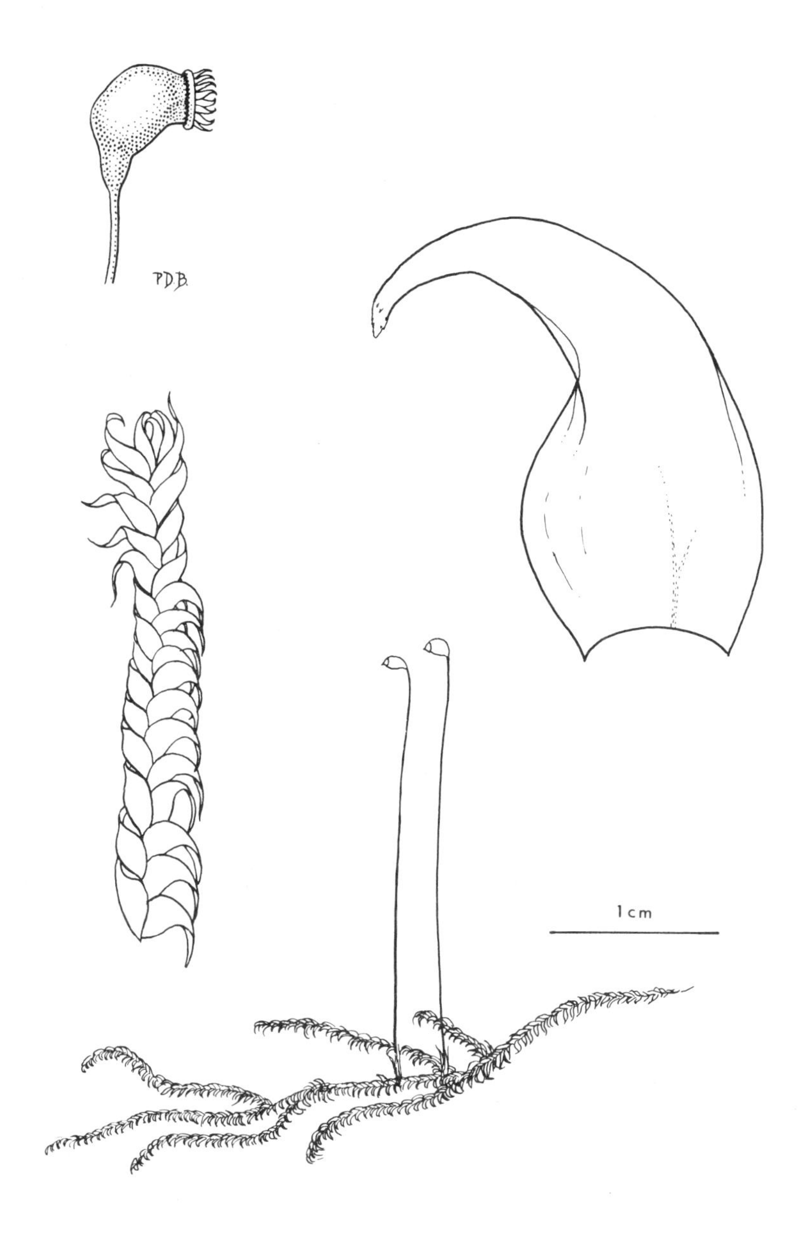

Hygrohypnum ochraceum

locomium splendens (Hedw.) B.S.G.

Name: Genus name meaning woodland inhabitant. Species name reflecting the intricate beauty of the plants.
Habit: Dull, glossy green to brownish-green plants with red-brown stems, forming loose carpets of interwoven much-branched arching shoots. Annual growth usually marked by an abrupt angle from the preceding year's feather-like, flattened branch system on the main shoot.
Habitat: Usually terrestrial or on decaying logs in forest but also on cliff shelves, from sea level to alpine elevations, rarely epiphytic on tree branches and trunks. Often forming extensive carpets on coniferous forest floor.
Reproduction: Sporophytes infrequent, but sometimes locally abundant, maturing in spring to summer.
World Distribution: Widely distributed in cooler parts of the Northern Hemisphere but also in to the Southern Hemisphere in New Zealand.
B.C. Distribution: Map 58, page 348.
Distinguishing Features: The elegantly feathery branched plants, with arching main shoots bearing feathery side branches, all arranged in a horizontal plane, coupled with the red-brown stems furry with paraphyllia, and the elongate sinuous tip of the stem leaves, are usually distinctive.
Similar Species: Species of the genus *Thuidium* have a similar branching pattern but the plants are not glossy, tend to be yellow-green rather than brownish-green, and the stems are not conspicuously red. *Kindbergia praelonga* is often elaborately branched but lacks the regular arching shoots and the single flattened plane of the annual branch system is not apparent; stem leaves have one (rather than two) midrib and lack the sinuous apex. *Kindbergia* is most frequent in swampy sites in forests, while the *Hylocomium* is in well-drained sites. *H. umbratum* lacks the regularly arched main branch system and the whole plant tends to be golden brownish-green.
Comments: This species is often called the "stair-step moss" because of the distinctive branching pattern. Another common name is "mountain fern moss", a name evoked by its distinctive branching.

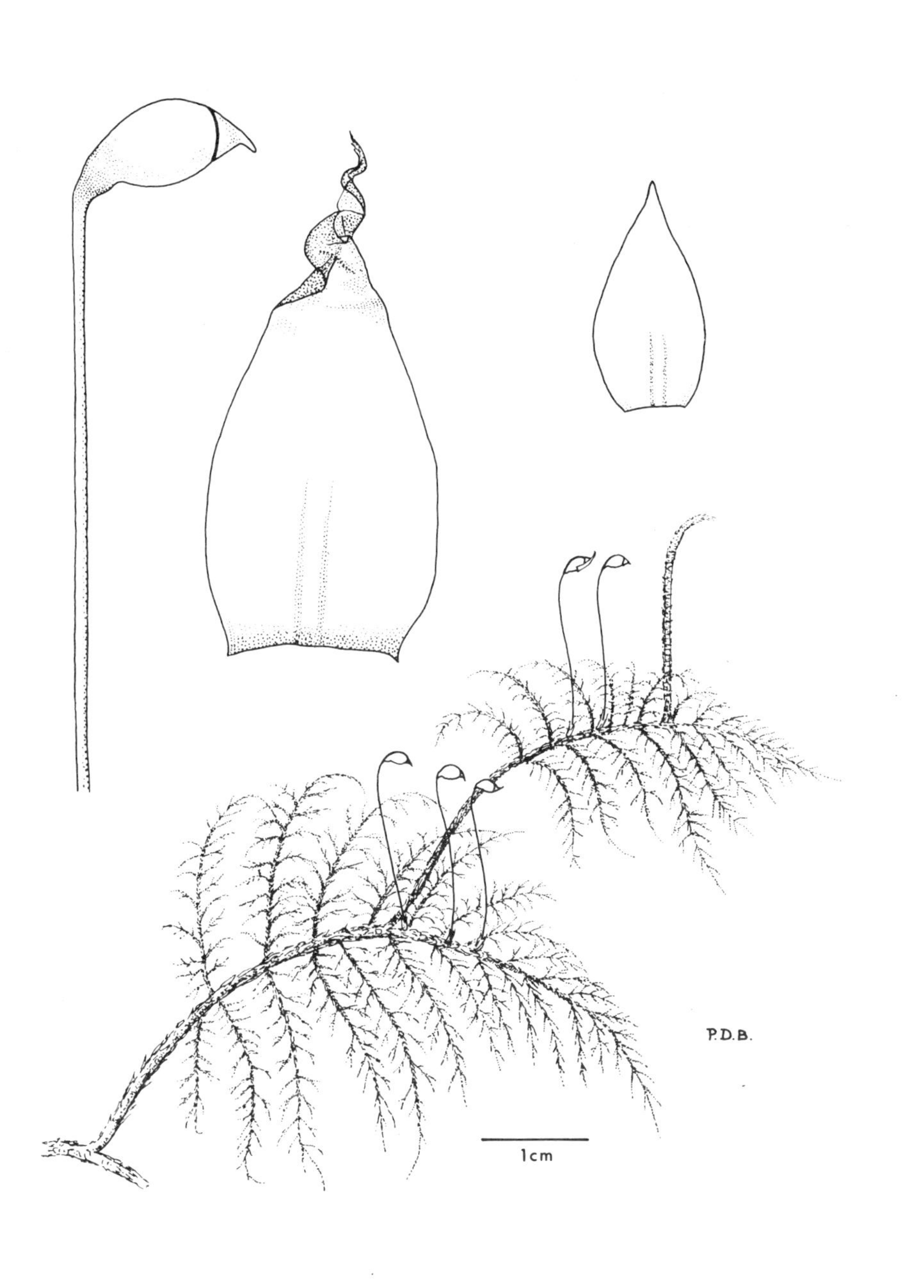

Hylocomium splendens

um circinale Hook.

.. Genus name derived from a Greek word meaning sleep, apparently reflecting the ancient use of some mosses as medicinal ingredients. Species name describing the strongly curved leaves.

Habit: Forming glossy, pale green to yellow-green mats or rarely, turfs.

Habitat: On tree trunks, logs in early stages of decomposition, and rocks. Occasionally on humus in subalpine forest areas. From sea-level to subalpine, most frequently in humid forest.

Reproduction: Sporophytes frequent, maturing in late winter to early spring; the small stout sporangia are characteristic.

World Distribution: Confined to western North America from southeastern Alaska southward to California and eastward to western Montana.

B.C. Distribution: Map 59, page 348.

Distinguishing Features: The short, stout, inclined, red-brown sporangia, associated with the pale glossy green, creeping shoots with facate-secund to strongly curved leaves that have attenuate tips and lack a midrib, usually separate this species.

Similar Species: *H. subimponens* often grows in similar habitats but the plants are larger, more regularly branched and the sporangia are suberect and cylindric. Some forms of *Drepanocladus uncinatus* also occur in the same habitats, but the leaves have a midrib and the sporangium is cylindric and usually curved. *Pseudotaxiphyllum elegans* may resemble *H. circinale*, but the plants are usually terrestrial, the shoots are somewhat flattened, and sporophytes, although not common, are clearly nodding.

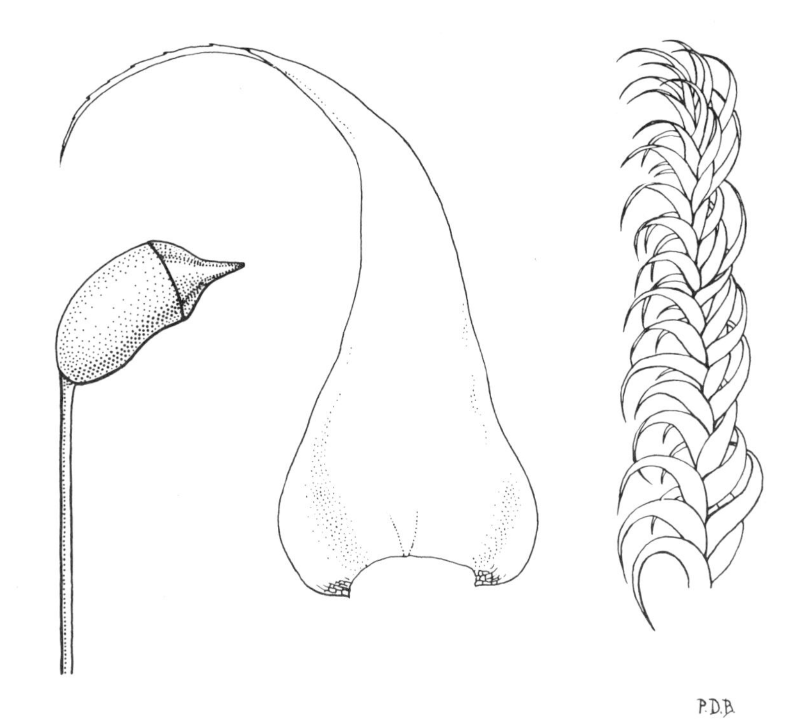

Hypnum circinale

Hypnum revolutum (Mitt.) Lindb.

Name: Species name describing the strongly recurved margins of the leaves.

Habit: Golden green to brownish-green, creeping, often regularly branched mats usually loosely attached to substratum.

Habitat: Usually on rock (especially in calcium-rich areas) but also terrestrial and on logs, occasionally ascending tree bases. Rare near the coast but frequent in the interior and in the northern third of the province.

Reproduction: Sporophytes occasional to rare; the plants fragment readily when dry and the fragments probably serve in propagation. Sporophytes are red-brown when ripe; sporangia are inclined to suberect, cylindric and curved.

World Distribution: Widely distributed in the Northern Hemisphere in both boreal and arctic regions, but known also from Antarctica and New Zealand.

B.C. Distribution: Map 60, page 349.

Distinguishing Features: The strongly revolute leaf margins and the seeming absence of midribs, combined with the dry substratum and the frequent regular branching of the golden brownish-green plants are useful features.

Similar Species: *H. cupressiforme* and *H. vaucheri* grow in similar habitats and are often the same colour as *H. revolutum*. They, however, lack regular branching and also lack revolute leaf margins. Microscopically, they have a triangular area of small alar cells that separate them readily from *H. revolutum*. *Drepanocladus uncinatus*, of similar habitats, has pleated leaves with a midrib; the leaves are very strongly curved. *H. recurvatum* is much more slender and is usually firmly affixed to its rock substratum.

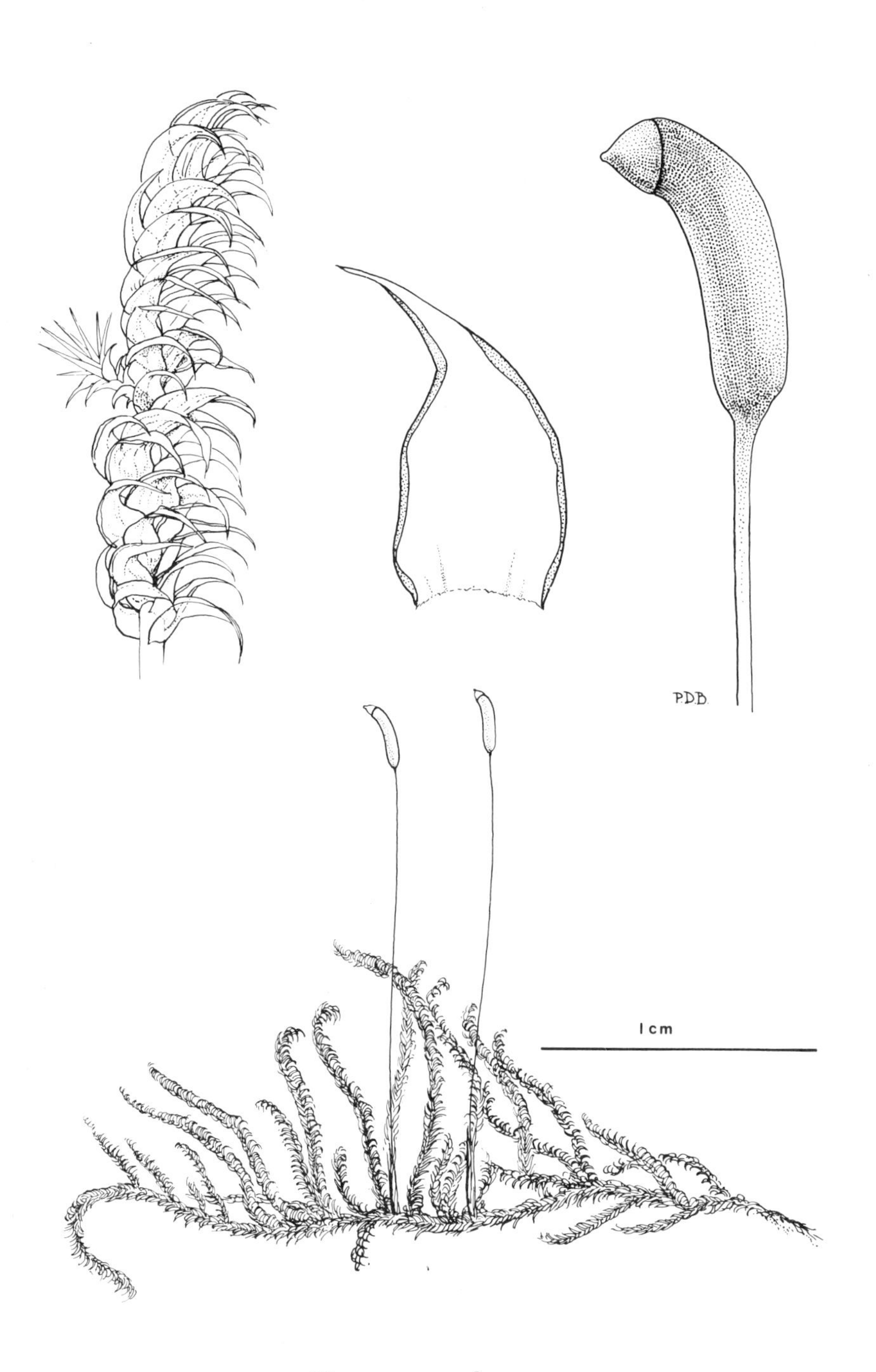

Hypnum revolutum

Hypnum subimponens Lesq.

Name: Species name meaning somewhat like *imponens*, referring to its resemblance to *H. imponens*. The name *imponens* means imposter, indicating that this species strongly resembles, at times, other species of *Hypnum*.
Habit: Forming reclining, glossy, pale yellow-green mats of regularly pinnate plants loosely attached to the substratum.
Habitat: On living trees, usually in open forest, on cliffs and boulders, occasionally on rotten logs or terrestrial.
Reproduction: Sporophytes frequent, erect to suberect, especially when unripe (green), reddish-brown when ripe.
World Distribution: Southeast Asia and western North America from southeastern Alaska to California and eastward to western Montana and Alberta, also noted from arctic North America and Greenland.
B.C. Distribution: Map 61, page 349.
Distinguishing Features: The yellow-green colour, abundant regular pinnate branching, coupled with the usually non-terrestrial habitat and the erect to suberect sporangia are valuable features.
Similar Species: See notes under *H. circinale*. *Ptilium crista-castrensis* is of a similar colour but the plants are very regularly pinnate, tend to be erect to suberect, rather than reclining, and are generally terrestrial within forests. The leaves of *Ptilium* are pleated, a feature absent in the *Hypnum*, and shoots of the *Hypnum* frequently grow with the apex pointing downward rather than consistently upward as in *Ptilium*. Sporangia of *Ptilium* are strongly inclined.

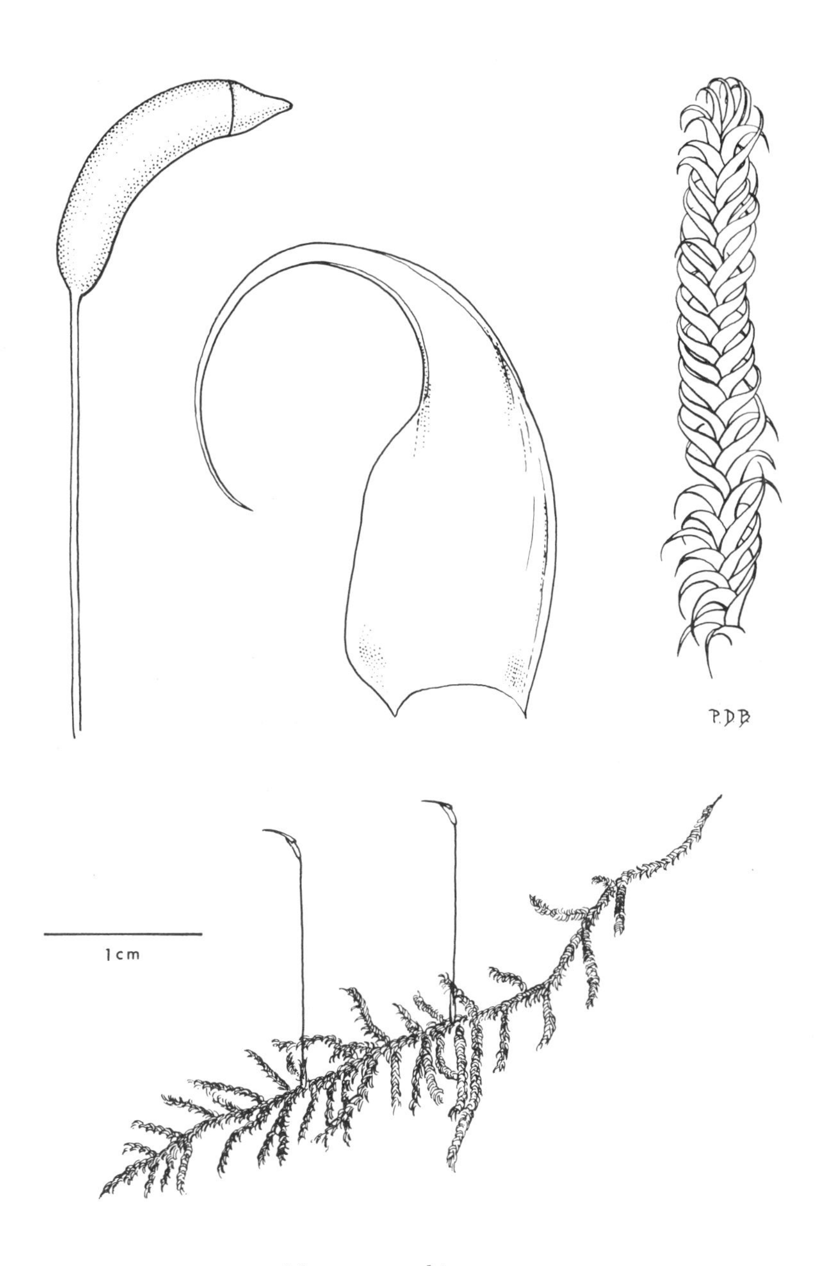

Hypnum subimponens

Isothecium stoloniferum Brid.

Name: Genus name referring to the symmetric sporangia, in spite of the fact that the sporangia are often asymmetric. Species name describing the creeping shoots characterizing some plants.
Habit: An extremely variable moss, as shown by the illustrations (pages 180 and 181). Plants are usually much-branched, occasionally very regularly, usually glossy, pale to dark green (occasionally golden brownish-green). Attenuate shoot-tips and branches are frequent in plants of shaded sites, especially on perpendicular surfaces and on tree or shrub branches. See also comments below.
Habitat: Most frequently epiphytic, up tree bases, on tree trunks and branches, and on shrubs, also on rock, rotten logs, commonly in forests, especially humid coniferous forests, from sea level to subalpine elevations. Occasionally on rooftops. Plants up bases of rocks and near tree bases are often coarse and regularly branched; those on shaded tree trunks, especially in coniferous forests, often produce slender branches that hang downward forming a loose mass of silky shoots; on tree branches the plants sheathe the upper surface of the branch and produce branches hanging downward; on sunny rock surfaces, especially near watercourses, the plants are golden brownish-green, the shoots are often very glossy and worm-like in appearance. Rarely, in extremely humid coastal forest, the epiphytic shoots hang downward and bear very regular lateral branches that emerge almost at right angles.
Reproduction: Sporophytes often produced in abundance, maturing in early spring, red-brown and with the sporangia suberect to inclined when mature. Probably fragmented branches and shoots can be blown from one site to another and establish new populations.
World Distribution: Confined to western North America from southeastern Alaska to central California and extending eastward to eastern British Columbia and western Montana. Reports from eastern North America and Europe need verification.
B.C. Distribution: Map 62, page 350.
Distinguishing Features: Although an extremely variable moss, this species can be readily recognized with some experience. Hand lens features are needed: the sharply toothed leaf-apices that are usually not narrowly pointed, the obvious midrib, the usually untidy pinnate arrangement of branches that emerge from the main stem at an acute angle, and the frequent presence of attenuate shoots or main branches are useful features.
Similar Species: Most species of *Brachythecium* can be distinguished

from *Isothecium* by noting that their leaves lack the coarse marginal teeth. *I. cristatum* is difficult to distinguish from julaceous forms of *I. stoloniferum*. The two species sometimes grow on the same tree or rock. *I. cristatum* is generally rusty green, shoots are strongly worm-like when dry and, in microscopic features, has a distinct and large group of small alar cells. All species of *Homalothecium* have pleated leaves, while the leaves of *Isothecium* lack pleats. Species of *Sceleropodium* also lack marginal teeth on leaves. *Eurhynchium* often has toothed leaf margins but the leaves tend to be blunt.

Comments: This species has been called *I. spiculiferum* but *I. stoloniferum* is the correct name. This moss is very common in humid coastal forests and is especially impressive when it festoons branches of trees and shrubs.

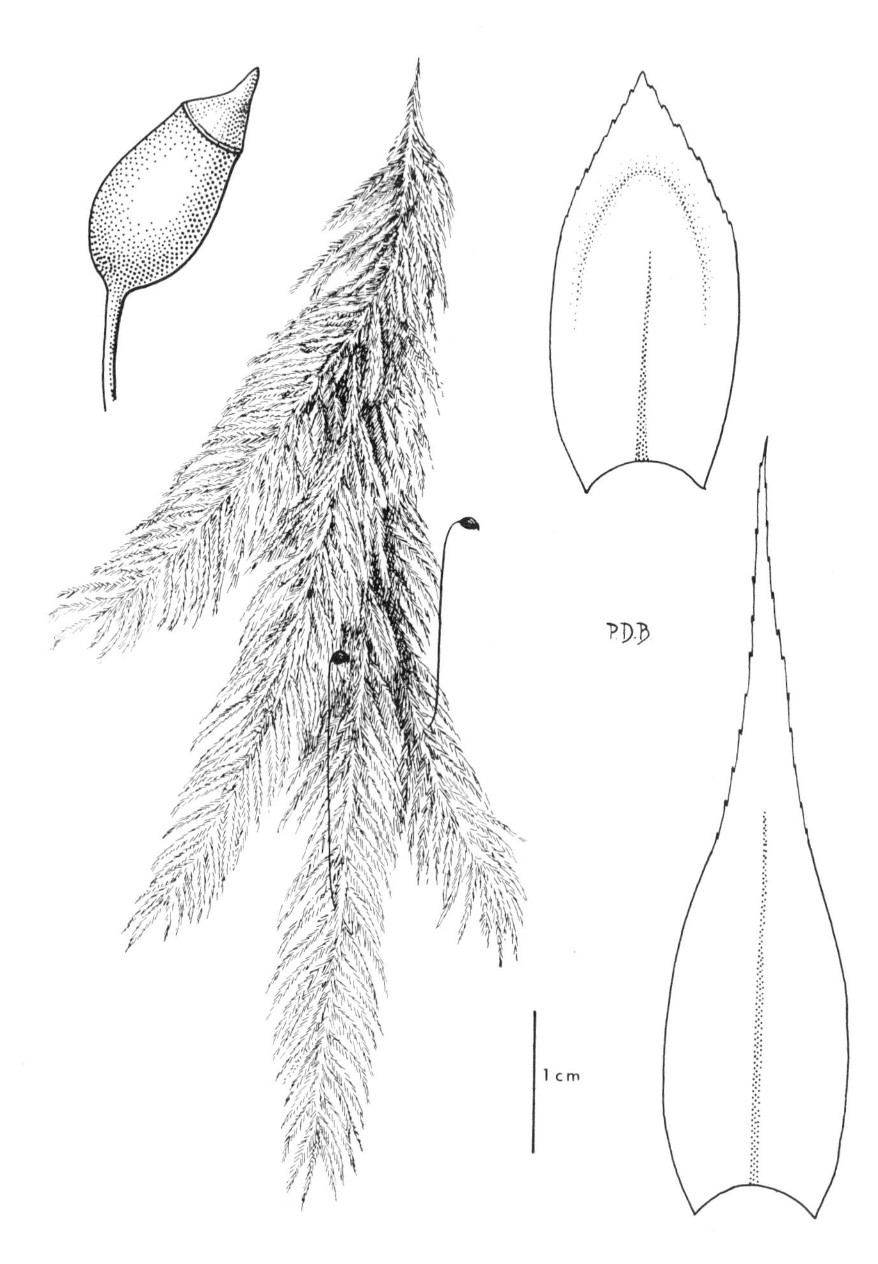

Isothecium stoloniferum

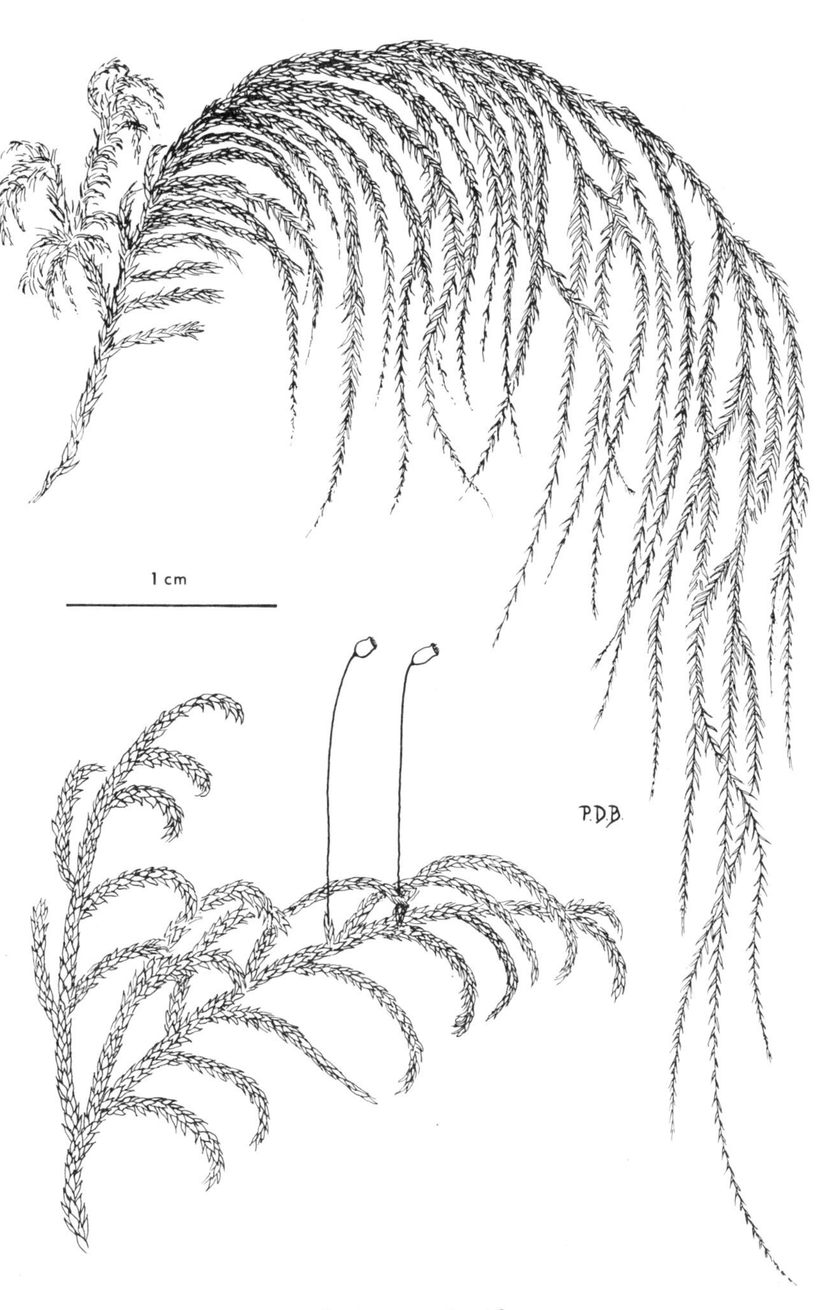

Isothecium stoloniferum

Kiaeria starkei (Web. & Mohr) I. Hag.

Name: Genus named in honour of the 19th-century Norwegian bryologist, F.C. Kiaer. Species named in honour of the original collector, V.C. Starke.
Habit: Forming short turfs of densely to loosely attached light green to yellow-green plants in which leaves become strongly curled at the tips when dry.
Habitat: Usually on siliceous rock surfaces, sometimes on soil, at subalpine to alpine elevations.
Reproduction: Sporophytes frequent, grooved when ripe, maturing in early summer or late spring when the snow melts to expose the populations.
World Distribution: Circumpolar in the Northern Hemisphere, mainly in the mountains; in the Western Hemisphere from Greenland to Nova Scotia in the east and from Alaska to California and inland to western Alberta and Montana in the west.
B.C. Distribution: Map 63, page 350.
Distinguishing Features: Considering only vegetative characteristics, this is a troublesome moss. However, the sporangia are usually present and, when mature, are grooved, which separates it from most mosses that resemble *K. starkei* vegetatively. The species frequently occurs on perpendicular rock surfaces and has usually falcate-secund leaves that are curled at tips when dry.
Similar Species: *K. falcata* usually forms tight turfs on more-or-less horizontal rock surfaces in late-summer snow-melt areas; *K. starkei* is usually on perpendicular surfaces, forms loose turfs and release from snow is early in the season. *K. falcata* also has sporangia without grooves when mature. *K. blyttii* is difficult to distinguish on vegetative features but is frequently terrestrial, is usually weakly falcate-secund and sporangia lack grooves when mature. From *Dicranoweisia crispula*, *K. starkei* differs in the non-falcate leaves of the *Dicranoweisia* that occurs mainly in rock crevices; in *Dicranoweisia* sporangia are straight and erect rather than curved and inclined. *Oncophorus wahlenbergii* is superficially similar, but the sporangia have a conspicuous knob at the neck, lacking in *Kiaeria*, and *Oncophorus* occurs on humus or wood, not on rock. *Dicranella heteromalla* is also a terrestrial species; the sporangia, although grooved, have the grooves obliquely arranged while in *K. starkei* the grooves are mainly longitudinal; in *Dicranella* the leaves are dark green and almost hair-like, while those in *Kiaeria* are yellowish-green and are not hair-like.

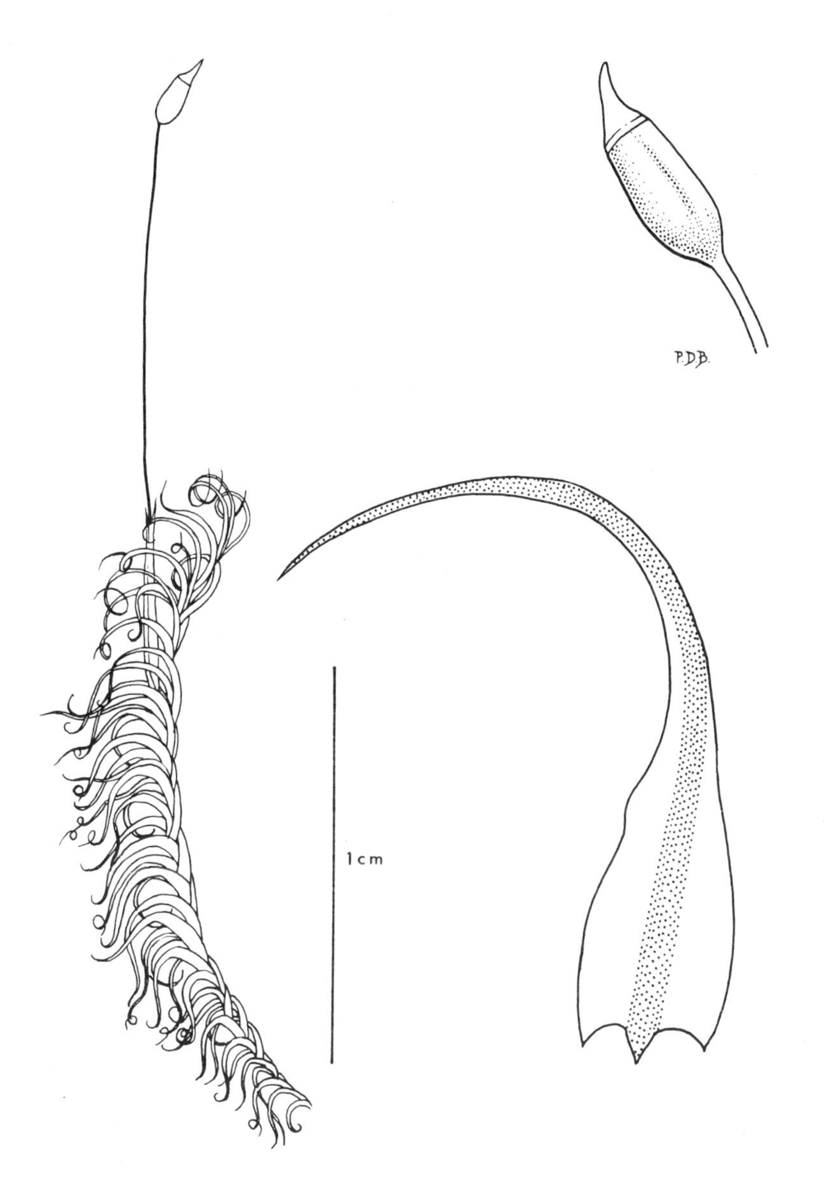

Kiaeria starkei

Kindbergia oregana (Sull.) Ochyra

Name: Genus named in honour of the 19th-century Swedish bryologist, N.C. Kindberg. Species named for the state of Oregon where the original specimen was collected.
Habit: Forming loose mats of interwoven light green to dark green to yellow-green, pinnately branched, creeping plants with the branches arising in a single horizontal plane.
Habitat: Terrestrial, on humus, on tree trunks, rotten logs and rocks, usually in humid coniferous forests, frequent in all vegetation types near the coast, but confined to humid forests in the interior. Mainly at lower elevations.
Reproduction: Sporophytes occasional to frequent, maturing in early spring, red-brown when ripe and with very beaked operculum.
World Distribution: Confined to western North America from southeastern Alaska to California, eastward to Idaho.
B.C. Distribution: Map 64, page 351.
Distinguishing Features: The regularly pinnate, large terrestrial plants that occur in near-coastal or humid interior forests are valuable features in combination with the broadly heart-shaped and sharply pointed stem leaves, and sporangium that bears a sharply snouted operculum. The rough seta is another useful character.
Similar Species: Some specimens of *K. praelonga* are difficult to distinguish from *K. oregana* but the occurrence in moister sites, the more complex branching pattern, and the usually more slender aspect of *K. praelonga* are usually diagnostic. *Eurhynchium pulchellum* is generally much smaller and the leaves tend to be blunt; in *Kindbergia* they are sharply pointed. *E. pulchellum* var. *barnesiae* also strongly resembles *K. oregana*, but this plant has blunt leaves and is a calcicole.
Comments: Sometimes named *Eurhynchium oreganum* or *Stokesiella oregana*; these are synonymous.

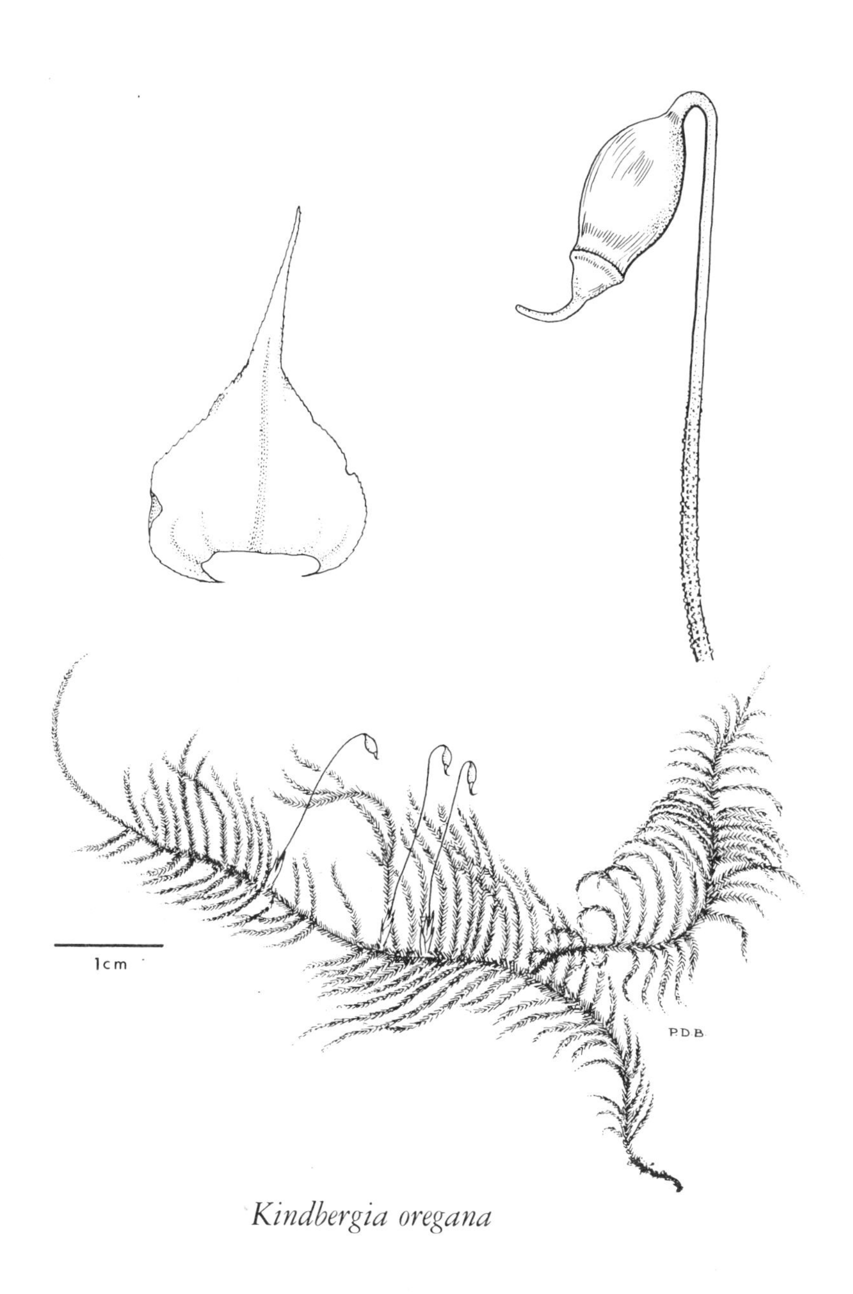

Kindbergia oregana

rgia praelonga (Hedw.) Ochyra

pecies name referring to the very long trailing plants.

Habit: Loose to dense mats of light to dark green, somewhat arched, interwoven, much-branched shoots, sometimes densely branched with the main branches also bearing slender branchlets.

Habitat: Most frequent in damp, swampy forest sites, around springs and in swampy depressions from sea level (indeed, on maritime rocks) to subalpine elevations; also on rotten logs, epiphytic on tree trunks and on rock.

Reproduction: Sporophytes frequent, reddish-brown, maturing in early spring.

World Distribution: Circumpolar in the Northern Hemisphere, also in the Southern Hemisphere, in Australasia and South America; in eastern North America from southern Newfoundland, Ontario and Nova Scotia (where rare); in the west from southeastern Alaska to California and eastward to western Montana and Nevada, where often abundant.

B.C. Distribution: Map 65, page 351.

Distinguishing Features: The slenderly branched, multipinnate plants that occur frequently in swampy depressions in forests and bear heart-shaped, sharply pointed, somewhat squarrose stem leaves are distinctive in most populations but the wide habitat tolerance of the species is often reflected in considerable variation in its growth form. The long-snouted operculum and rough seta are also useful characters.

Similar Species: See also notes under *K. oregana* and *Hylocomium spendens*. From species of *Brachythecium*, the more complex branching of *Kindbergia*, as well as the squarrose stem leaves are usually enough to separate it. From *Eurhynchium pulchellum* the blunt leaves of the *Eurhynchium* are usually diagnostic. *Bryhnia hultenii*, extremely rare in British Columbia is similar but occurs on logs subject to flooding; the leaves, however, are abruptly narrowed to a point, are never squarrose and possess rather swollen, thin-walled alar cells, features absent in the *Kindbergia*.

Comments: Sometimes called *Stokesiella praelonga*, *Eurhynchium stokesii* or *Eurhynchium praelongum*, all of which are synonyms.

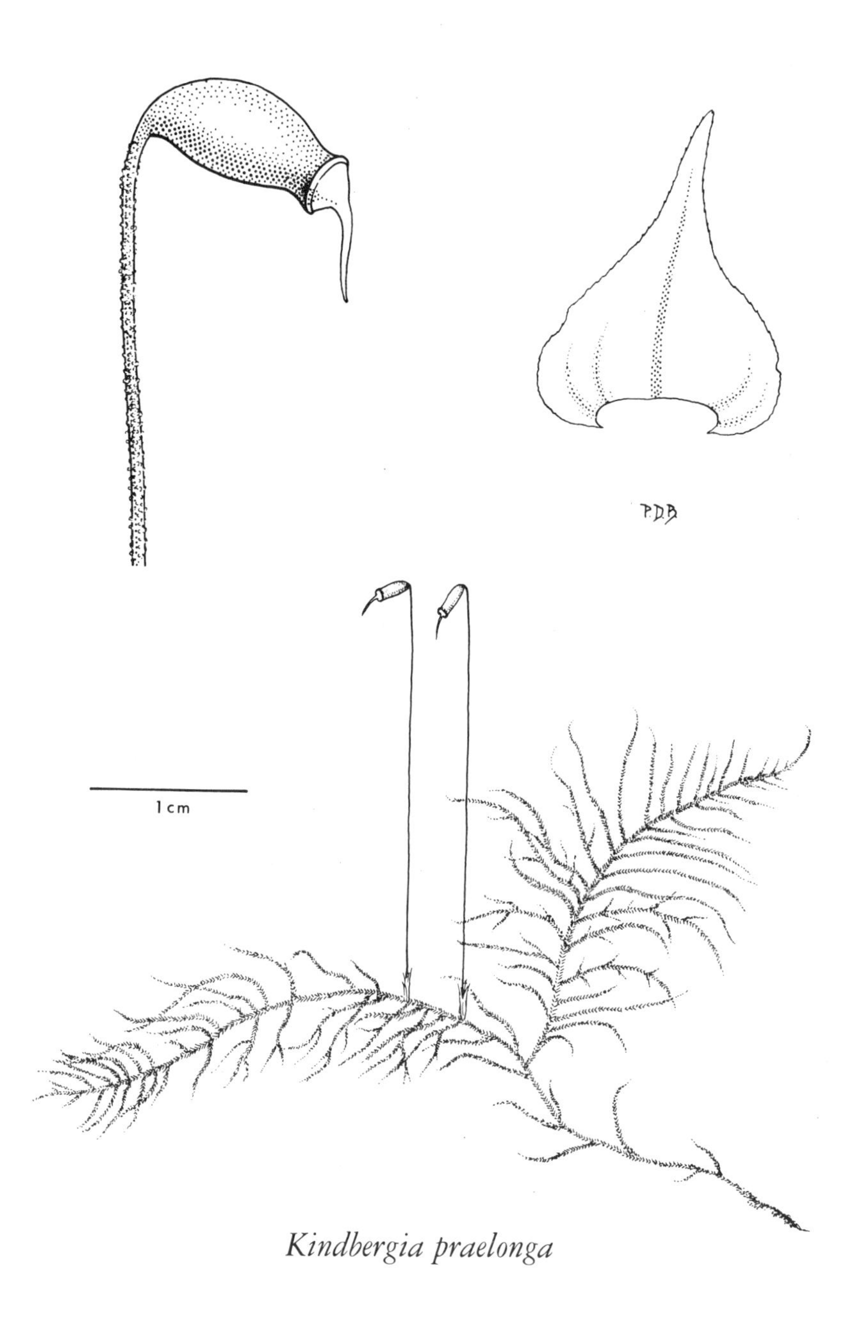

Kindbergia praelonga

m pyriforme (Hedw.) Wils.

us name meaning delicate or slender *Bryum*, probably de- the slender leaves as well as the slender, shiny seta. Species name describing the pear-shaped sporangium.

Habit: Silky, small, light green to yellow-green plants forming loose to dense short turfs; leaves almost hair-like when dry, strongly divergent when humid.

Habitat: Disturbed earth, especially in burned-over forests but also in gardens, as a greenhouse weed, and on concrete of walls.

Reproduction: Sporophytes common, the shiny golden-brown, nodding, pear-shaped sporangium is diagnostic.

World Distribution: Nearly cosmopolitan but more frequent in temperate climates; found on all continents except Antarctica.

B.C. Distribution: Map 66, page 352.

Distinguishing Features: The slender silken leaves, broadly divergent from a subclasping base, combined with the glossy, golden brown, pear-shaped, nodding sporangia are distinctive.

Similar Species: From species of *Bryum* and *Pohlia*, the slender, silken, shiny leaves are enough to separate the *Leptobryum*. No other moss in British Columbia has nodding, pear-shaped sporangia on a similar leafy plant with slender silken leaves.

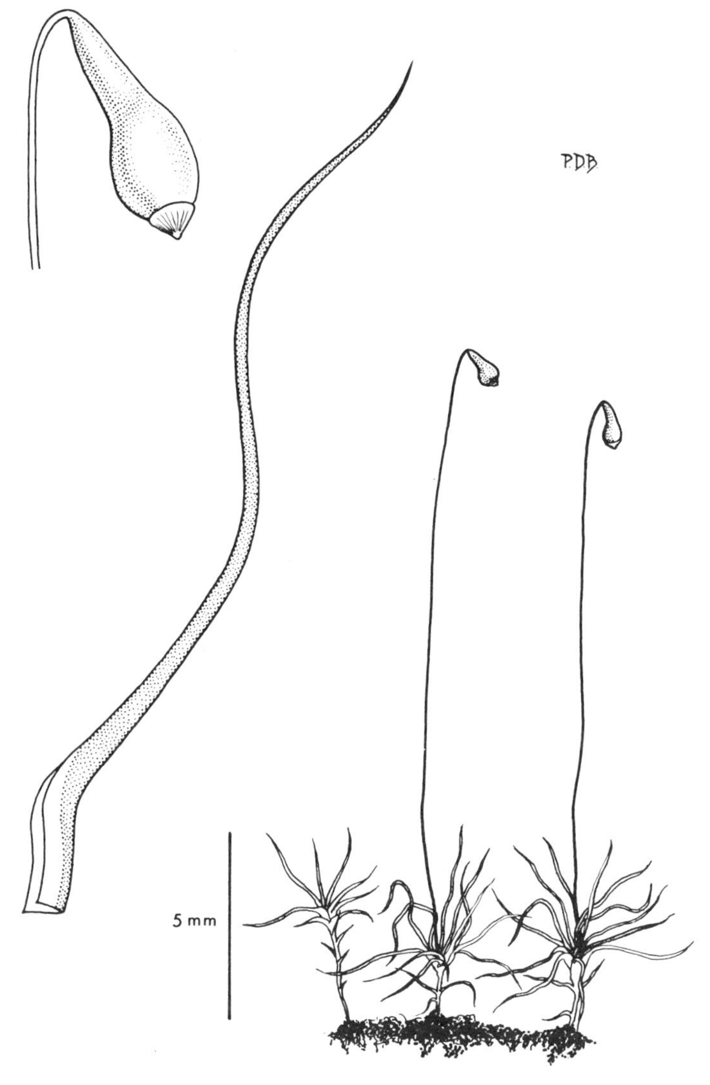

Leptobryum pyriforme

...ɔis acanthoneuron (Schwaegr.) Lindb.

...enus named for the white, scale-like leaves on the main stem. Species name meaning spiny nerve, probably in reference to the teeth on the back of the midrib.

Habit: Forming loose to tall turfs of minature, tree-like, dark green plants, superficially resembling tiny palm trees.

Habitat: Usually in dampish rich soil of shaded sites, especially near water courses in forests, but also an epiphyte (especially on broad-leafed maple), on rotten logs and soil over rock.

Reproduction: Sporophytes frequent, maturing in spring, with red-brown sporangia and seta when ripe.

World Distribution: Confined to western North America from south-eastern Alaska to central California and eastward to Idaho.

B.C. Distribution: Map 67, page 352.

Distinguishing Features: The leafy plants that resemble small trees, the white, narrowly triangular leaves on the main stem and the nodding sporangia are useful features. In male plants, the rosette of leaves surrounding the sex organs is distinctive in a dwarf tree-like plant; no other moss in British Columbia resembles it.

Similar Species: Possibly *Climacium dendroides* might be considered similar, but the broad heart-shaped stem-leaves, the more shiny appressed leaves and the usual presence of a subterranean creeping stem separate *Climacium* from *L. acanthoneuron*. *Pleuroziopsis ruthenica* has plants in which the apical mass of branches forms a horizontally flattened, intricately branched system. *Thamnobryum neckeroides* also possesses somewhat flattened branched systems but the leafy branches are decidedly swollen compared to *Leucolepis*. *Hypopterygium fauriei* also has somewhat flattened but tiny, tree-like shoots but the leafy shoots are also conspicuously flattened with the leaf on the underside of the branch much smaller than the lateral ones; this moss is a calcicole.

Comments: Sometimes called *Mnium menziesii* or *L. menziesii*, both of which are synonyms. Sometimes called "palm tree moss".

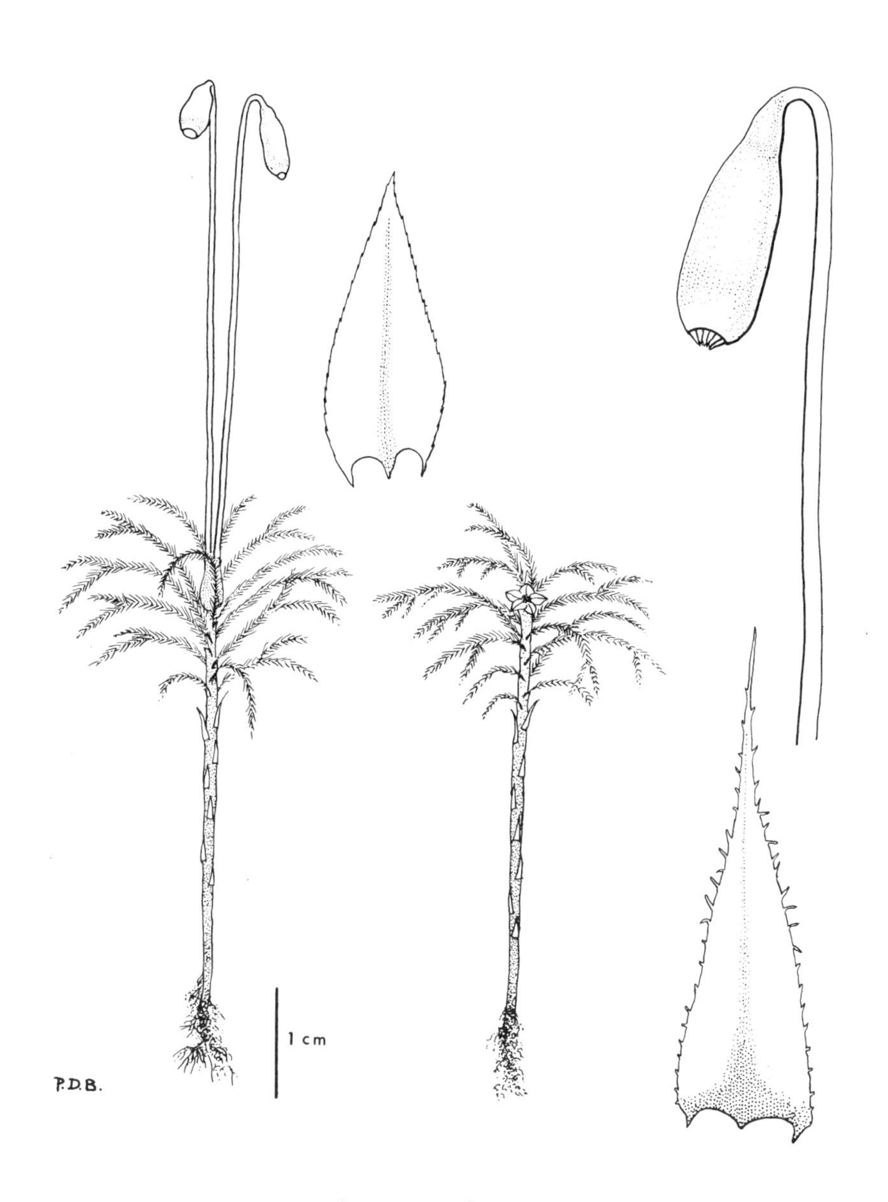

Leucolepis acanthoneuron

Meesia triquetra (Richt.) Aongstr.

Name: Genus named in honour of a Dutch gardener, D. Meese. Species name describing the leaf arrangement in three neat rows (three-angled).

Habit: Tall to short turfs of dense to loosely compacted, unbranched, shoots in which the dark green leaves are strongly divergent when wet and are in three rows, making leafy shoots appear three-angled when viewed from above.

Habitat: Occasional in calcium rich areas in wetlands or on wet cliff ledges.

Reproduction: Sporophytes infrequent, red-brown when ripe.

World Distribution: Circumpolar in temperate and arctic regions of the Northern Hemisphere; in North America across the northern portion of the continent, extending southward in the east to the Great Lakes area and in the west to California.

B.C. Distribution: Map 68, page 353.

Distinguishing Features: The most striking feature is the regularly three-rowed arrangement of the leaves that are dark green and strongly divergent from a somewhat sheathing base. These features, plus the wet calcareous habitat are usually enough to separate this moss.

Similar Species: Some specimens of *Dichodontium pellucidum* have leaves that are structurally similar to the *Meesia* but they are never in three distinct rows, as in *Meesia*. *Oxystegus tenuirostris* and *Oncophorus virens* may also superficially resemble *M. triquetra* but their leaves are never in three rows.

Meesia triquetra

Metaneckera menziesii (Hook. *ex.* Drumm.) Steere

Name: Genus name meaning substituted for *Neckera*, the genus in which the species was formerly placed. Species name in honour of A. Menzies, ship's surgeon and naturalist on Capt. George Vancouver's expedition to western North America in the late 18th century. Menzies collected the specimen from which the species was described.
Habit: Rusty red-brown to golden-brown to pale brownish-green, pinnately branched, flattened shoots that have the apex of the shoot pointing downward. Loosely attached when forming mats, usually producing attenuate, brittle branchlets as well as larger lateral branches.
Habitat: Epiphytic, especially on tree trunks of broad-leafed maple but also on other trees; frequent on shaded, dry rock surfaces.
Reproduction: Sporophytes abundant on epiphytic plants but infrequent on plants of rock surfaces; immersed and on the undersurface of the main shoots.
World Distribution: From central Europe to the Mediterranean area of western Asia and North Africa; reappearing disjunctively in western North America from southeastern Alaska to California and eastward to western Alberta, Montana and South Dakota.
B.C. Distribution: Map 69, page 353.
Distinguishing Features: The flattened leafy shoots, which are usually golden brownish-green and with regular lateral branching, the frequent presence of slender, fragile branchlets, the undulate leaves, and the immersed sporangia on the underside of the shoots are all extremely useful characters. Male plants produce numerous tiny, distinctive, bulb-like branches on the lateral shoots.
Similar Species: *Neckera pennata* plants are similar in some respects but the leaves are glossy, while those of *Metaneckera* are somewhat dull; *N. pennata* also lacks fragile slender branchlets. *N. douglasii* also lacks the branchlets and tends to be pale yellow-green with sporophytes on a conspicuous elongate seta, unlike the short seta of *N. pennata* and *Metaneckera*. Species of *Plagiothecium* are all irregularly branched, produce sporophytes on an elongate seta that emerges on the upper side of the shoot, and lack brittle lateral branchlets.
Comments: The name *Neckera menziesii*, considered a synonym, also has been applied to this species. The distribution of this moss is especially interesting representing, as it does, a group of species otherwise known only from the Mediterranean region.

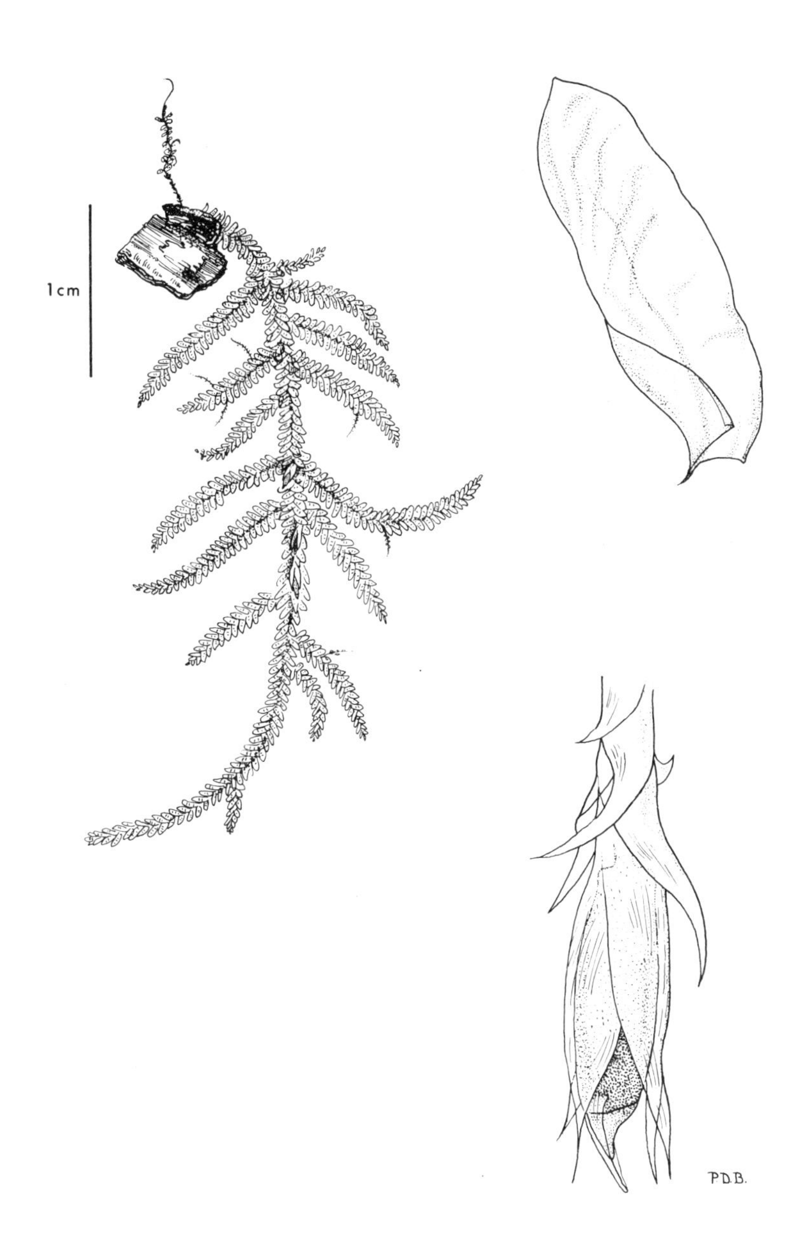

Metaneckera menziesii

Mnium spinulosum B.S.G.

Name: Genus name an ancient Greek name for moss. Species name referring to the spiny teeth of the leaf margins.
Habit: Loose, bluish-green to dark green turfs of unbranched shoots attached to the substratum by red rhizoids, the upper leaves often forming a rosette.
Habitat: Frequent in coniferous forests, usually on humus or at tree bases but also on rock and rotten logs.
Reproduction: Sporophytes frequent, maturing in late spring to summer; seta red-brown, sporangium light brown when mature, peristome mouth and teeth red-brown.
World Distribution: Circumboreal; in North America across the continent, mainly in boreal coniferous forests but extending southward from Labrador to Maryland in the east and from Alaska to California in the west.
B.C. Distribution: Map 70, page 354.
Distinguishing Features: This *Mnium* usually has sporophytes, is usually on humus or at tree bases and the leaves differ little whether wet or dry. Under the hand lens the pairs of teeth on the leaf margin are distinctive. The mouth of the sporangium also turns dark brown well before the sporangium ripens, a feature unusual in *Mnium*. The red-brown peristome teeth are also characteristic.
Similar Species: The double-toothed, differentiated margins of the leaves separate the genus *Mnium*; of the local species of this genus, *M. ambiguum*, *M. arizonium*, *M. marginatum* and *M. thomsonii* are possible to confuse with *M. spinulosum*. *M. arizonicum* and *M. thomsonii* are usually less than half the size of *M. spinulosum* and are normally not forest-floor species. Peristome teeth are yellow in *M. thomsonii*; they are red-brown in *M. spinulosum*. *M. ambiguum* and *M. marginatum* tend to have the leaves contorted when dry, are more frequent on rock rather than humus and are found in more humid sites than *M. spinulosum*, a species of well-drained humus or logs. Microscopic features are more reliable but these pose some difficulties to a beginner.

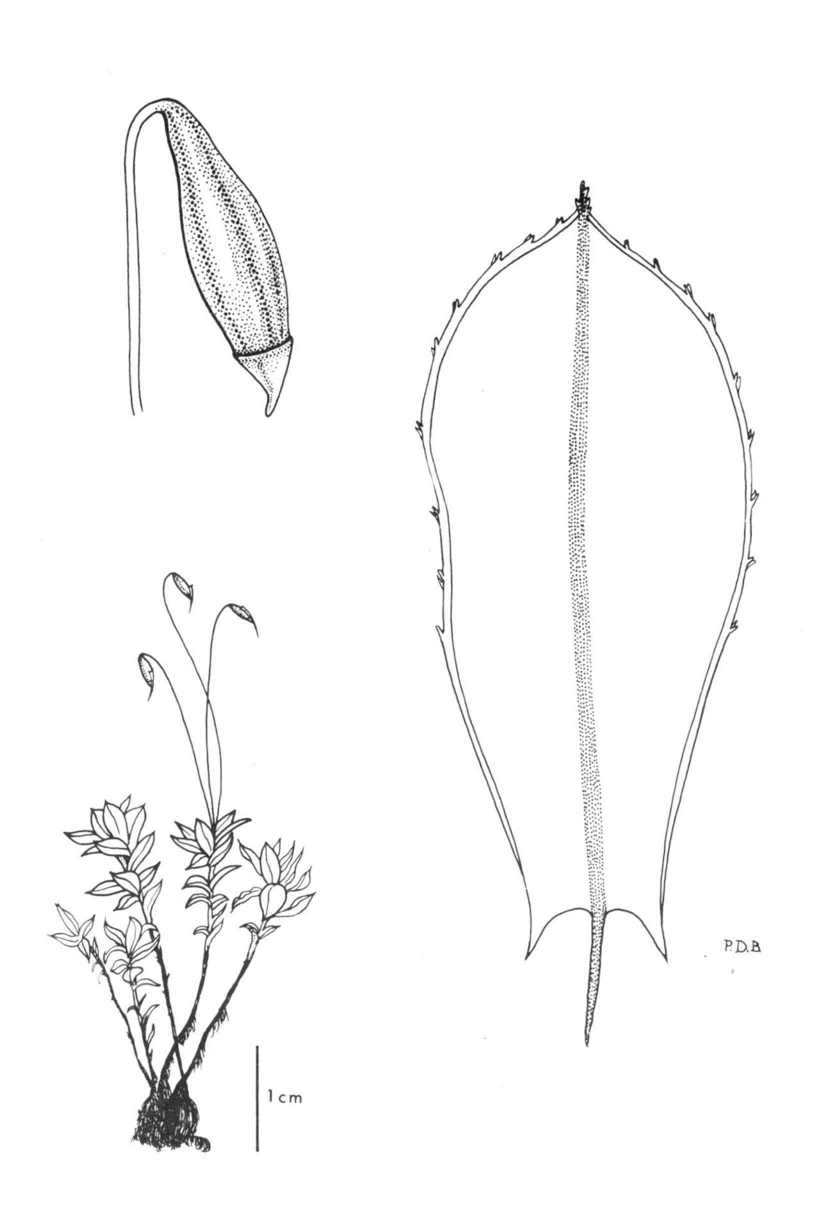

Mnium spinulosum

Neckera douglasii Hook.

Name: Genus named in honour of N.J. de Necker, an 18th-century German botanist. Species named in honour of David Douglas, a well-known, 19th-century Scottish plant explorer of western North America.
Habit: Pale yellow-green, regularly to irregularly branched flattened shoots, hanging down on perpendicular sites and pendent on horizontal sites, sometimes forming dense mats.
Habitat: Most commonly epiphytic on trunks and branches of living trees in open coniferous or broad-leafed forests, also on cliffs and rocks.
Reproduction: Sporophytes occasional to locally abundant, pale red-brown and hanging downward from the underside of the leafy shoot.
World Distribution: Confined to western North America from south-eastern Alaska to California and extending eastward to western Montana.
B.C. Distribution: Map 71, page 354.
Distinguishing Features: The usually soft, glossy, yellow-green, flattened shoots with undulate leaves and shining, relatively regular pinnate branching serve as useful features, especially when sporophytes are present.
Similar Species: See notes under *Metaneckera menziesii*.
Comments: This species is sometimes extremely abundant in humid coastal forests where it festoons tree branches and sheathes trunks of small trees and shrubs.

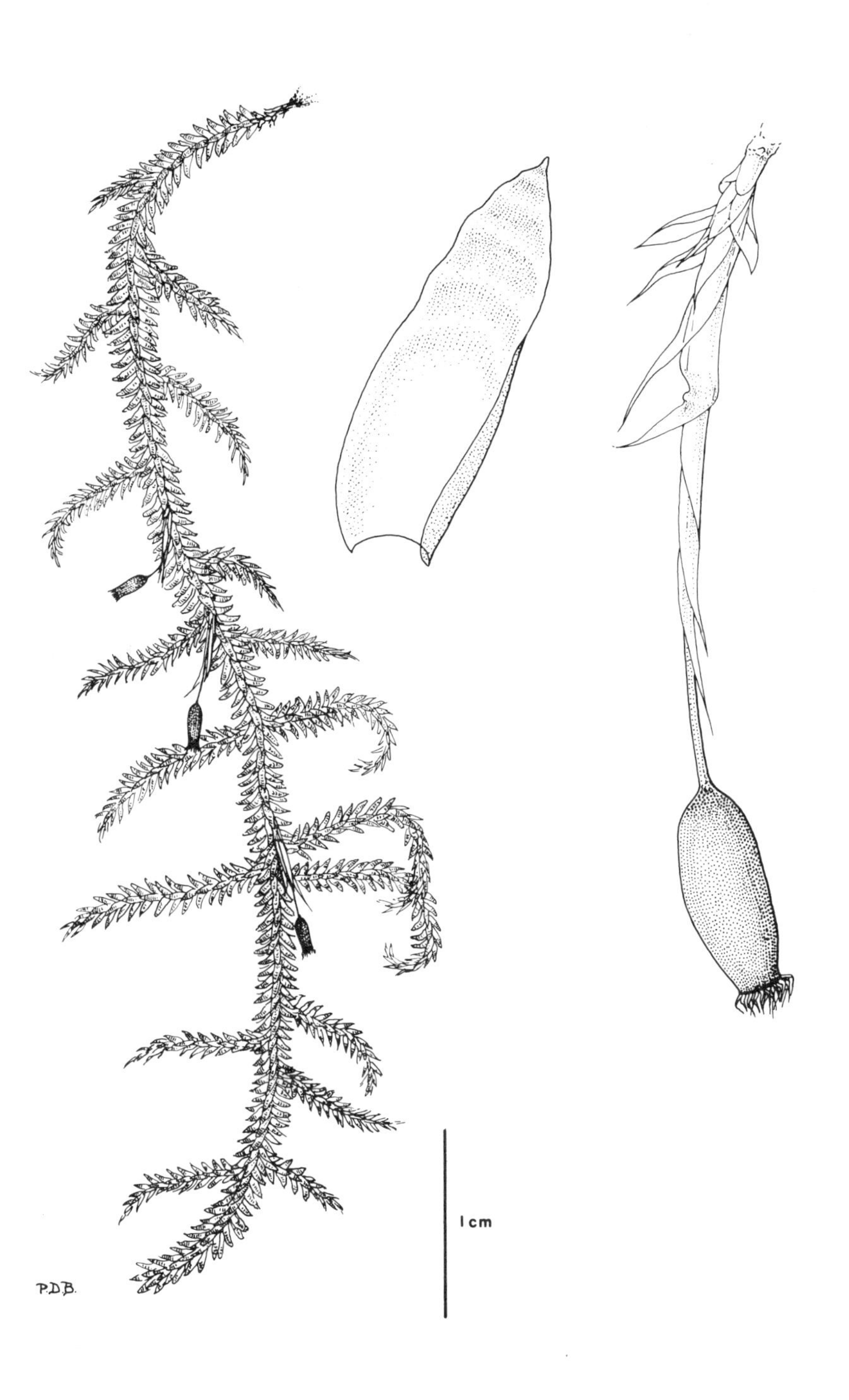

Neckera douglasii

Oligotrichum aligerum Mitt.

Name: Genus name meaning few hairs, in reference to the few hairs on the calyptra. Species name meaning winged, probably in reference to the ridges on the back of the midrib.
Habit: Loose to dense, dark-green to reddish-brown turfs with the leaves divergent when humid but imbricate when dry.
Habitat: Frequent on the disturbed earth of banks and trail margins, also along water courses, from sea level to subalpine elevations.
Reproduction: Sporophytes frequent, maturing in summer, dark brown when mature. Male plants are usually in separate colonies, the bracts around the sex organs forming a reddish-brown cup.
World Distribution: In western North America from southeastern Alaska to California and eastward to Idaho, also in Mexico; in eastern Asia from Japan, Taiwan and the Philippines.
B.C. Distribution: Map 72, page 355.
Distinguishing Features: The very small, often dark green to wine-red tinted, rigid plants with leaves showing ridge-like lamellae on both surfaces, combined with the disturbed mineral soil habitat are characteristic.
Similar Species: *O. hercynicum* is very similar but lacks the ridge-like lamellae on the back of the leaves. *O. hercynicum* is also alpine. *Psilopilum cavifolim* also lacks the ridges on the back of the leaf and the plants are extremely small (usually less than ¼ the size of *Oligotrichum*). *Psilopilum* is extremely rare and is known from only the northermost part of the province.

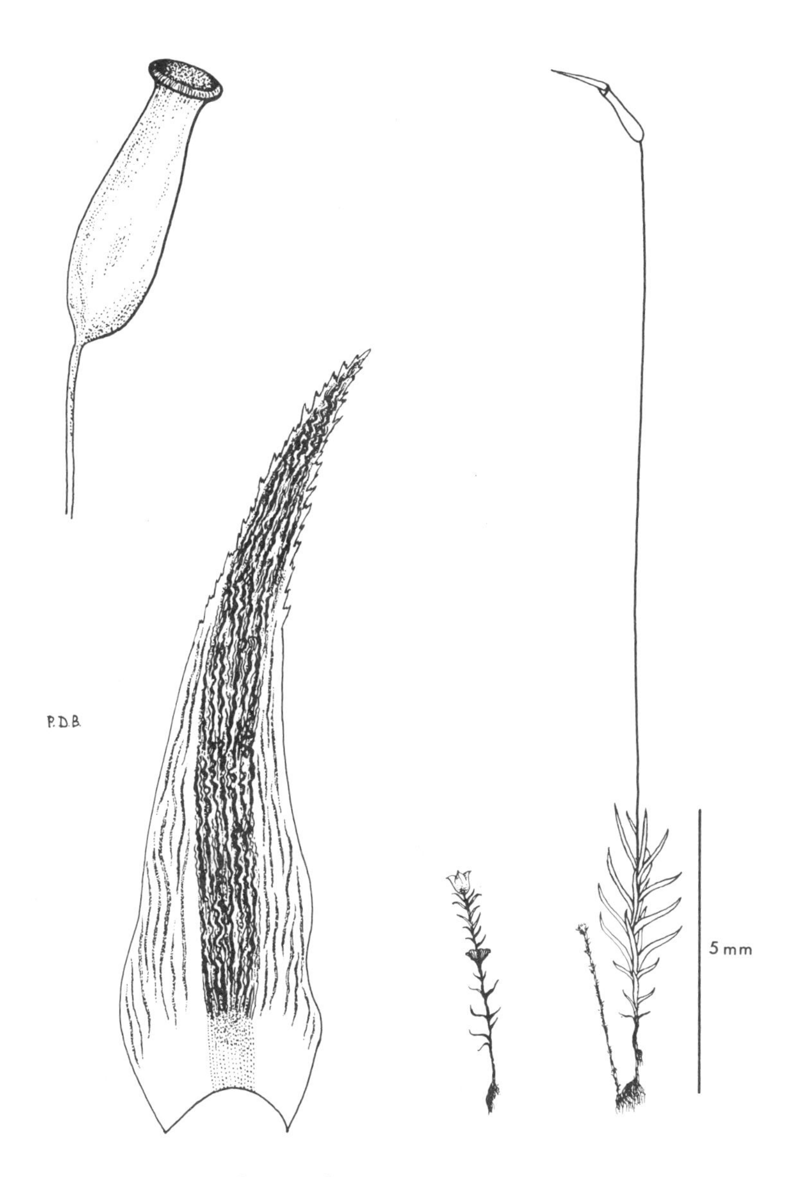

Oligotrichum aligerum

Oncophorus wahlenbergii Brid.

Name: Genus name referring to the Adam's apple-like swelling at the neck of the sporangium. Species named in honour of G. Wahlenberg, a Swedish botanist of the 19th century who collected the original specimen.
Habit: Forming pale green, yellow-green to brownish-green, short turfs. Leaves sheathing at base and contorted when dry.
Habitat: Rotten logs and rock crevices, especially in areas where there is a fluctuation in water level; from near sea level to subalpine forests.
Reproduction: Sporophytes frequent, reddish-brown when mature, with bright red-brown peristome teeth.
World Distribution: Circumpolar in the Northern Hemisphere; in North America across the continent in arctic latitudes, southward in the east to the New England and Great Lakes states and in the west to California.
B.C. Distribution: Map 73, page 355.
Distinguishing Features: The subsheathing leaves, their remarkable contortion when dry, the sporangia with very red-brown peristome teeth and distinct knob in the neck, and the usual habitat on logs in perennially flooded areas are useful features.
Similar Species: *O. virens* has recurved leaf margins and enlarged cells at the lowermost margin of the leaf while *O. wahlenbergii* lacks these features visible with a hand lens. *O. wahlenbergii* usually has somewhat clasping leaf bases while *O. virens* lacks them. See also notes under *Kiaeria starkei*.

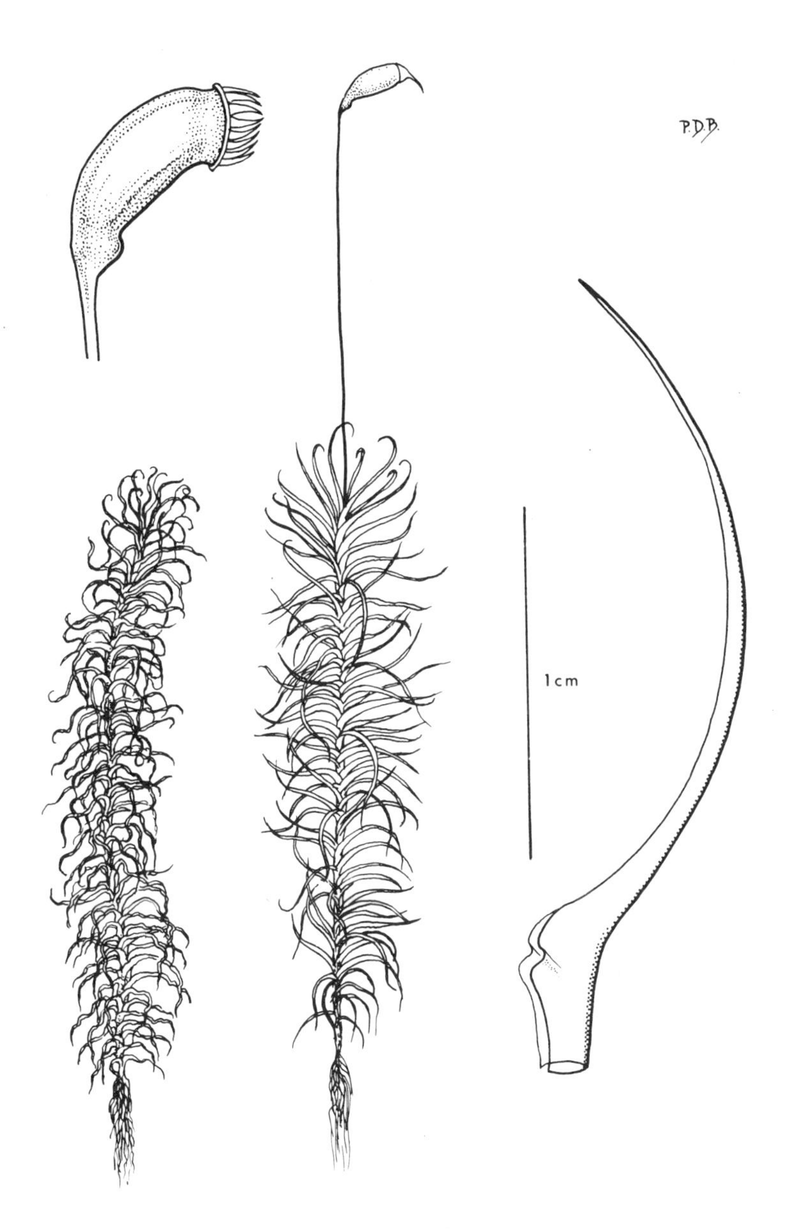

Oncophorus wahlenbergii

›trichum *lyellii* Hook. & Tayl.

..ıe: Genus name denoting the straight, erect hairs on the calyptra of many species. Species named in honour of C. Lyell (1767–1849), a British botantist and father of the famous geologist of the same name.
Habit: Forming tufts of dark green plants attached at the base by rhizoids.
Habitat: Epiphytic on living trees, especially alder, maple and oak, but also on spruce, hemlock, yew and yellow cedar, most frequent near sea level on the coast, but rare in humid forests of the interior.
Reproduction: Sporophytes frequent, pale brown to dark brown when maturing in spring; also producing tiny gemmae on the leaves.
World Distribution: Scattered through Europe southward to the Mediterranean area of North Africa and western Asia; in North America confined to west of the Rocky Mountains from southeastern Alaska to Baja California, eastward to Idaho.
B.C. Distribution: Map 74, page 356.
Distinguishing Features: The open tufts, the epiphytic habitat, the grooved (when mature) sporangia that are barely emergent from perichaetial leaves, the white peristome teeth, and the long narrow leaves are useful features.
Similar Species: There are many species of *Orthotrichum* in British Columbia, and many are epiphytic. In *O. lyellii*, however, male and female plants are separate and the male plants bear distinctive bulb-like branches. Male plants are usually present near female plants. *O. striatum*, of similar habitats, has sporangia that are not grooved when mature.

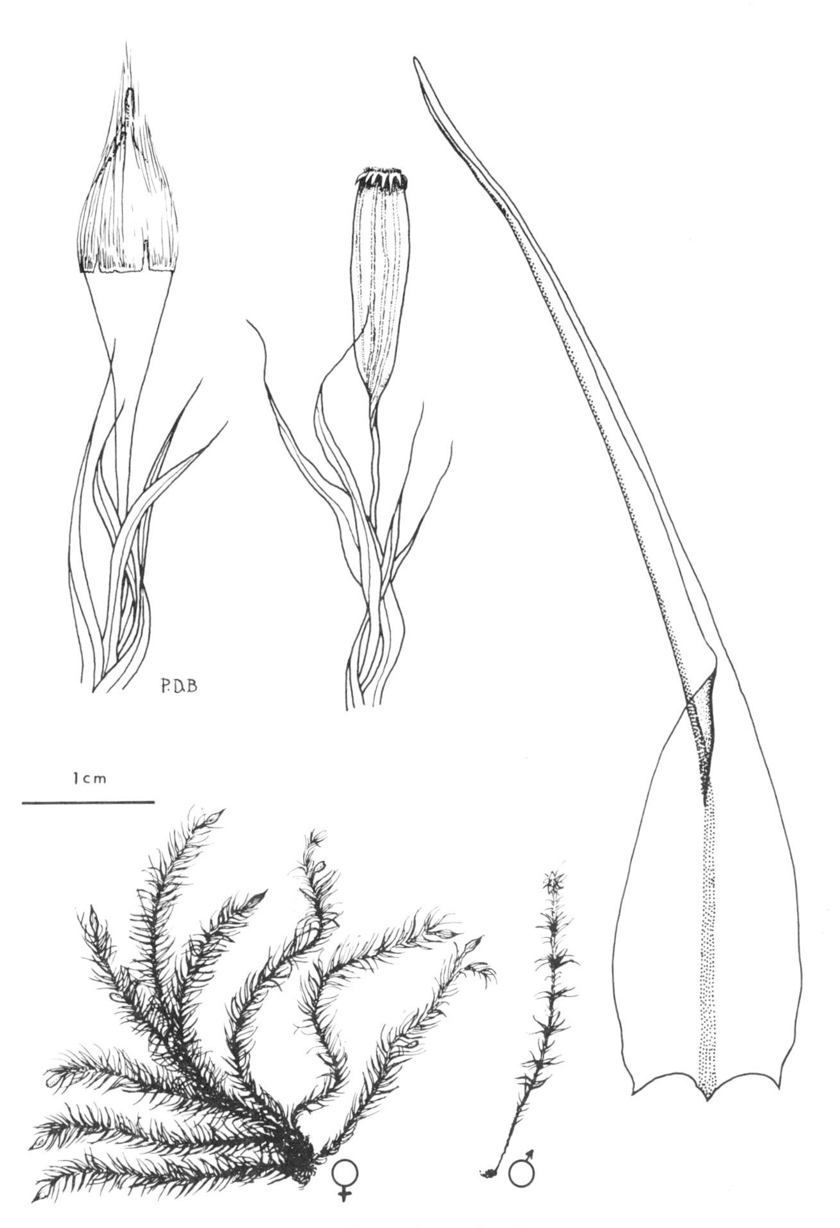

Orthotrichum lyellii

Paludella squarrosa (Hedw.) Brid.

Name: Genus name denoting a little marsh in reference to the habitat. Species name describing the leaf arrangement where the upper part of the leaf curves abruptly downward (see illustration).
Habit: Forming dense, tall turfs of yellow-green to pale green plants in which the leaves are arranged in five neat rows. The shoot, when viewed from above, has the leaves radiating like the five arms of a star.
Habitat: In calcium-rich wetlands.
Reproduction: Sporophytes rare, pale brown when mature.
World Distribution: Circumpolar in the Northern Hemisphere; in North America extending southward in the east to the Great Lakes states and in the west to Wyoming.
B.C. Distribution: Map 75, page 356.
Distinguishing Features: The regular, five-rowed arrangement of the extremely squarrose leaves, plus the moist calcareous habitat make this moss strikingly distinctive in British Columbia.
Similar Species: None.

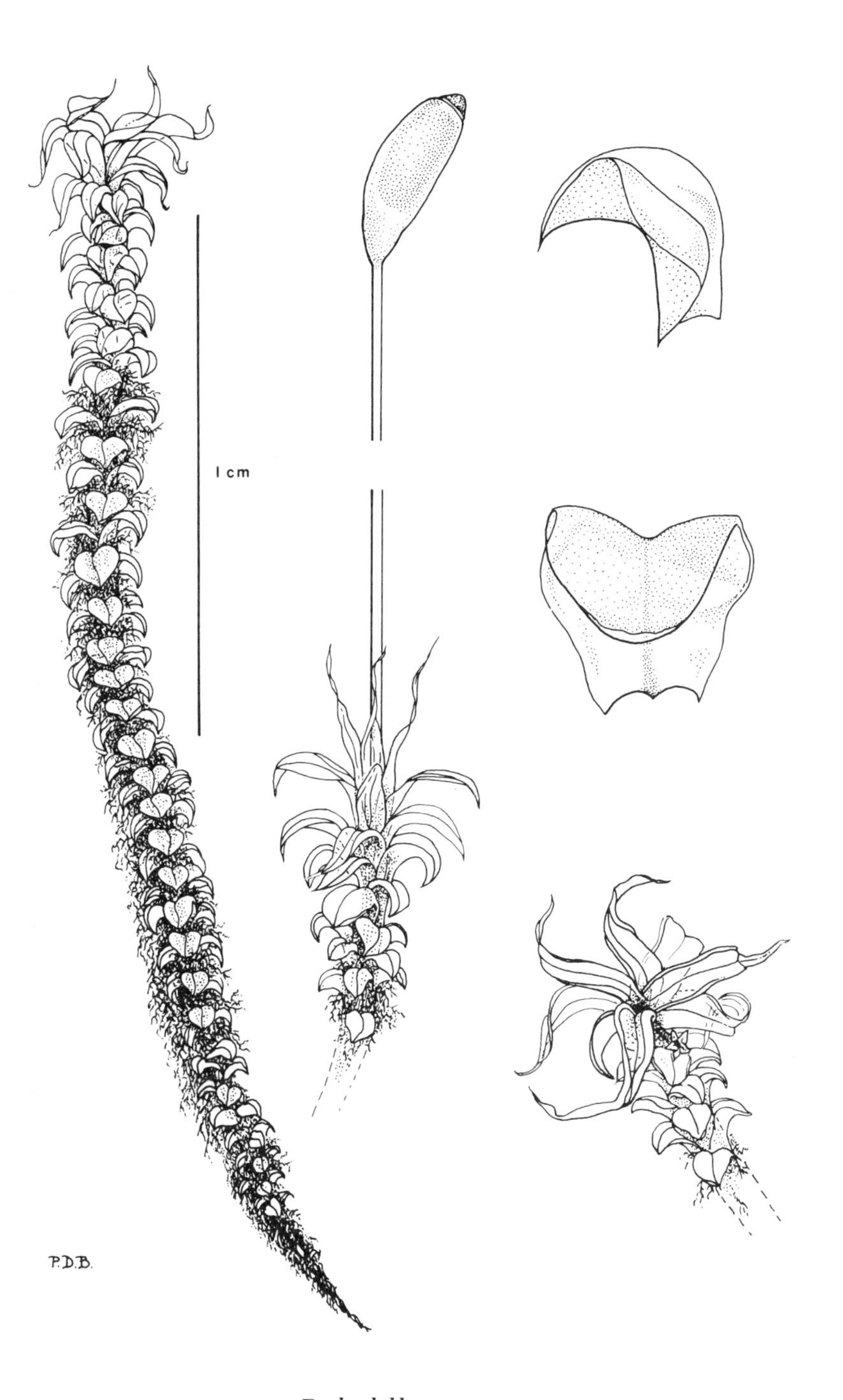

Paludella squarrosa

Philonotis fontana (Hedw.) Brid.

Name: Genus name meaning lover of moisture, in reference to the wet habitat. Species name also referring to the habitat of springs and fountains.

Habit: Forming dense, tall, bright yellow-green to golden-green turfs densely tangled with red rhizoids.

Habitat: Wet, seepage or springy sites over rock, gravels or cliff shelves, often terrestrial, from sea level to alpine elevations, always in open sites; sometimes forming extensive turfs along streams in alpine areas and covering wet cliffs and banks.

Reproduction: Sporophytes occasional to locally abundant, reddish-brown when mature; the sporangium erect, green and subspherical when immature, but red-brown, inclined and regularly grooved when ripe. *P. fontana*, especially in humid, shaded sites, often produces masses of deciduous branches that serve in reproduction.

World Distribution: Circumpolar in the Northern Hemisphere; throughout North America.

B.C. Distribution: Map 76, page 357.

Distinguishing Features: The tall, turf-like plants in wet habitats, tightly compacted with red-brown rhizoids, the male shoots terminated by a flower-like rosette of leaves and often bearing short lateral branchlets below this rosette, plus the round sporangia, grooved and inclined when dry, are useful features.

Similar Species: *P. capillaris* is less than ¼ the size of *P. fontana* and grows on disturbed mineral soil. Other species of *Philonotis* are impossible to distinguish on field characters and are troublesome even with microscopic features. *Conostomum*, similar in some respects to *Philonitis*, tends to form hard, rounded tufts in well-drained sites and the stiff, leafy shoots have the leaves in five neat rows, thus differing from the soft mats of *Philionitis*.

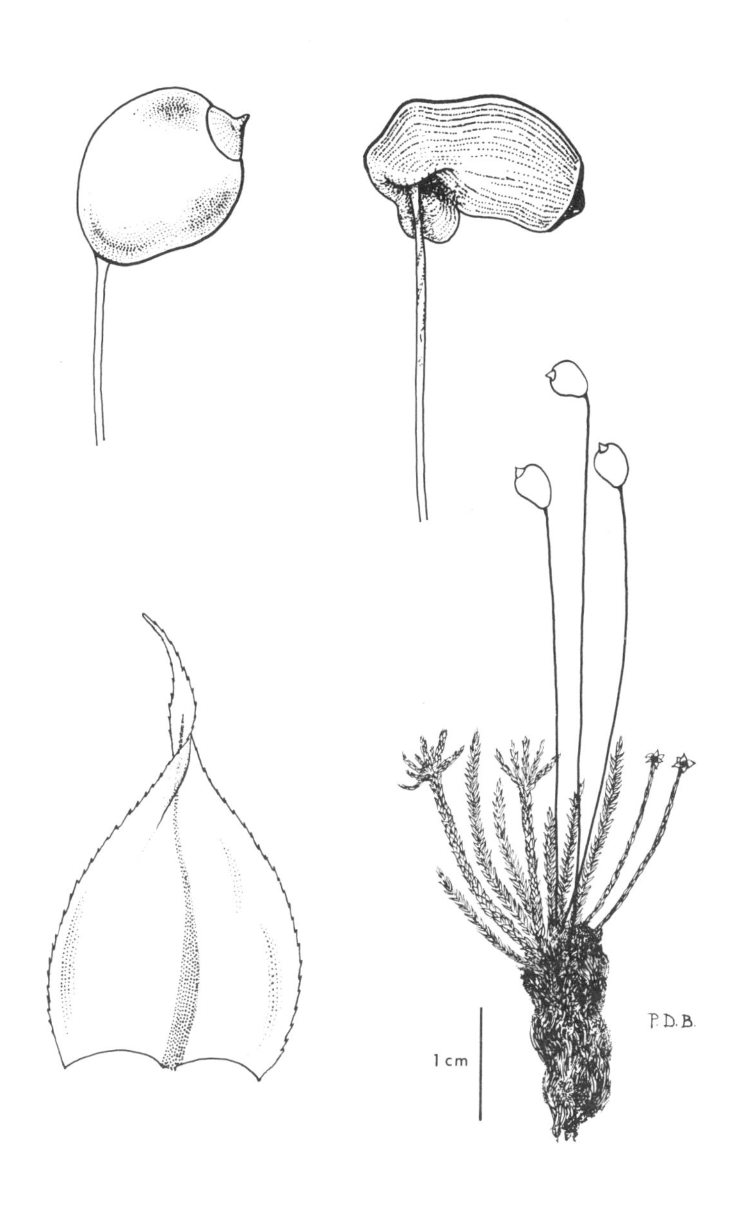

Philonotis fontana

,iomnium insigne (Mitt.) Kop.

Name: Genus name meaning oblique *Mnium*, probably in reference to the obliquely arching shoots of many species. Species name meaning notable, possibly in reference to the remarkable size of the plants or perhaps to the apical rosette of leaves in male shoots that resemble a decorative medal.
Habit: Forming loose, tall turfs of translucent, dark green to light green plants attached to the substratum by dark brown rhizoids; leaves strongly contorted when dry.
Habitat: Somewhat shaded banks and swampy areas or alluvial sites, occasionally in lawns, especially in shaded sites, most frequent near sea level near the coast where it is most common in floodplain forests, but extending to subalpine elevations.
Reproduction: Sporophytes frequent, usually several arising in a cluster on each shoot, maturing in spring, light brown when ripe, later turning dark brown. Male plants with flower-like cluster of leaves around the male sex organs; sexes usually in separate colonies.
World Distribution: Confined to western North America from southeastern Alaska to California and eastward to western Montana.
B.C. Distribution: Map 77, page 357.
Distinguishing Features: This is the largest *Plagiomnium* in British Columbia. It is usually terrestrial or on rotten logs in broad-leafed forests of humid climates. The strongly decurrent leaves with pointed tips are distinctive, as is the flower-like rosette of leaves at the apex of the male shoots. The presence of several sporophytes from each female plant characterizes many species of *Plagiomnium*.
Similar Species: *P. ciliare*, although with leaves bearing decurrent bases, regularly produces a single sporophyte on each shoot. In *P. cuspidatum* and *P. drummondii*, the marginal teeth of the leaves are above the leaf middle only (to the base in *P. insigne*); *P. venustum* is less than half the size of *P. insigne* and confined to well-drained sites (especially frequent as an epiphyte) and separate male plants are absent; *P. venustum* is also yellow-green rather than translucent dark green. *P. medium*, although having decurrent bases, has older shoots heavily clothed with rhizoids; old shoots of *P. insigne* are nearly naked. *P. ellipticum* lacks decurrent leaf bases.
Comments: This species also has been treated under the name *Mnium insigne*, a synonym.

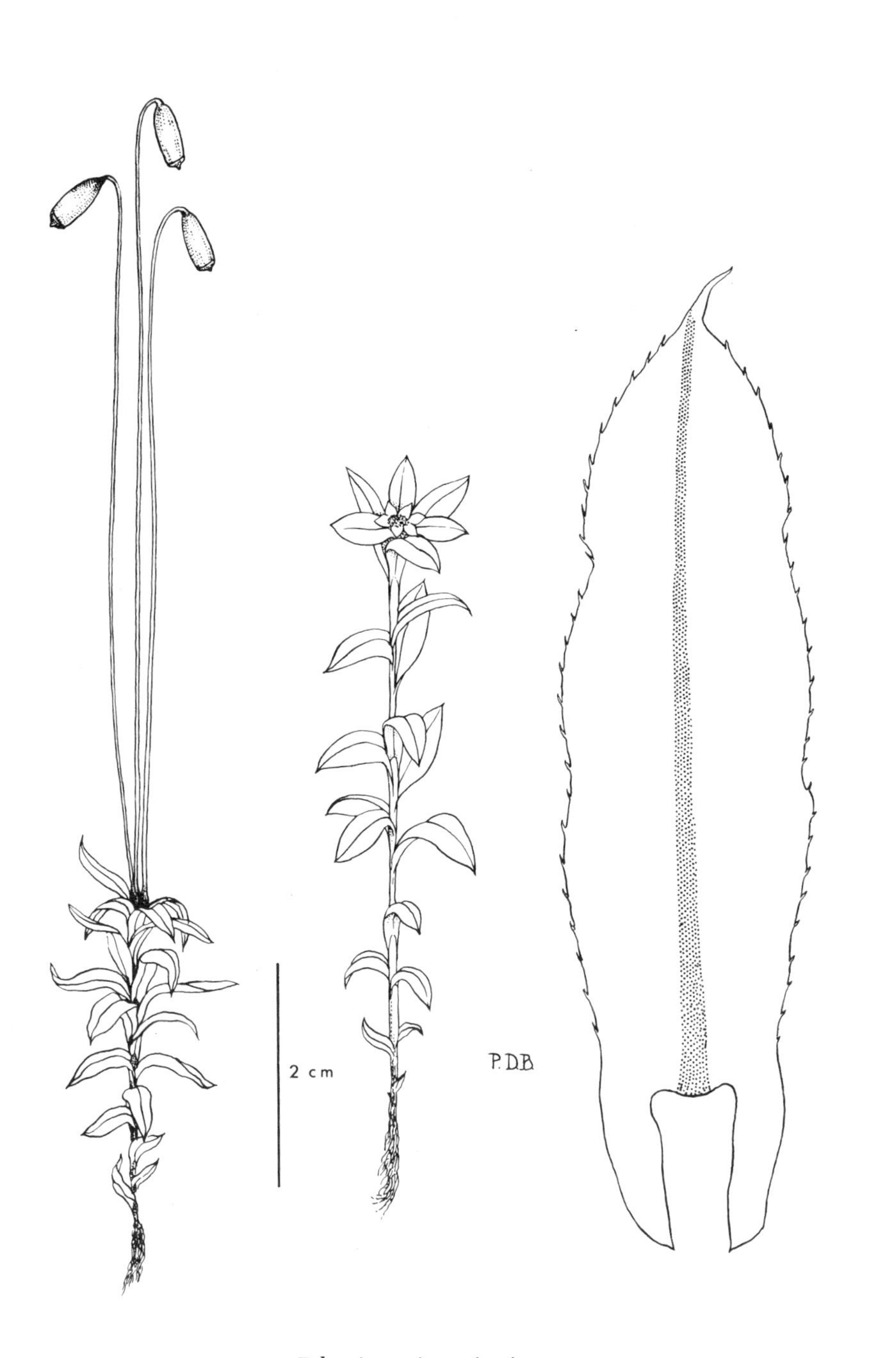

Plagiomnium insigne

Plagiopus oederi (Brid.) Limpr.

Name: Genus name meaning oblique footed, perhaps in reference to the seta and its connection to the sporangium in the original specimen upon which the name is based. Species named in honour of G.C. von Oeder (1728–1791), a Danish botanist who first recognized this moss as distinctive.
Habit: Forming short to tall, dark green turfs of unbranched plants, sometimes with notably three-ranked leaf arrangement (the stem is triangular in cross section).
Habitat: On humid calcium-rich cliff shelves.
Reproduction: Sporophytes frequent, maturing in spring; dark brown when ripe.
World Distribution: Circumpolar in the Northern Hemisphere; in North America extending southward in the east to Virginia and Iowa, in the west to Oregon and eastward to Colorado and Utah.
B.C. Distribution: Map 78, page 358.
Distinguishing Features: The roughly three-ranked leaf arrangement, the dark green leaves that are not glossy, the calcareous habitat and spherical sporangia are useful features.
Similar Species: All species of *Bartramia* have somewhat glossy, yellowish-green, rather than dull dark green leaves as in *Plagiopus*. In *Anacolia menziesii* the plants are heavily invested with rhizoids and the leaves are little altered when dry and golden yellow-green compared to the dark green of *Plagiopus*.

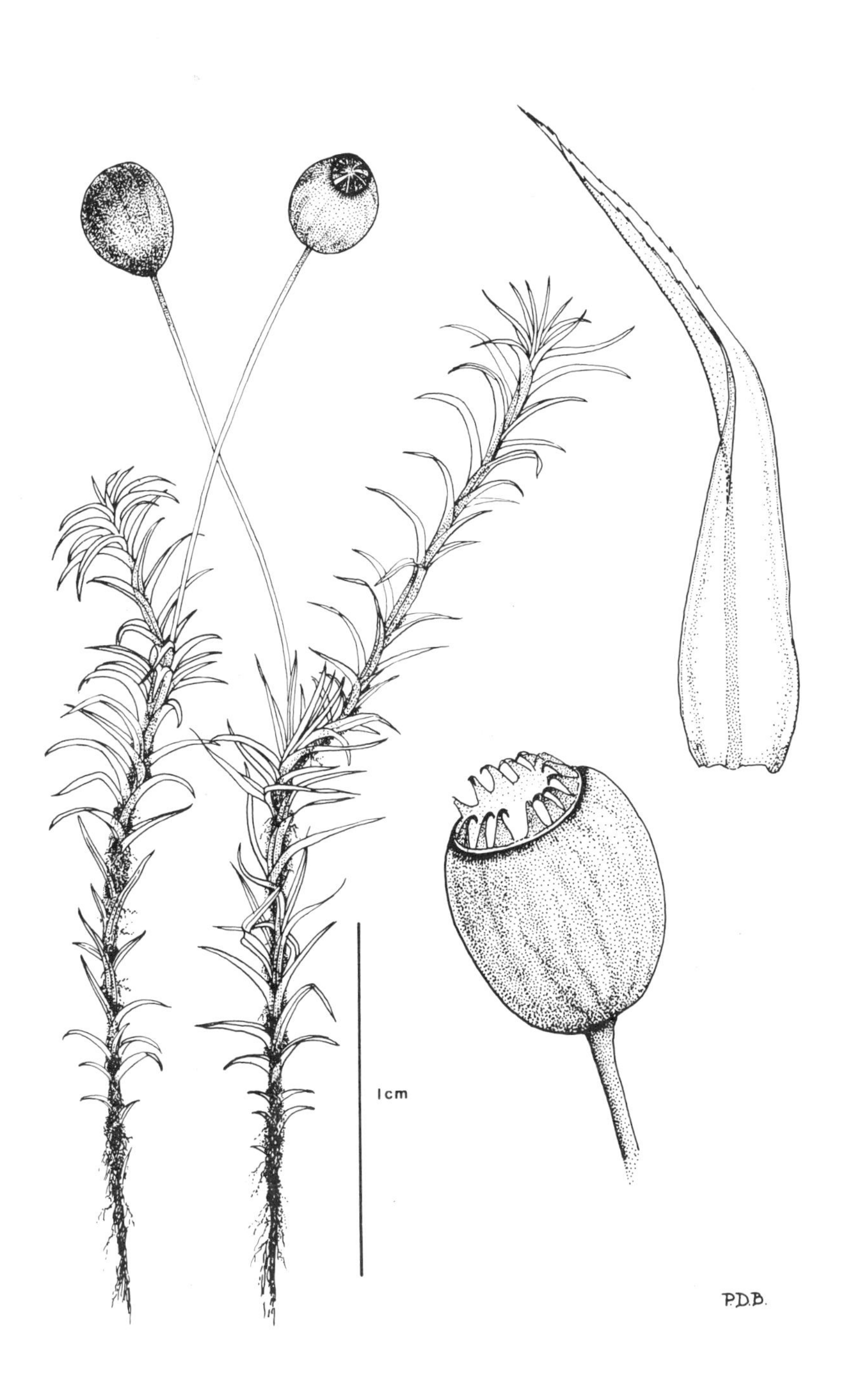

Plagiopus oederi

Plagiothecium denticulatum (Hedw.) B.S.G.

Name: Genus name describing the obliquely oriented sporangia of many species. Species name referring to the teeth at the leaf apex.
Habit: Forming glossy, flattened, pale green to dark green, creeping shoots in which leaves are slightly or not undulate.
Habitat: Most frequent on humid, shaded cliffs but also terrestrial and on tree bases; rarely on rotten logs.
Reproduction: Sporophytes occasional, light brown when ripe; maturing in spring.
World Distribution: Circumpolar in the Northern Hemisphere, also in Africa and Australasia; widely distributed in North America.
B.C. Distribution: Map 79, page 358.
Distinguishing Features: The flattened, glossy, pale to dark green plants, the leaves that lack an obvious midrib (the double midrib is usually obscure), the abruptly pointed leaf apex and the elongate, slightly curved sporangium are useful distinguishing features.
Similar Species: *Plagiothecium laetum* closely resembles *P. denticulatum* but is usually much smaller and found most frequently on shaded, rotten logs; *P. denticulatum* is frequently on humid rock surfaces. Microscopic details, especially the cells of the leaf base as they extend down the stem, are the most useful identification features: in *P. denticulatum* these cells are thin-walled and swollen; in *P. laetum* they are not swollen. *P. cavifolium* usually has somewhat turgid, instead of flattened, shoots and is mainly terrestrial or in cliff crevices. *Porotrichum bigelovii* also has flattened leafy shoots, like *Plagiothecium denticulatum*, and occurs on damp cliffs, but the leaves have a strong midrib, tend to diverge almost at right angles to the stem (those of the *Plagiothecium* diverge at an acute angle toward the shoot apex), and the shoots are wiry, compared to the soft shoots of *Plagiothecium*. *Hookeria lucens* and *H. acutifolia* show the outlines of the leaf cells clearly under 10X magnification and lack the decurrent bases present in *Plagiothecium denticulatum*. *Rhynchostegium serrulatum*, extremely rare in British Columbia, has a midrib and the leaves have toothed margins. See also notes under *Neckera*, *Metaneckera* and *Plagiothecium undulatum*.

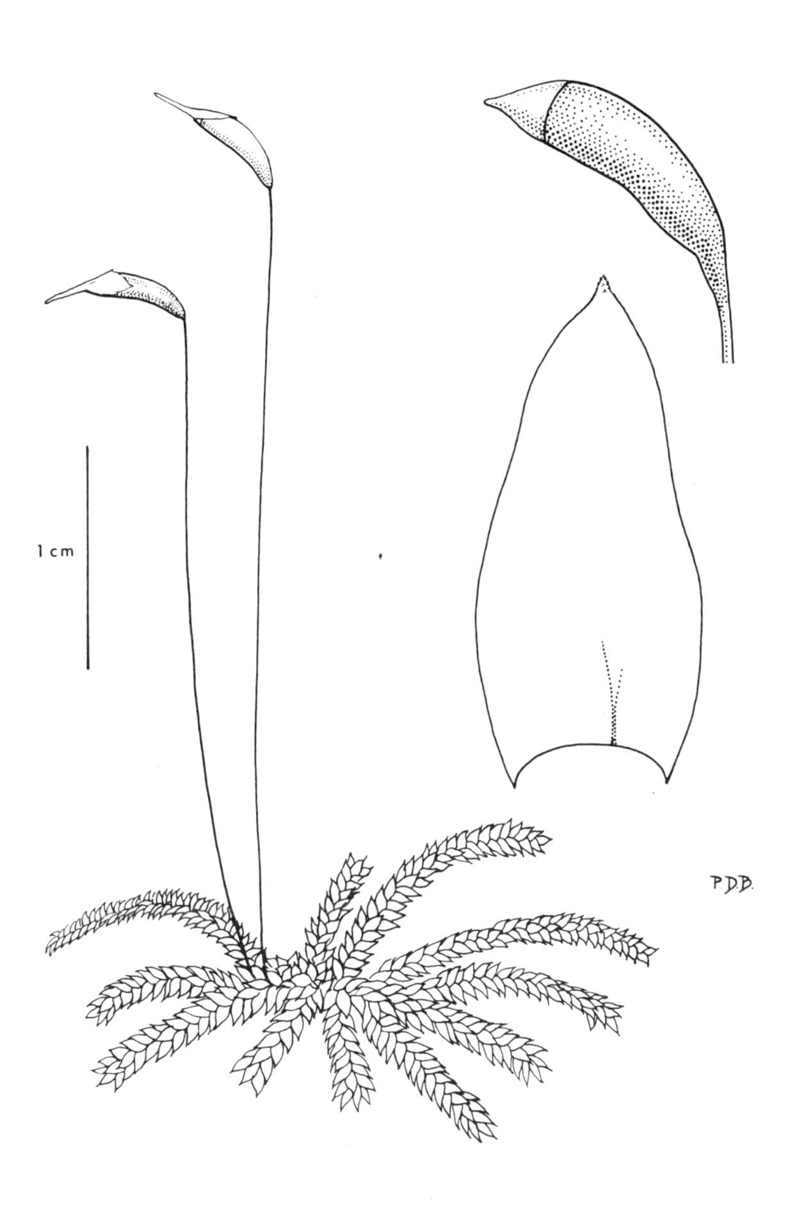

Plagiothecium denticulatum

:ium undulatum (Hedw.) B.S.G.

ies name describing the undulations of the leaves.
ning pale, yellow-green to whitish-green mats of undulate-
leafed shoots.

Habitat: Rotten logs, humus and rocks in humid, shaded forest sites near the coast and mainly at lower elevations.

Reproduction: Sporophytes occasional, red-brown and grooved when ripe in late spring.

World Distribution: Widespread in Europe and western North America; in North America from southeastern Alaska to northern California and eastward to Idaho.

B.C. Distribution: Map 80, page 359.

Distinguishing Features: The habitat on soil, humus or rotten logs combined with the pale, yellow-green to whitish-green colour, the flattened leafy shoots in which the leaves are undulate, plus irregular branching, lack of obvious glossiness and presence of grooved sporangia on a long seta make this moss distinctive.

Similar Species: From other species of *Plagiothecium*, the strongly undulate, pale yellow-green to whitish-green, barely glossy leaves are usually enough to separate this species. See, however, notes under *Neckera* and *Metaneckera* and *Plagiothecium denticulatum*.

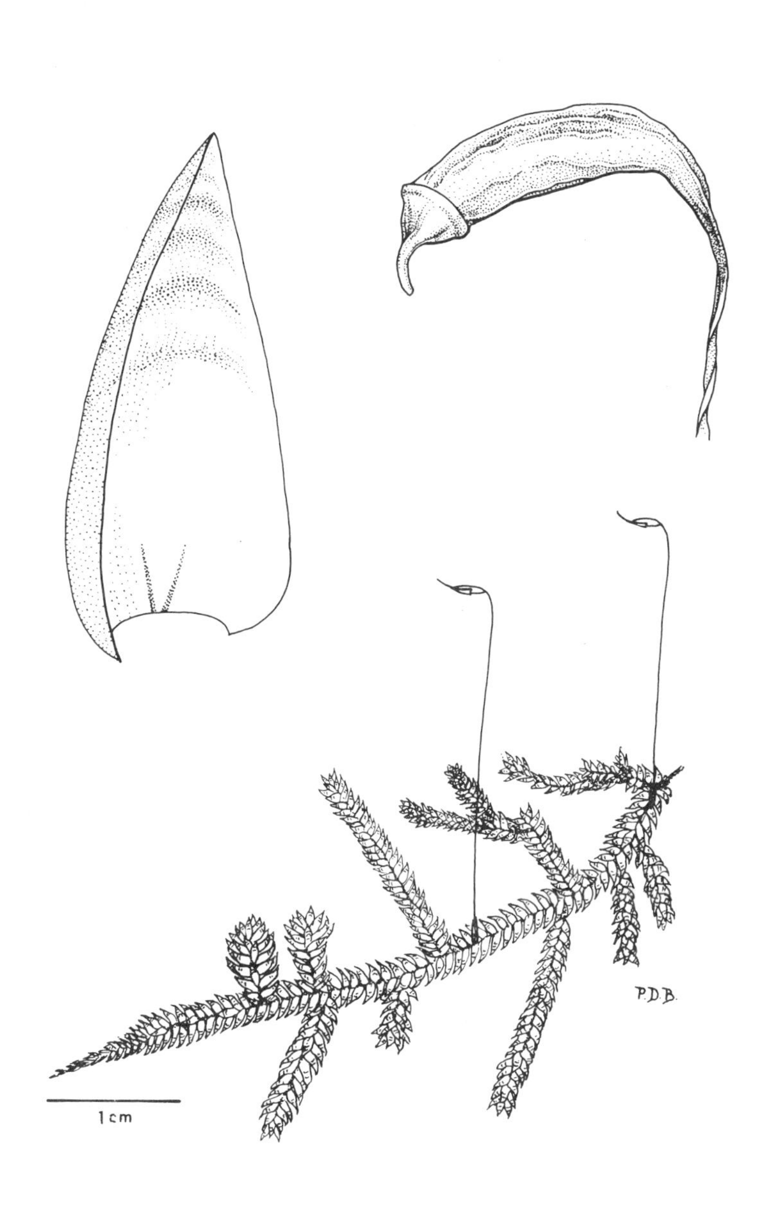

Plagiothecium undulatum

Pleurozium schreberi (Brid.) Mitt.

Name: Genus name presumably referring to the loosely pinnate branching of the plants. Species named in honour of J.C. Schreber (1739–1810) a German botanist.
Habit: Thick loose mats of semi-erect to reclining, glossy yellow-green, golden-green to dull green, regularly branched shoots with conspicuously red shoots.
Habitat: On sterile litter, over rock and on cliff ledges, occasionally ascending tree bases and in bogs. Mainly at lower elevations but ascending to subalpine forests and occasionally into alpine elevations.
Reproduction: Sporophytes occasional to locally abundant, dark red-brown when mature, usually maturing in summer to autumn.
World Distribution: Circumpolar in the Northern Hemisphere, south to northern South America. Widespread across North America, extending southward in the east to North Carolina and Arkansas and in the west to Oregon and eastward to Colorado.
B.C. Distribution: Map 81, page 359.
Distinguishing Features: The red stem, the pinnate branching, the broadly ovate leaves with an apiculate tip, and the short double midrib usually separate this species of drier, usually forested habitats.
Similar Species: *Pseudoscleropodium purum*, a moss of city lawns, has green, rather than red, stems and leaves with a single midrib. *Calliergonella cuspidata*, a plant of wet sites, including lawns, also has green rather than red stems, and the leaves show no obvious midrib, but show swollen alar cells (alar cells in *Pleurozium* are small and angular). Some specimens of *Hylocomium splendens* from alpine areas may resemble *Pleurozium*, but the stems of *Hylocomium* are furry (under 10X hand lens) with paraphyllia.
Comments: Sometimes popularly called red-stem moss.

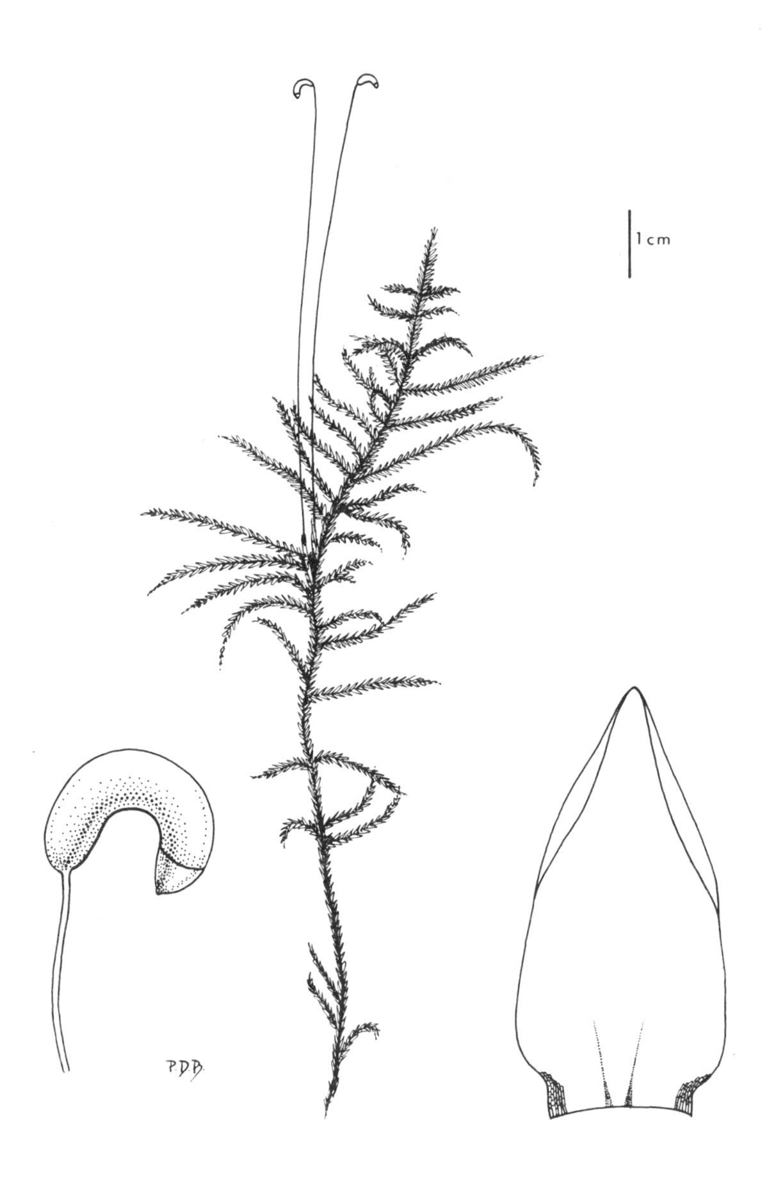

Pleurozium schreberi

ɪmatum contortum (Brid.) Lesq.

Name: Genus name meaning born with a beard, in reference to the hairy calyptra. Species name describing the leaves that become contorted when dry.
Habit: Tall, loose to dense turfs of dark green, unbranched plants with leaves strongly divergent when humid, much contorted when dry. When young, a green webbing of protonema stabilizes the sandy substrata and scattered small shoots are produced. This protonema disappears when the turfs become continuous.
Habitat: Commonly on disturbed mineral soil of shaded banks, especially in forest areas, but from sea-level to open alpine elevations.
Reproduction: Sporophytes frequent, maturing in early spring, with the red-brown sporangia sheathed by a whitish, hairy calyptra.
World Distribution: Southeast Asia and western North America; in North America from southeast Alaska to California eastward to eastern British Columbia.
B.C. Distribution: Map 82, page 360.
Distinguishing Features: The dark green plants with leaves diverging outward exposing a wide flat surface, the many teeth on leaf margins, the strong contorting on the leaf margins, the strong contorting of the leaves, when dry, and the sporangia (circular in cross section) all serve as useful distinguishing features.
Similar Species: See also notes under *Polytrichum commune*, *P. alpinum* and *Atrichum selwynii*.

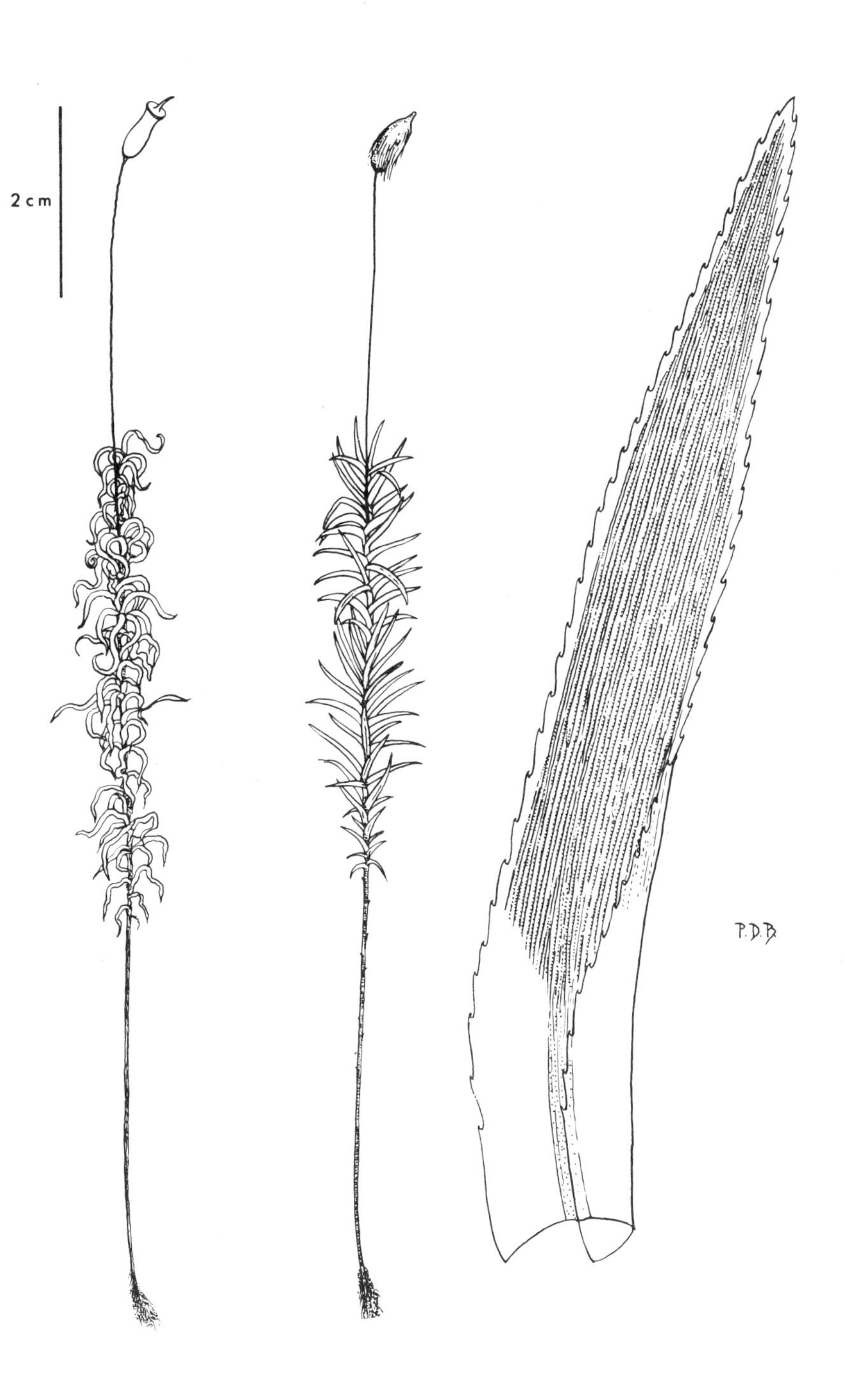

Pogonatum contortum

Pogonatum urnigerum (Hedw.) P. Beauv.

Name: Species name meaning urn-bearing, in reference to the sporangia.
Habit: Dense to loose, short, bluish-green turfs in which the stems are wine-red and attached at the base to the substratum by many rhizoids.
Habitat: Common in open areas on mineral soil and over rock, especially near water courses, from sea level to alpine elevations.
Reproduction: Sporophytes common, maturing in summer to autumn but persisting sometimes through two years.
World Distribution: Circumpolar in the Northern Hemisphere; in North America south to the New England states and Great Lakes in the east and to northern California in the west.
B.C. Distribution: Map 83, page 360.
Distinguishing Features: The powdery bluish-green plants with wine-red stems, the leaves occasionally tinged with red, with toothed margins, combined with the stiff leaves with lamellae on the upper surface, and the autumn maturing sporangia that are round in cross section, are all useful characters. Often the shoots are branched and each branch may bear a sporophyte.
Similar Species: *Pogonatum dentatum* cannot be readily distinguished on hand lens characters. *Polytrichum alpinum* of high elevations lacks the bluish-green leaves and may resemble *Pogonatum*. *Polytrichum sexangulare* has glossy, dark green leaves that lack marginal teeth.

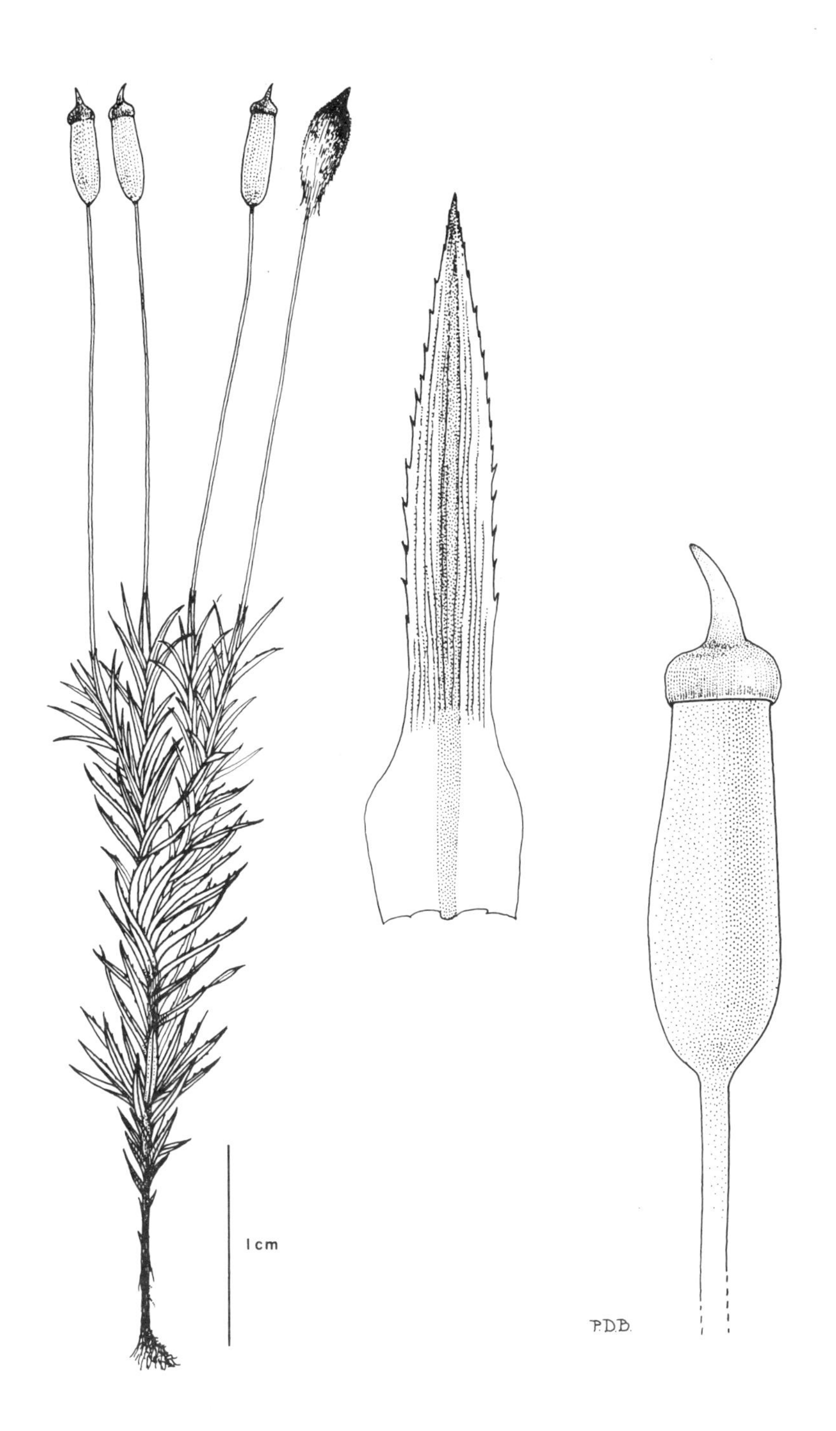

Pogonatum urnigerum

Pohlia annotina (Hedw.) Lindb.

Name: Genus name honouring J.E. Pohl, a German physician. Species name referring to last year's growth, the significance of which is not clear.
Habit: Loose, glossy, pale green to yellow-green turfs with reddish stems, at least in lower parts.
Habitat: Disturbed banks, especially along water courses, trail margins and in neglected gardens.
Reproduction: Clusters of gemmae are abundant and pale green, especially from autumn until spring; sporophytes are relatively frequent, red-brown when mature in late spring.
World Distribution: Interruptedly circumtemperate in the Northern Hemishere; in North America, from southern Newfoundland southward to Georgia and westward to the Great Lakes in the east, and from southeastern Alaska to northern California eastward to Idaho in the west.
B.C. Distribution: Map 84, page 361.
Distinguishing Features: When bearing gemmae, these quickly identify *P. annotina*. The sporangia lack any hint of glossiness, a feature of the genus. The widely spaced marginal teeth also characterize the genus.
Similar Species: Many other species of *Pohlia* resemble *P. annotina*, and gemmae are frequently needed to distinguish them. Gemmae differ in colour and form among the species but need microscopic examination for discrimination. From *Bryum*, the toothed leaves and non-glossy sporangia are usually enough to separate these genera.

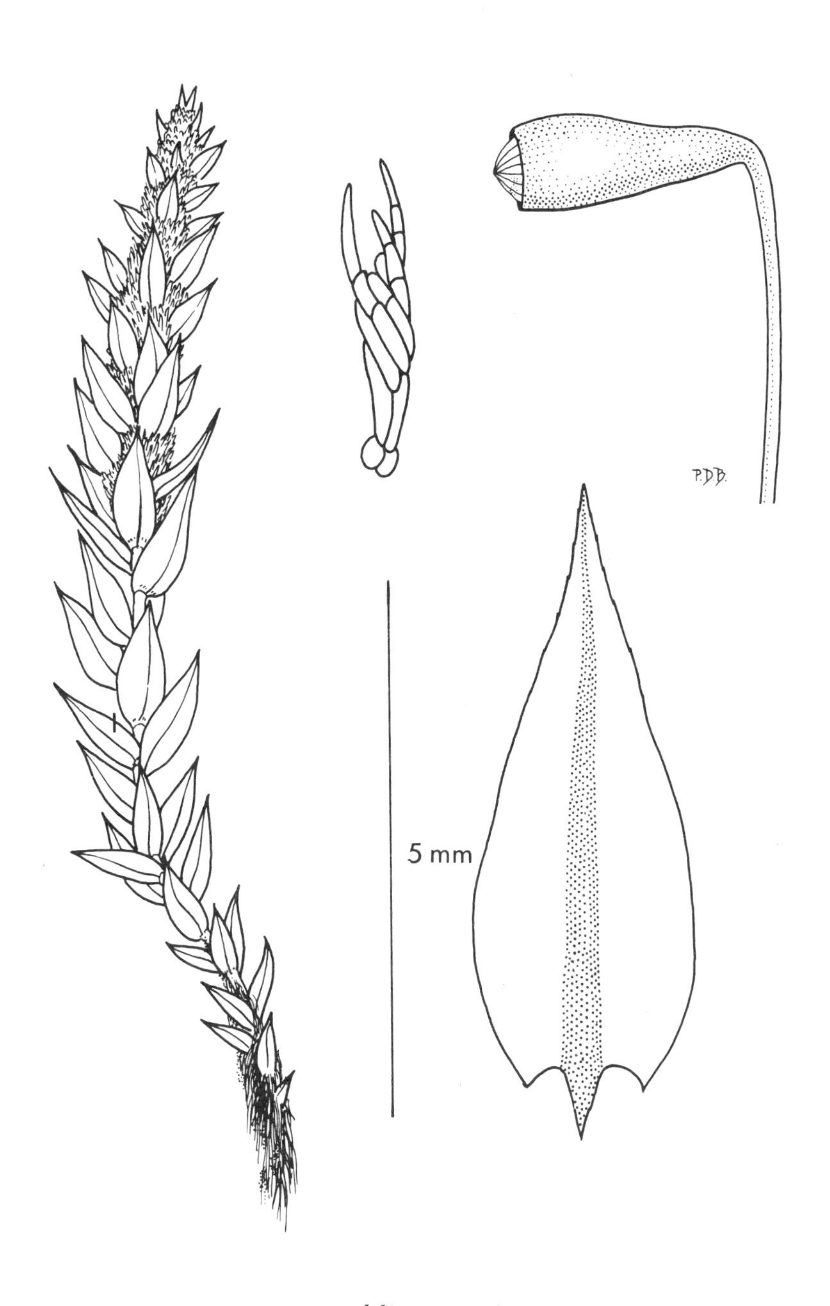

Pohlia annotina

Pohlia cruda (Hedw.) Lindb.

Name: Species name denoting raw or uncooked, presumably in reference to the opalescent sheen of the leaves that sometimes characterizes uncooked meat.
Habit: Forming loose, opalescent, glossy short turfs.
Habitat: Shaded humus of tree bases and cliff crevices from sea level to alpine elevations.
Reproduction: Sporophytes occasional, maturing in spring, light red-brown when ripe, varying from erect to nodding.
World Distribution: Circumpolar in the Northern Hemisphere, also in southern South America and Autralasia; in North America widely distributed, extending southward to Mexico.
B.C. Distribution: Map 85, page 361.
Distinguishing Features: The toothed, opalescent, glossy leaves, the occasionally erect to suberect sporangia and the shaded terrestrial or cliff crevice habitat are all useful features.
Similar Species: In *Pohlia* only *P. cruda* shows the glossy, opalescent leaves.

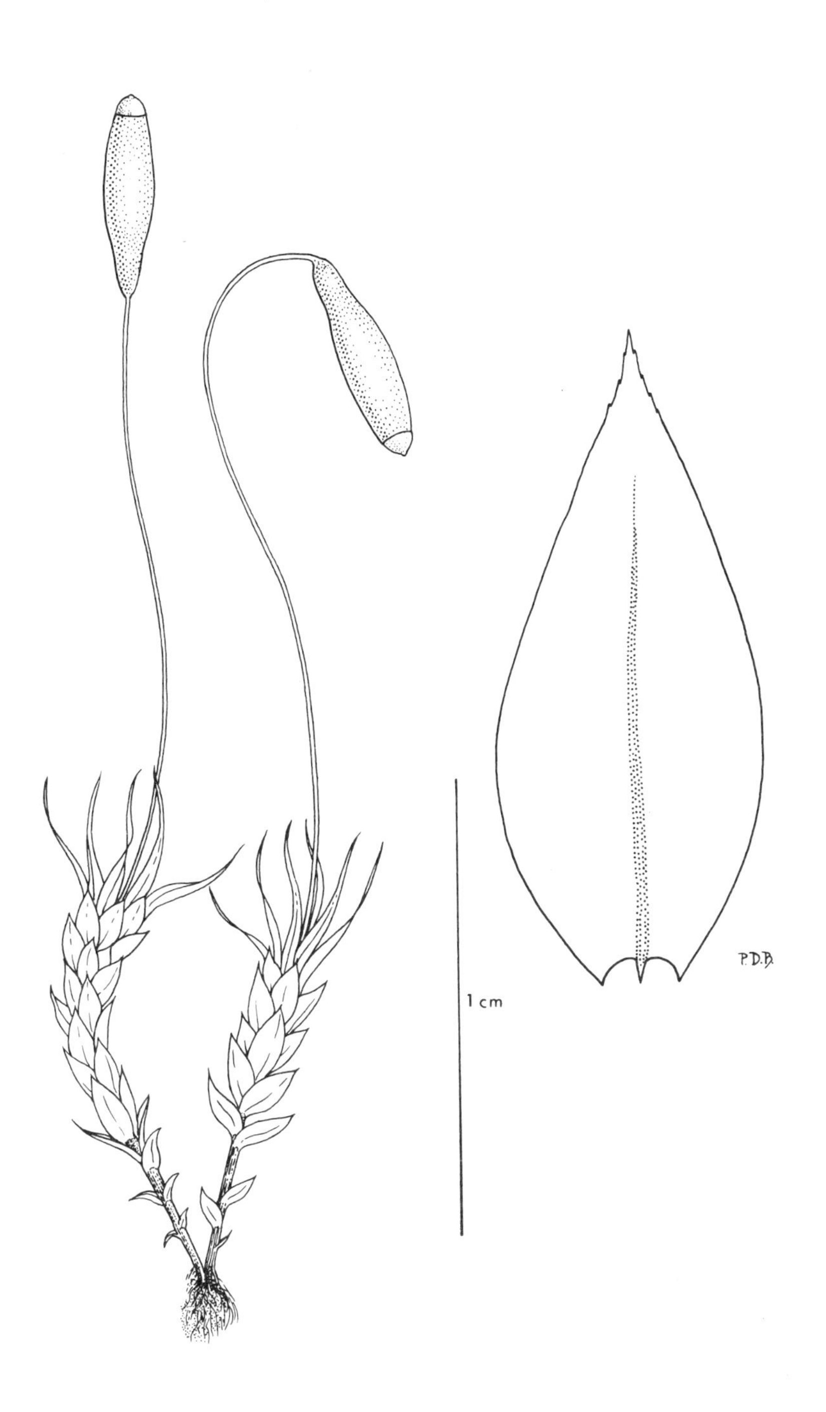

Pohlia cruda

Pohlia nutans (Hedw.) Lindb.

Name: Species name describing the nodding sporangium.
Habit: Forming short, dense, light green to dark green turfs.
Habitat: Mineral soil banks, humus soil and rotten logs in somewhat shaded to sunny sites from sea level to alpine elevations.
Reproduction: Sporophytes common; sporangia light red-brown, not glossy, maturing in summer.
World Distribution: Circumpolar in the Northern Hemisphere, extending in North America southward to Mexico (where rare).
B.C. Distribution: Map 86, page 362.
Distinguishing Features: This is an extremely common and widespread species throughout British Columbia, especially frequent at high elevations and in higher latitudes. There is no reliable field character, however, to separate it from many species of *Pohlia* and *Bryum*.
Similar Species: (See Distinguishing Features above.)

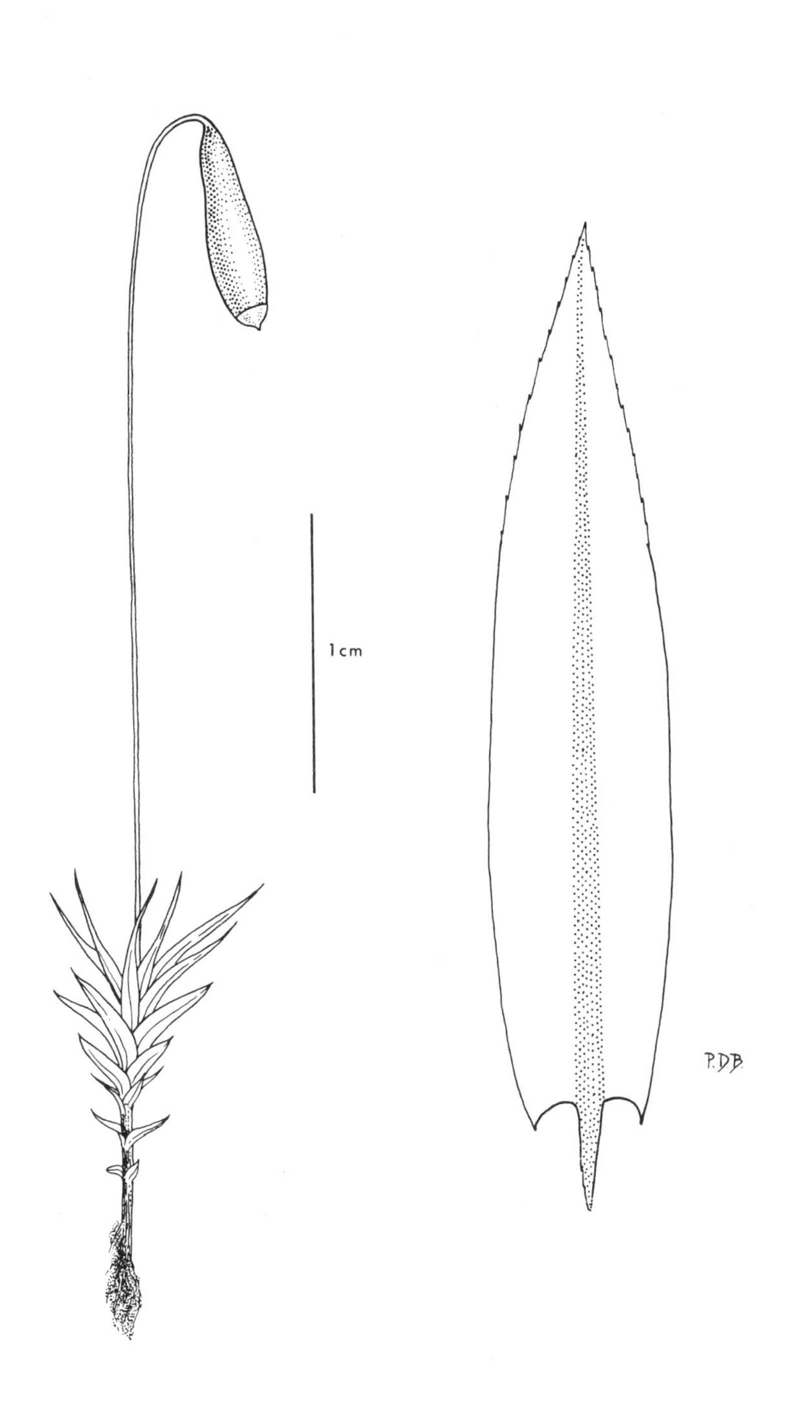

Pohlia nutans

Pohlia wahlenbergii (Web. & Mohr) Andr.

Name: Species named to honour G. Wahlenberg (1780–1851), a Swedish botanist.
Habit: Forming loose, tall to short turfs of whitish-green to yellow-green, unbranched plants with red stems.
Habitat: On mineral soils, frequently calcareous, in damp ditches, stream and lake margins, around springs, forming extensive turfs along alpine streams near late-snow areas; from sea level to alpine elevations.
Reproduction: Sporophytes occasional to locally abundant, red-brown, maturing from late spring to late summer. Sporangia shrinking to half their length from moist to dry conditions.
World Distribution: Circumpolar in the Northern Hemisphere, also in South America and Australasia; widespread in North America except in arctic and interior grassland areas.
B.C. Distribution: Map 87, page 362.
Distinguishing Features: The whitish, yellow-green turfs, the reddish stems, the rosette-tipped male shoots, the nodding sporangia that shrink to half their length when dry, and the wet terrestrial habitat are all useful characters.
Similar Species: When without sporophytes this species may resemble *Philonotis* which grows in the same habitats, but sporophytes separate them (see *Philonotis*). *Pohlia columbica* is similar but smaller (in most cases), and the peristome teeth are yellow rather than red-brown, a feature of *P. wahlenbergii*. *Pohlia longibracteata* is similar, but is a species of sandy banks and the male plants have very long leaves around the sex organs, compared to the bulbiform bracts of *P. wahlenbergii*.
Comments: In mossy meadows around late snow-melt streams, dense turfs of this moss and *Philonotis* often retain glistening spheres of dew or rain for extended periods.

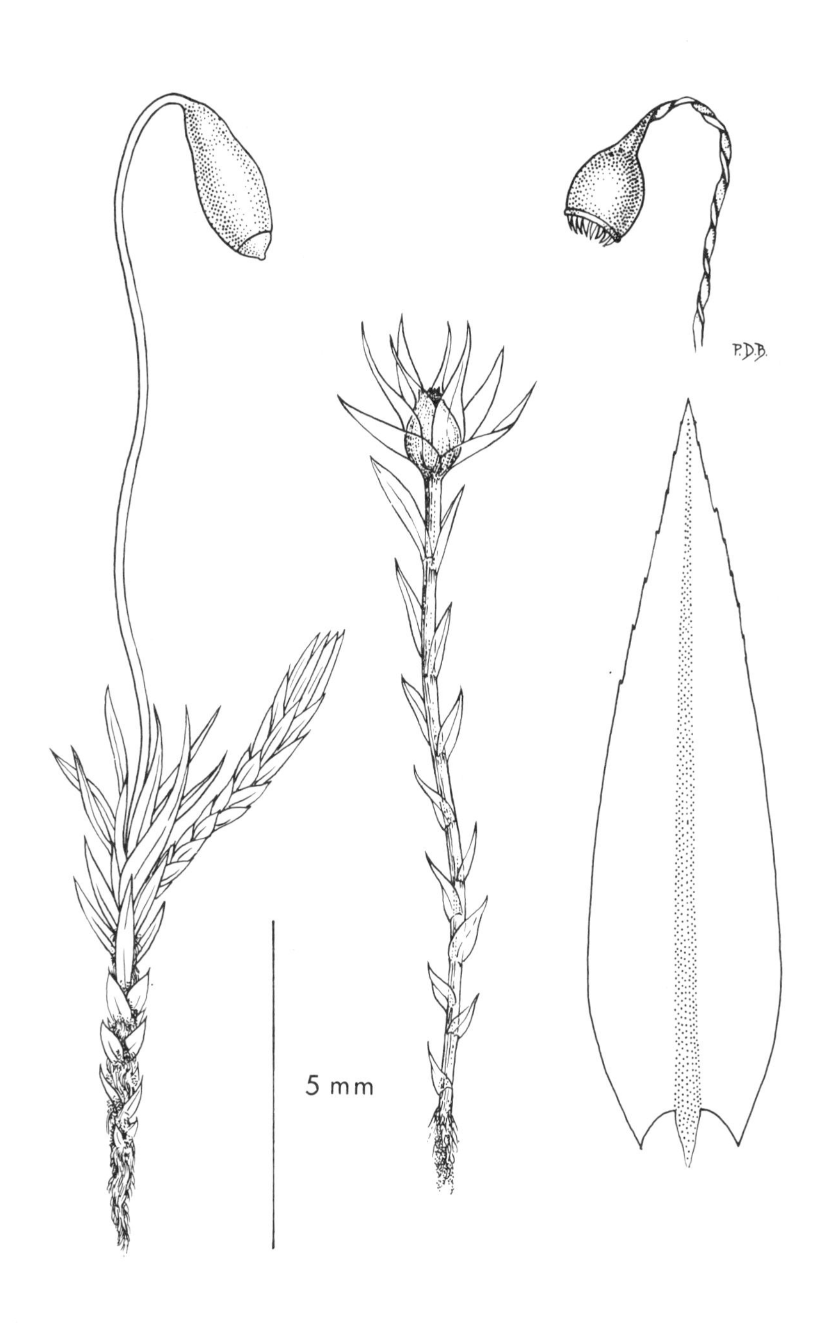

Pohlia wahlenbergii

richum alpinum Hedw.

Name: Genus name referring to the many hairs on the calyptra. Species name indicating the Alps, where the species was first collected.
Habit: Loose, dark green, turfs 20–150 mm tall, attached to the substratum by rhizoids at the base of the shoots.
Habitat: Frequent on earth of banks, cliff crevices, on overturned tree roots, and forest floor, occasional on rotten logs, frequent in coniferous forest; extending from sea level to alpine elevations.
Reproduction: Sporophytes frequent, the setae light brown to straw-coloured when mature, 10–50 mm tall; sporangia pale green when mature in late spring to summer, turning brown with age. Sporangia vary considerably in length; large plants of humid lower elevations have elongate, somewhat curved sporangia while plants of alpine sites sometimes have short, stout, almost spherical sporangia.
World Distribution: Widespread in the Northern Hemisphere, also in mountains of Africa and in southern South America and Australasia.
B.C. Distribution: Map 88, page 363.
Distinguishing Features: Useful features include the toothed leaves with numerous lamellae on the upper surface, the lower portion of the stem with extremely reduced leaves that lack lamellae and the usually elongate sporangium that is rounded in cross section.
Similar Species: From species of *Pogonatum*, *Polytrichium alpinum* differs conspicuously in the time of maturation of the sporangia, in spring and summer instead of autumn. From other *Polytrichum* species, the sporangium that lacks angles, and the lack of a conspicuous bulge at the apex of the seta where it joins the sporangium, are usually diagnostic.
Comments: This species has been treated under the names *Polytrichastrum alpinum*, *Pogonatum alpinum*, *Pogonatum macounii* and *Pogonatum alpinum* var. *sylvaticum*, all of which are considered synonyms.

Polytrichum alpinum

ytrichum commune Hedw.

Name: Species name presumably in reference to its common occurrence.

Habit: Forming dense, tall, dark green turfs 50–500 mm tall, in which the leaves are strongly divergent from sheathing bases when humid, becoming incurved to somewhat contorted when dry.

Habitat: Common on organic soils of sites that retain moisture longer than the surrounding terrain, reaching their greatest size in swamp margins, usually in sunny sites.

Reproduction: Sporophytes common, seta red-brown and wiry, calyptra whitish to light brown, sporangium erect until mature, then inclined. Sporangia maturing in summer.

World Distribution: Cosmopolitan, but more abundant in temperate and frigid climates.

B.C. Distribution: Map 89, page 363.

Distinguishing Features: The toothed leaves with many lamellae on the upper surface, the sheathing leaf bases, the angular sporangia, when mature, and the generally large size (up to 200 mm or more long) usually separate this species.

Similar Species: From *P. juniperinum* and *P. piliferum*, *P. commune* is readily distinguished on the basis of its toothed leaf margins (teeth are lacking the first two species). From *P. alpinum*, the most useful character is the sporangium angled in *P. commune* and not angled in *P. alpinum*. From *P. formosum*, a species of forested habitats, *P. commune* differs in its more obvious shiny sheathing bases of the leaves. In *P. formosum*, the sporangia contract conspicuously below the mouth when dry and the mouth flares outward; in *P. commune* this is not apparent. Young mature sporangia of *P. formosum* are greenish to pale brown; those in *P. commune* are brown. In *P. lyallii*, the sporangium tapers to the mouth and usually has only one longitudinal ridge; that of *P. commune* is four-angled in cross section and has four longitudinal ridges.

Comments: *P. commune* was used to make mattresses in early times. The stems, stripped of leaves, were also bound together to make brooms. An infusion made of this moss has been used as a hair restorative but there is no evidence that it is effective. Small rodents and birds often consume the sporangia. Commonly called hair-cap moss or pigeon wheat.

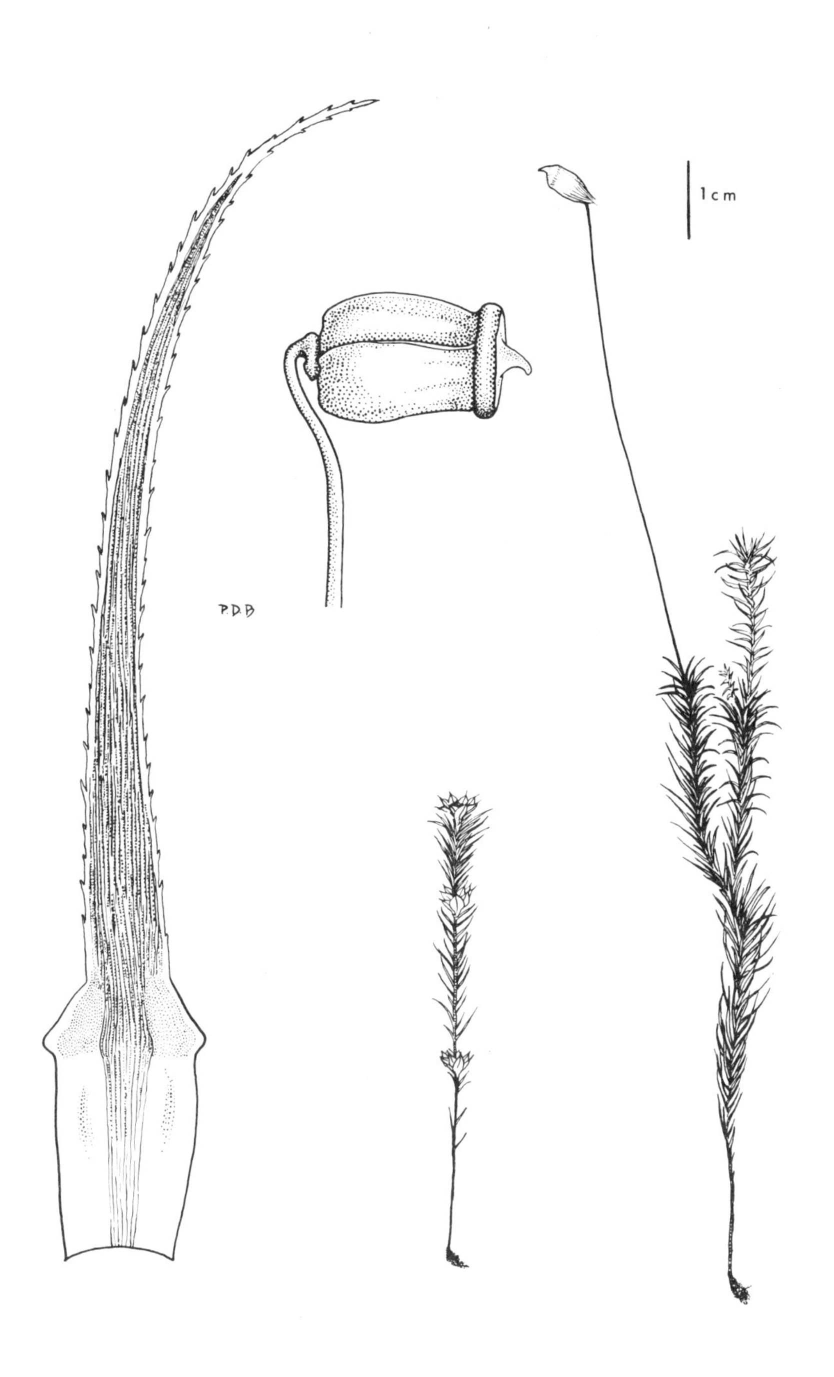

Polytrichum commune

olytrichum juniperinum Hedw.

Name: Species name probably denoting the resemblance of the plants to juniper seedlings.
Habit: Forming short to tall turfs 10–100 mm tall, of dusty blue-green unbranched shoots with leaves strongly divergent when dry and with reddish points.
Habitat: Open dry sterile soil and within forest, also on road banks and cliff ledges, from sea level to alpine elevations.
Reproduction: Sporophytes frequent, the seta wine-red and glossy and the sporangium dull red-brown when mature in summer; erect when young, inclined when ripe.
World Distribution: Cosmopolitan, but more abundant in temperate and frigid climates.
B.C. Distribution: Map 90, page 364.
Distinguishing Features: The most useful features are the red, gradually tapering hair points of the leaves and the incurved, clear leaf blade that overlaps the lamellae on the upper surface of the leaves; this gives the leaves a powdery, bluish-green colour.
Similar Species: *P. piliferum* resembles *P. juniperium* in many respects. In *P. piliferum*, the white hair points are usually much longer (about ¼ the length of the leaf) and taper abruptly from the body of the leaf; those of *P. juniperinum* are red, usually less than ⅛ of the leaf length and taper gradually from the body of the leaf. *P. strictum* is a miniature version of *P. juniperinum*, differing from it in its abundant felting of rhizoids along much of the stem length, its leaves that are generally half the size of those of *P. juniperinum* and in its usually boggy habitat.

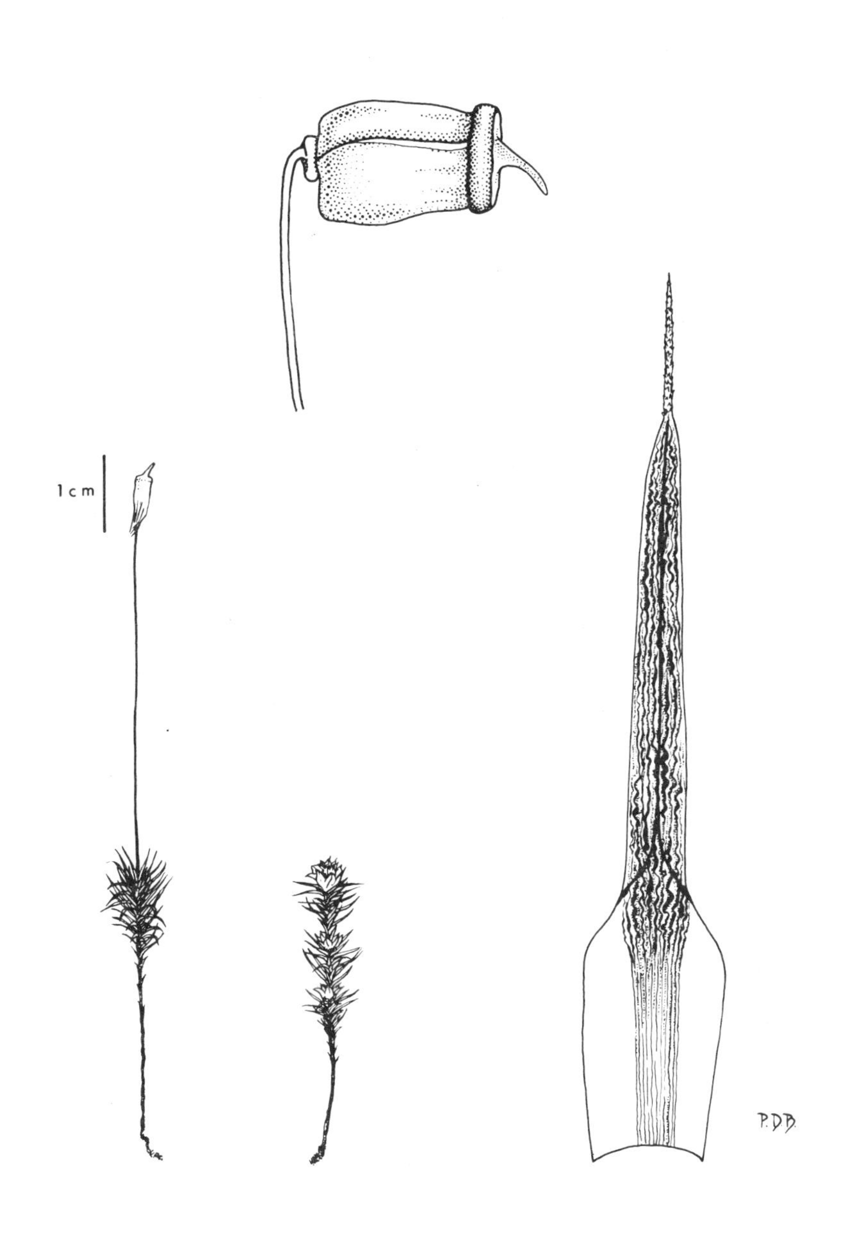

Polytrichum juniperinum

richum piliferum Hedw.

...ne: Species name referring to the white hair points of the leaves.
Habit: Plants usually bluish-green, forming relatively short turfs 10–20 mm tall with wine-reddish stems; sometimes the whole plant, except the long, white hair points of the leaves, tinged with red.
Habitat: Shallow soil over outcrops, open sandy soil of banks and disturbed areas from sea level to alpine elevations.
Reproduction: Sporophytes frequent, red-brown, maturing in summer.
World Distribution: Cosmopolitan, more frequent in temperate and frigid climates.
B.C. Distribution: Map 91, page 364.
Distinguishing Features: The very long, white hair point and incurved blade of the leaf over the lamellae, mark this species of sterile habitats.
Similar Species: See notes under *P. juniperinum*.
Comments: *P. piliferum* is extremely tolerant of long periods of dryness and, in the dry condition, of extremely high temperatures in its open sunny habitat. It, like *P. juniperinum*, is a reliable indicator of nutrient-poor soil.

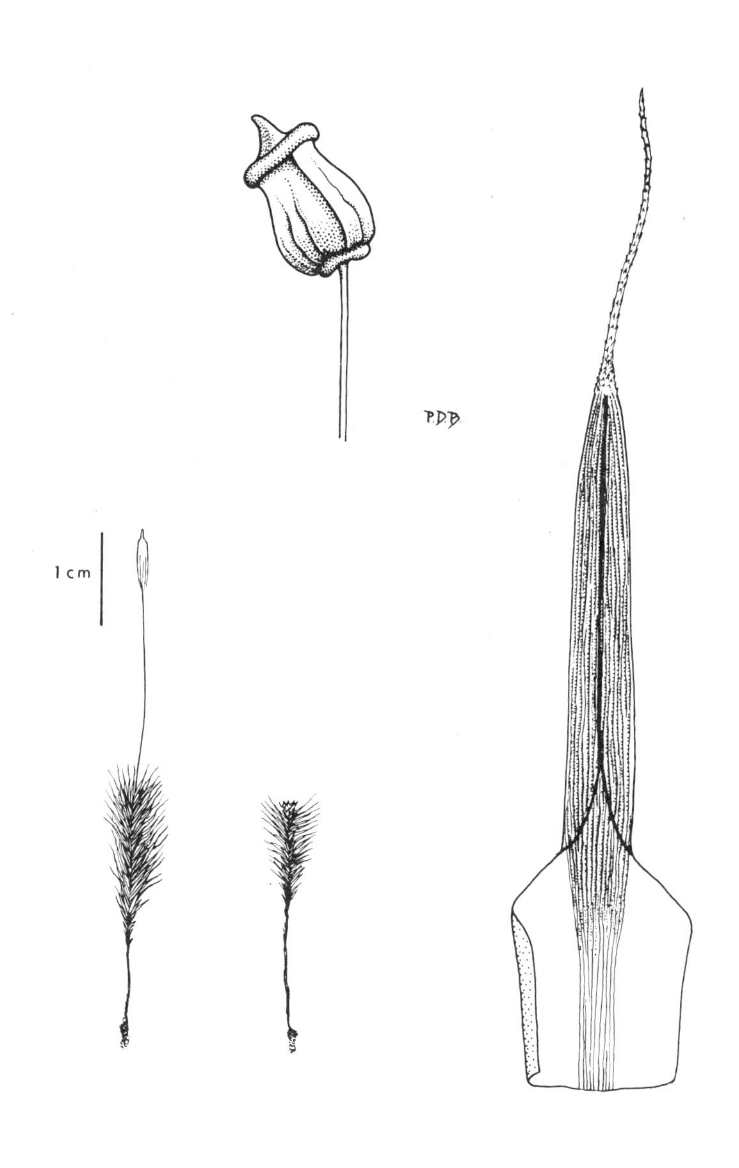

Polytrichum piliferum

Pseudotaxiphyllum elegans (Brid.) Iwats.

Name: Genus name indicating that the genus resembles the genus *Taxiphyllum*, a name meaning that the leaves appear to be in two rows because the plants are somewhat flattened. Species name reflecting the elegance of the species, perhaps indicating the original author's (Bridel) affection for the species.
Habit: Creeping, shiny, pale green to yellow-green, usually irregularly branched, somewhat flattened plants with many leaves diverging at right angles from the reclining stem and branches when the plants are moist; there is little change in appearance when the plants dry except that the leaves tend to curve downward at the points.
Habitat: In usually somewhat shaded sites on earth of banks in forests, humus, rotten logs, rock and up bases of tree trunks from sea level to subalpine forest and rarely at alpine elevations.
Reproduction: During winter, plants, especially on banks, often produce masses of tiny, brittle, yellow-green branches that serve as reproductive bodies; in late spring sporangia are produced occasionally; these are strongly nodding and red-brown when ripe.
World Distribution: Western Asia to eastern Europe, also in North Africa; in eastern North America from southern Labrador to northern Georgia westward to the Great Lakes, Missouri and Arkansas; in the west, from southeastern Alaska to central California, most frequent near the coast but also found in eastern British Columbia.
B.C. Distribution: Map 92, page 365.
Distinguishing Features: The somewhat flattened, pale green plants with many leaves diverging at right angles to the stem, the usually terrestrial habitat, coupled with the decidedly nodding sporangia and the masses of tiny attenuate reproductive gemmae, usually separate this species.
Similar Species: See notes under *Hypnum circinale*. From species of *Plagiothecium*, *Pseudotaxiphyllum elegans* is distinguished by the narrower, more sharply pointed leaves, the gemma branches and the nodding sporangium. The leaves of *Pseudotaxiphyllum* often point downward toward the substratum when dry, a feature absent in *Plagiothecium*.
Comments: This moss has suffered from many name changes that are related to an improved understanding of its relationship to similar mosses. It has been called *Plagiothecium elegans*, *Isopterygium elegans* and *I. borrerianum*, all of which are synonymous.

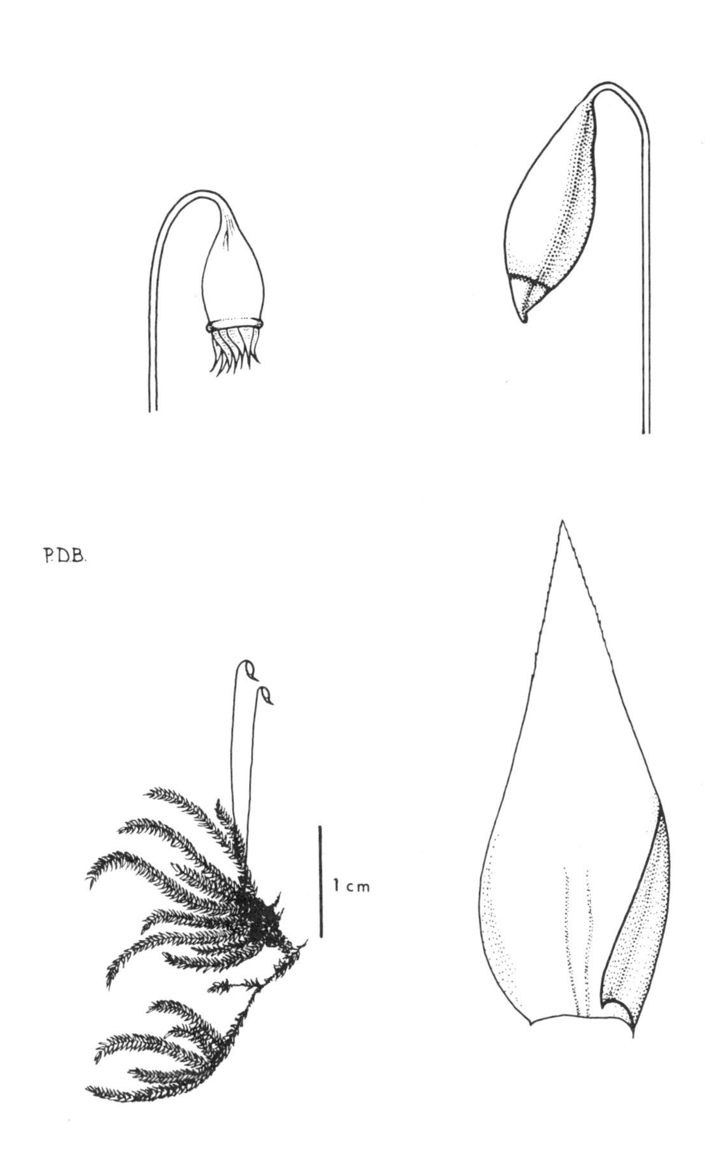

Pseudotaxiphyllum elegans

Ptilium crista-castrensis (Hedw.)

Name: Genus name referring to the feather-like growth form. Species name meaning military plume, in reference to its resemblance to the feather on a soldier's hat.

Habit: Forming mats of pale, yellow-green to golden green, regularly branched shoots that may be suberect or reclining.

Habitat: Coniferous forest floors and occasionally on bog margins, from sea level to alpine elevations, occasionally in tundra.

Reproduction: Sporophytes occasional, maturing in summer, red-brown when mature.

World Distribution: Circumtemperate in the Northern Hemisphere; widespread in northern North America, extending in the east southward to the southern Appalachian Mountains, and in the west southward to Wyoming.

B.C. Distribution: Map 93, page 365.

Distinguishing Features: The regular, feather-like plants with the branches almost at right angles to the main shoot, the pale yellow-green colour tending to golden in sunny sites, and the pleated falcate-secund leaves that point toward the base of the main shoot, are distinctive.

Similar Species: Within the genus *Hypnum* there are several species superficially similar to *Ptilium*. *H. subimponens* and *H. callichroum* are often regularly branched and yellow-green but the falcate-secund leaves are not pleated and point to the underside of the stem rather than to the stem base as in *Ptilium*. *H. procerrimum* sometimes shows the same regular branching but plants are not yellow-green but tend to reddish-brown. The regular branching of *Ctenidium schofieldii* is also similar to *Ptilium*, but *C. schofieldii* is reddish-brown to dark green and the main stem leaves are conspicuously squarrose. In calcareous marshy habitats *Cratoneuron commutatum* is sometimes regularly branched but forms dense mats of erect, rather than creeping plants; *Ptilium* is always in well-drained sites, usually in forests. Some specimens of *Drepanocladus uncinatus* may resemble *Ptilium* but the *Drepanocladus* has a single midrib and the branch leaves curve downward toward the soil surface rather than in the direction of the base of the main shoot, which is characteristic for *Ptilium*.

Comments: Sometimes called *Hypnum crista-castrenesis*, a synonym. Commonly called the knight's plume moss.

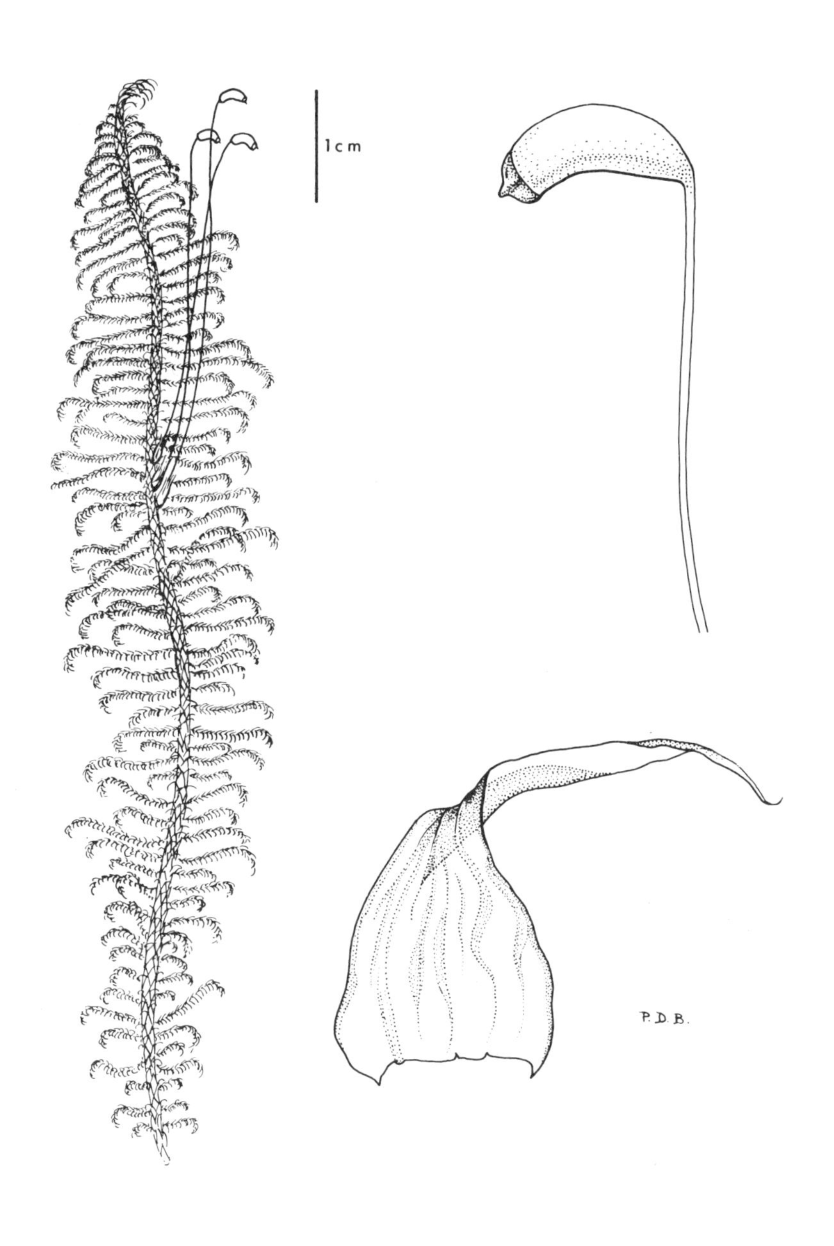

Ptilium crista-castrensis

Racomitrium aciculare (Hedw.) Brid.

Name: Genus name denoting a torn cap, in reference to the lacerate fringe at the base of the calyptra of many species. Species name referring to the pin-like beak of the operculum.
Habit: Dark green (sometimes almost black) to reddish-brown to orange-brown loose turfs to tufts, 20–50 mm tall, with leaves strongly divergent when wet and imbricate when dry; firmly affixed by dark brown to black rhizoids at the stem base.
Habitat: Generally on rocks wet from irrigation, flooding or the splash zone of watercourses and fresh-water bodies. From sea level to subalpine elevations.
Reproduction: Sporophytes frequent, dark brown to nearly black when ripe, maturing in spring.
World Distribution: Europe and western Asia, also in Japan; widespread in North America, occurring in the east southward to the southern Appalachian Mountains in Georgia and in the west southward to California; infrequent in the prairies and plains.
B.C. Distribution: Map 94, page 366.
Distinguishing Features: The habitat on irrigated rock surfaces combined with the blunt-tipped leaves that have distant blunt teeth on the margins, plus the black erect sporangia on a long seta (when mature and old) usually separate this species.
Similar Species: *Schistidium rivulare* and *Orthotrichum rivulare* grow in the same habitat and superficially resemble *R. aciculave* but their sporangia have an inconspicuous seta. *R. aquaticum* usually has yellow-green plants and the leaf apex is strongly tapered to a bluntish point and shows no blunt marginal teeth on the leaves. *Scouleria aquatica*, also of similar habitats, has sporophytes with very short setae and sporangia are subspherical rather than cylindric. The leaves of *Scouleria* are strongly divergent when wet; those of *R. aciculare* are slightly divergent. *R. pacificum* is a smaller species of open outcrops near the coast; it has narrow tapering leaves that are rather blunt-tipped but lack widely spaced marginal teeth and the peristome teeth radiate outward when dry; in *R. aciculare* they are erect.

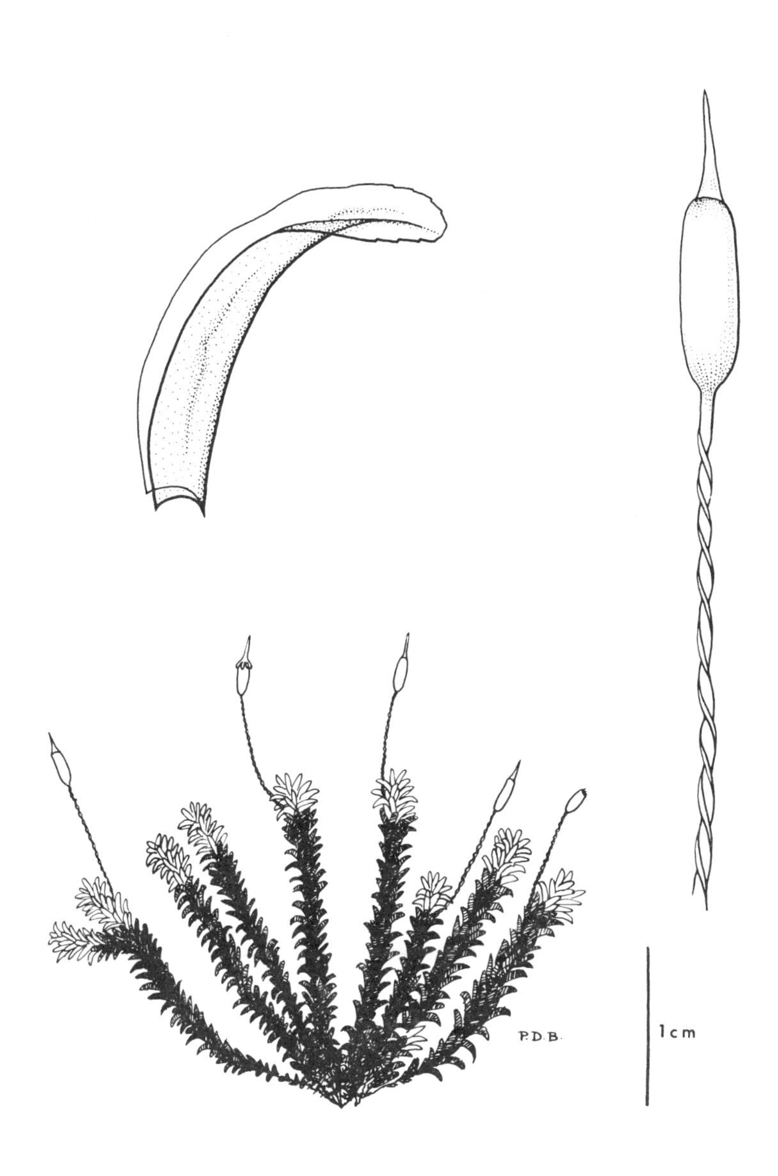

Racomitrium aciculare

꞉itrium canescens (Hedw.) Brid.

ɪɴame: Species name referring to the grayish appearance of dried plants.
Habit: Forming bright, yellow-green mats or rounded tufts of suberect to reclining plants, usually bearing many blunt lateral branches; plants whitish yellow to nearly white when dry, resulting from the white hair points as well as the ornamentation of leaf cell surfaces. Loosely affixed to substratum.
Habitat: Open, dry, sandy soil of banks, over rock surfaces, among stones and on boulders and outcrops subject to complete drying, also in open forests, especially pine. From sea level to subalpine elevations.
Reproduction: Sporophytes occasional, dark red-brown when mature. The plants are readily fragmented when dry; these fragments undoubtedly serve in propagation.
World Distribution: Europe westward to central Asia and in North America from Newfoundland southward to the New England states and westward to the Great Lakes, in the west from Alaska southward to Oregon and eastward into the Rocky Mountains as far south as Colorado.
B.C. Distribution: Map 95, page 366.
Distinguishing Features: The dull, yellow-green plants with regular, short, side branches and white hair points on the leaves, plus the well-lit, rapidly drying habitat make this a moss that can be recognized even from a rapidly moving vehicle. The extensive bright patches on rock and exposed open banks of roads are especially conspicuous after rainfall.
Similar Species: This name includes a complex of species that are not easily recognized on field characters. In British Columbia the common species is *R. elongatum* but *R. ericoides* is also frequent. At high elevations *R. muticum* is frequent but lacks the white hair points of *R. canescens*. *R. lanuginosum* is superficially similar but hand lens examination of the hair points shows them to lack acute teeth and to possess rather obtuse teeth. (See illustrations of *R. lanuginosum*) The short lateral branches in *R. lanuginosum* are also very few in most specimens and the plants, when moist, are not yellow-green but dark green to brownish-green.

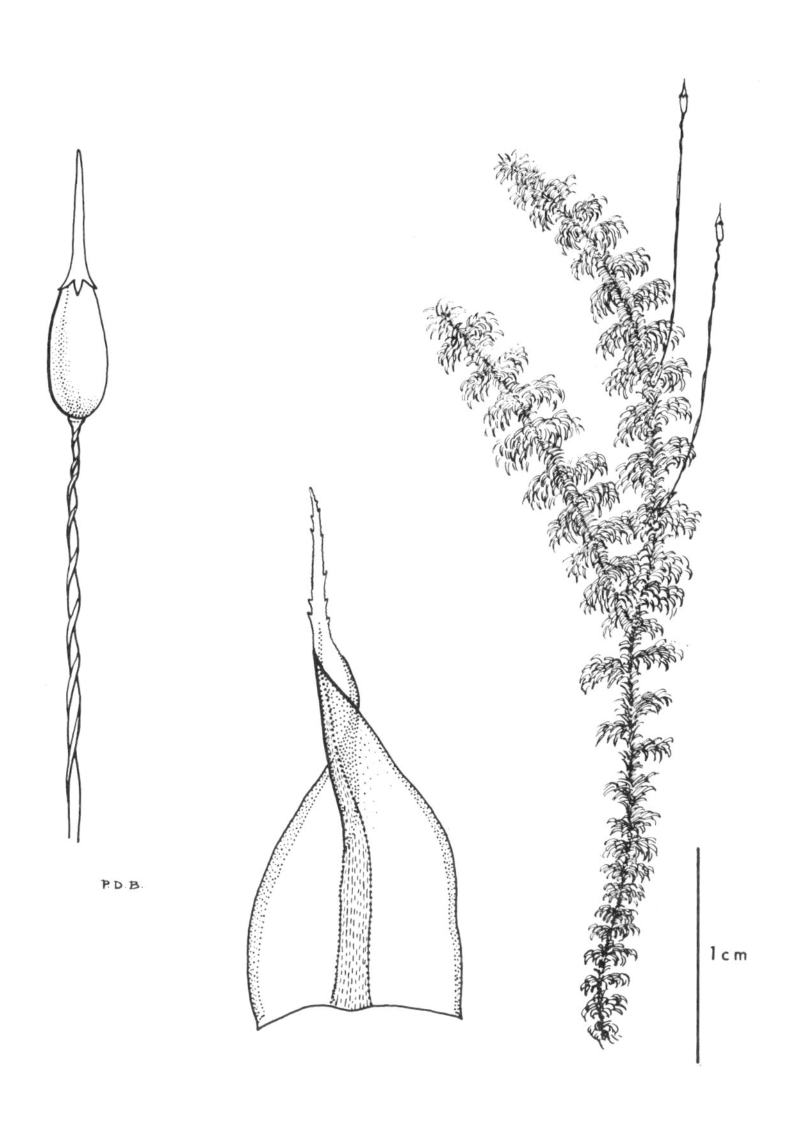

Racomitrium canescens

Racomitrium heterostichum (Hedw.) Brid.

Name: Species name referring to the somewhat one-sided arrangement of the leaves in some specimens.
Habit: Forming mats of reclining to suberect shoots strongly affixed to the substratum or loosely affixed in older plants. Plants dark green to brownish-green to nearly whitish-green when hair points are long; irregularly to regularly branched.
Habitat: Exposed rock surfaces of outcrops and boulder slopes, from sea level to alpine elevations.
Reproduction: Sporophytes common, reddish-brown to pale brown when mature; maturing in spring.
World Distribution: Widespread in temperate portions of the Northern Hemisphere; in North America southward to the southern Appalachian Mountains and westward to the Great Lakes states and northward across the continent to Alaska, in the west southward to California and eastward in the Rocky Mountains of Colorado.
B.C. Distribution: Map 96, page 367.
Distinguishing Features: This is the most variable of the species of *Racomitrium* in the province and is not easy to distinguish on hand-lens characters. Regrettably it shows considerable variation in growth form as well. Plants are never yellow-green, usually possess white hair points, and are usually firmly affixed to rock surfaces in dry sites, thus compressed tufts of irregularly branched plants are the rule.
Similar Species: As treated here, *R. heterostichum* includes a number of related species that are difficult to separate on hand-lens characters. In British Columbia the following species belong to this complex group: *R. affine*, *R. microcarpon*, *R. obesum* and *R. occidentale*, as well as *R. heterostichum*. Species of *Grimmia* with white hair points (e.g., *G. pulvinata*) and *Coscinodon calyptratus* can superficially resemble *R. heterostichum*. The *Grimmia* has curved setae, however; the *Coscinodon* has large sheathing calyptrae that differ from the small calyptrae of *Racomitrium*. Microscopic features (especially the sinuose walls of cells of *Racomitrium*) quickly separate this genus. *Hedwigia ciliata* lacks a midrib in the leaf and is usually whitish to yellow-green (see notes under *Hedwigia*). See also notes under *R. lanuginosum* and *R. canescens*. *R. varium* is difficult to distinguish on hand-lens characters. *Dryptodon patens* has two ridges on the back of the costa which are lacking in *Racomitrium*. See also *R. sudeticum*.

Racomitrium heterostichum

Racomitrium lanuginosum (Hedw.) Brid.

Name: Species name referring to the down-like appearance of dry mats as viewed from a distance.
Habit: Forming grayish-green to whitish-green, rounded masses or mats of loosely, but regularly short-branched plants that sometimes become almost white when dry.
Habitat: Exposed rock surfaces and among boulders, especially on boulder slopes, occasionally forming mounds in bogs and tundra, from sea level to alpine elevations.
Reproduction: Sporophytes occasional, sometimes locally abundant in populations near the coast, found most abundantly near bases of mounds on boulder slopes and sloping outcrops.
World Distribution: Circumpolar in the Northern Hemisphere, also scattered in mountains of the tropics and in the Southern Hemisphere widespread into subantarctic regions; in North America across the northern portion of Canada and extending southward in the east to Maine, and in the west to California and inland to the Rocky Mountains.
B.C. Distribution: Map 97, page 367.
Distinguishing Features: The usually distant, short, lateral branches, the white hair points with the teeth showing curved sinuses between them, and the sporophytes on short branches well below the stem apex are useful features. When moist, the plants are a dark grey-green.
Similar Species: See notes under *R. canescens* and *R. heterostichum*.

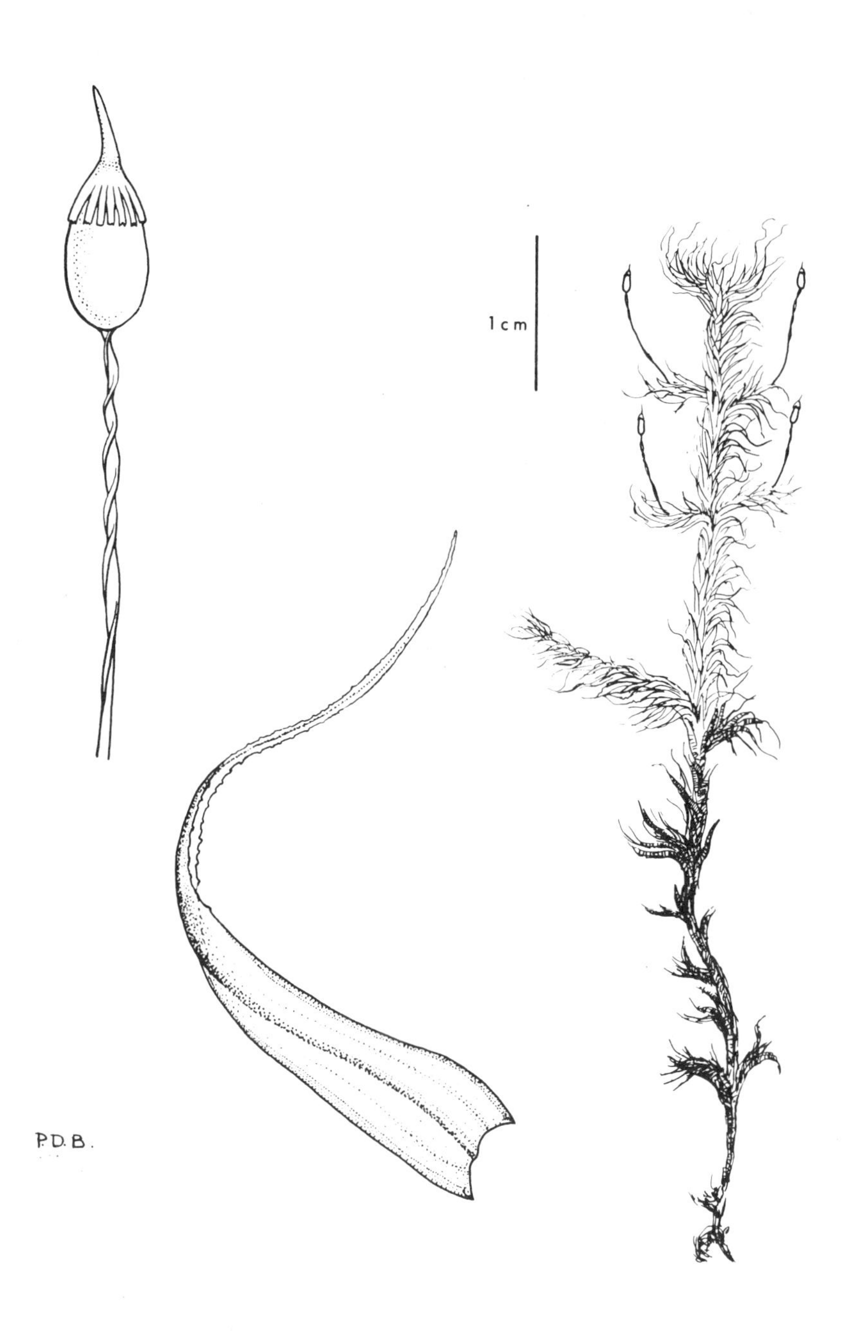

Racomitrium lanuginosum

Racomitrium sudeticum (Funck) B. & S. in B.S.G.

Name: Species name referring to the Sudeten area of Germany, from which specimens were collected upon which the name is based.
Habit: Forming dark green to nearly black turfs 20–100 mm tall, usually firmly attached to substratum.
Habitat: Usually on rock but also on soil among heaths in subalpine to alpine elevations.
Reproduction: Sporophytes frequent in populations on rock, especially in late-snow areas or somewhat shaded sites, dark brown when mature; maturing in late spring or when snow melts to expose the colony.
World Distribution: Circumpolar in the Northern Hemisphere, predominantly in mountains; in North America southward in the east to the northern Appalachian Mountains and around the Great Lakes and in the west to the Sierra Nevada of California.
B.C. Distribution: Map 98, page 368.
Distinguishing Features: The plants are usually very dark green to nearly black and are most frequent at subalpine to alpine elevations. Leaves usually lack white hair points or hair points are inconspicuous; branching is always irregular. Extremely variable in size.
Similar Species: *Dryptodon patens* is similar in general appearance but has two ridges on the back of the costa (visible at 10X); these are absent in the *Racomitrium*. *R. aciculare* has blunt leaf tips and blunt marginal teeth on the leaves, both features lacking in *R. sudeticum*. *R. macounii* is predominantly a terrestrial species and although usually larger than *R. sudeticum*, can be readily confused with it. *R. brevipes* is also similar, but the leaves are somewhat incurved at the margins while *R. sudeticum* margins are not. See also notes under *R. heterostichum*.

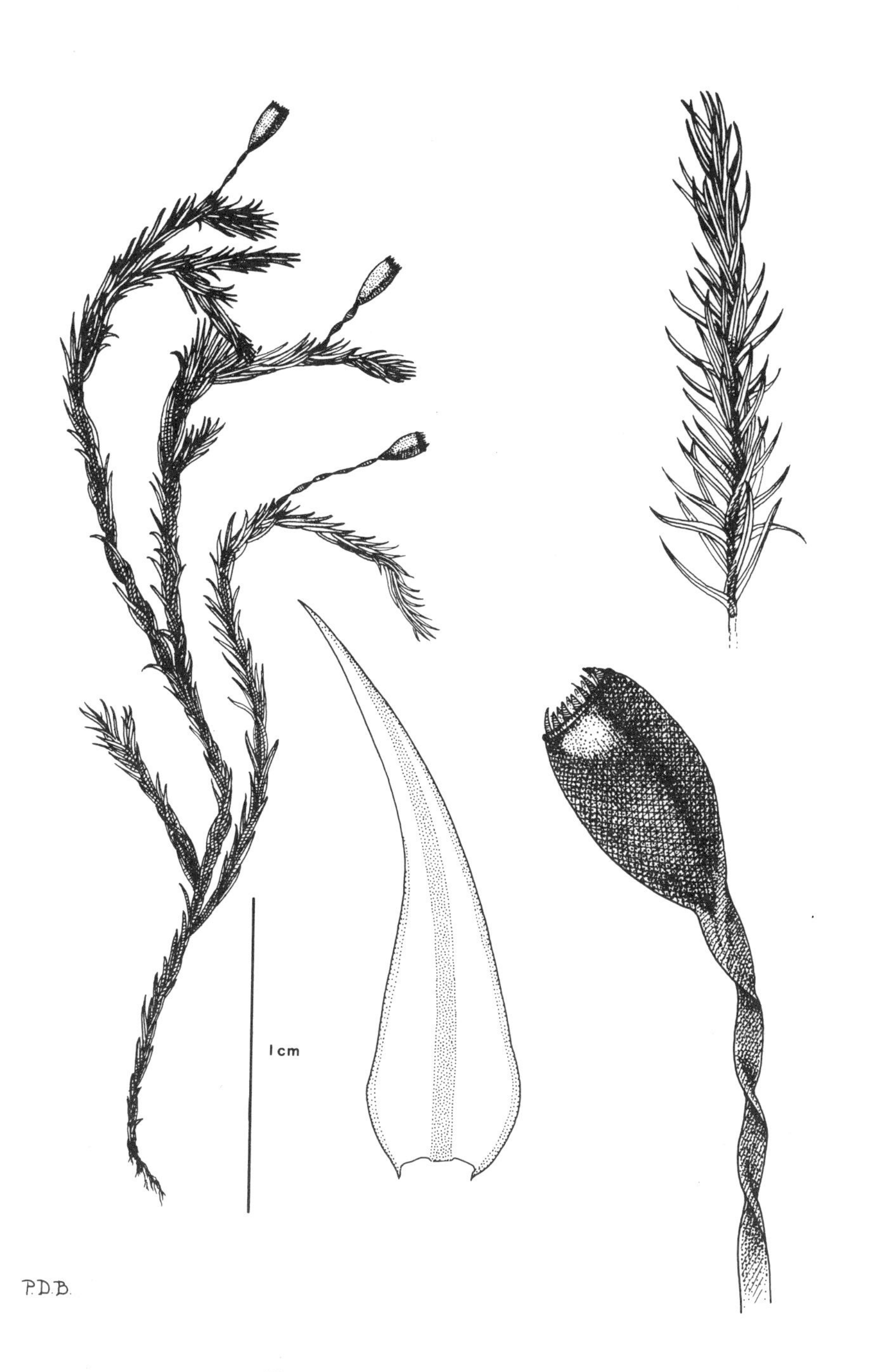

Racomitrium sudeticum

nium glabrescens (Kindb.) Kop.

Genus name meaning "the rhizoid-bearing *Mnium*" in reference to the numerous rhizoids on the stems of many species. Species name indicating that the stems of *this* species are usually smooth (or without rhizoids).

Habit: Forming relatively short, loose turfs of dark green to pale green plants tightly affixed to the substratum by reddish-brown basal rhizoids. The substratum in early colonization often matted with red rhizoids from which leafy plants arise.

Habitat: Most frequently on rotten logs and stones in coniferous forests, extending from sea level to subalpine elevations.

Reproduction: Sporophytes frequent, pale brown when mature, maturing in spring. Male plants in separate colonies or intermixed with sporophyte-bearing shoots, the male heads flower-like, and with the mass of antheridia turning red-brown when sperms have been released.

World Distribution: Confined to western North America from southeastern Alaska to central California and inland to western Montana.

B.C. Distribution: Map 99, page 368.

Distinguishing Features: The elliptic leaves that lack any suggestion of marginal teeth but possess a distinctive differentiated margin, the rosette-like apex of the male shoot, and the usual habitat on rotten logs in forest are useful features.

Similar Species: *R. gracile* and *R. pseudopunctatum* are both peatland species and less than half the size of *R. glabrescens*. *R. magnifolium* is also a species of wet sites, especially springy or seepage areas, and the stems are heavily clothed with a mat of rhizoids (*R. glabrescens* has few stem rhizoids). *R. nudum* is a terrestrial species of subalpine to alpine sites and the leaves are nearly circular in outline (rather than elliptic, as in *R. glabrescens*). When dry, *R. nudum* leaves are somewhat opalescent glossy and are little changed in form (in *R. glabrescens* leaves are dull, dark green and contorted when dry). *R. punctatum* is difficult to distinguish from *R. glabrescens* on field characters although the older stems of *R. puuctatum* tend to be red-brown rather than nearly black as in *R. glabrescens*.

Comments: Sometimes called *Mnium glabrescens*, a synonym.

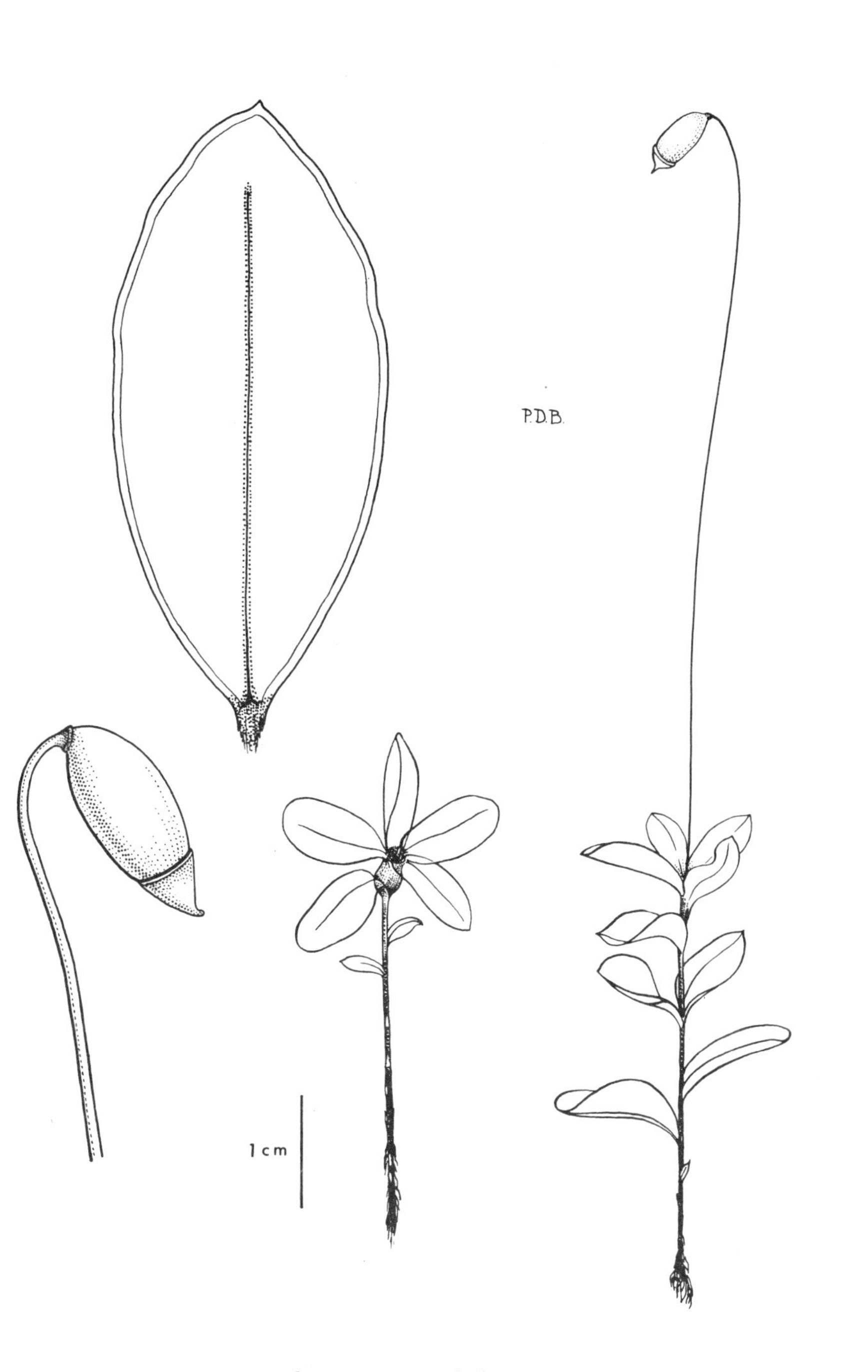

Rhizomnium glabrescens

diadelphus loreus (Hedw.) Warnst.

ne: Genus name meaning "brother of *Rhytidium*" referring to its resemblance to *Rhytidium*. Species name meaning striped, possibly referring to the striped appearance made by the pleats on the leaves.

Habit: Forming loose mats of interwoven, brownish-green to green tinged with red (from the red stems), loosely affixed to the substratum; leaves at the tips of shoots frequently curled mainly in one direction (= falcate-secund).

Habitat: On rotten logs, forest floors, rocks and occasionally epiphytic up tree bases, usually in humid coniferous forests from sea-level to subalpine elevations.

Reproduction: Sporophytes occasional, red-brown, ripe sporangia hard, glossy, subspherical and inclined.

World Distribution: Widespread, mainly in mountain forests, in Europe and on the east and west coasts of North America; in North America extending from Newfoundland southward to Maine in the east; in the west from southeastern Alaska southward to California and inland to western Montana.

B.C. Distribution: Map 100, page 369.

Distinguishing Features: The red-brown stems, the falcate leaves at the stem tips, the glossy leaves pleated at the base and lacking obvious midribs, plus the rather coarse, relatively regularly branched plants, are useful features. The species is predominantly on forest floors.

Similar Species: *R. squarrosus* is occasionally difficult to separate from *R. loreus*. *R. squarrosus*, however, regularly has squarrose leaves on the main stem and pleats are lacking or obscure in the leaves. *R. squarrosus* never possesses falcate-secund leaves at the main stem apex and is usually in open sites.

Comments: *R. loreus* is commonly used to decorate shop windows and decoratively in hanging baskets of plants. Small decorative animals, including bears, constructed mainly of this moss, are sometimes sold in shops.

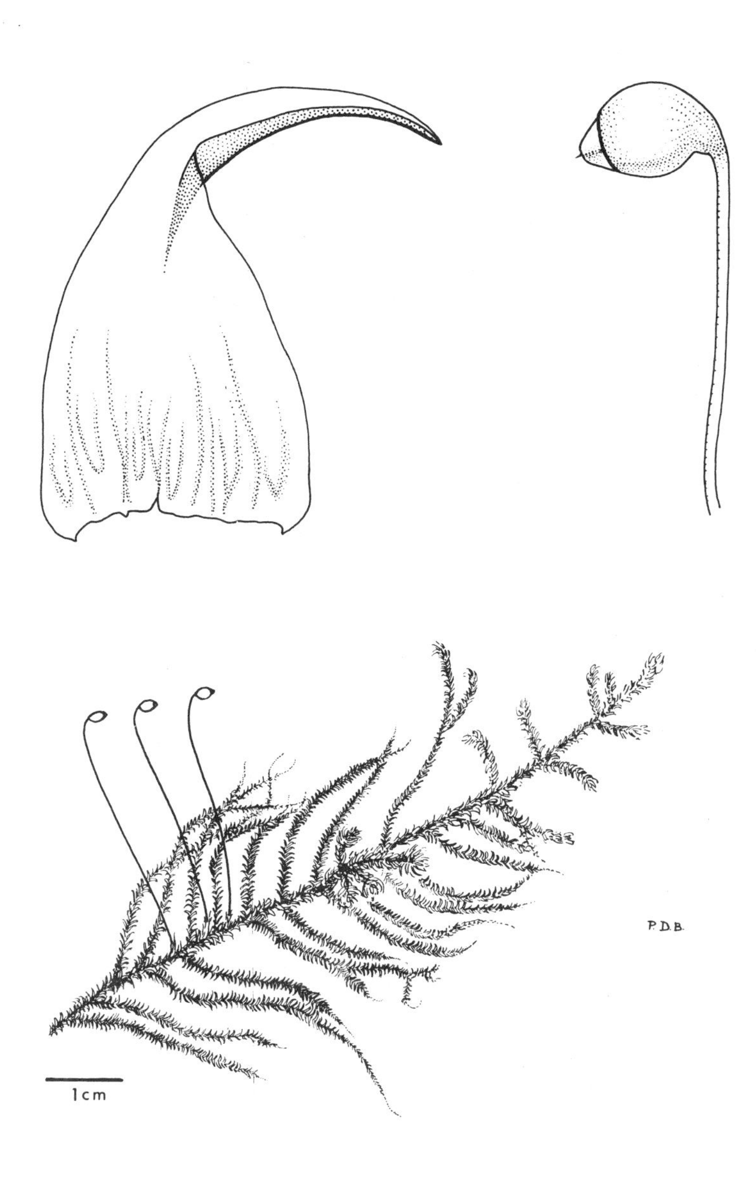

Rhytidiadelphus loreus

Rhytidiadelphus squarrosus (Hedw.) Warnst.

Name: Species name referring to the leaf form where the upper portion bends outward and downward at right angles to the body of the leaf.
Habit: Forming loose mats of suberect shoots with strongly squarrose leaves, giving the shoot apex a star-like appearance when viewed from above; light green in colour with a tinge of red brown, especially in open areas.
Habitat: A common "weed" in lawns, especially in areas where moisture persists in shade, also on stabilized sand near the sea, and among grasses. Along streams, plants and shoots are arched and reclining and frequently regularly branched. This variant is often treated as a separate species, *R. subpinnatus*; plants of subalpine and alpine elevations are in seepage sites and tend to be irregularly branched.
Reproduction: Sporophytes occasional, red-brown, maturing in spring to summer.
World Distribution: Circumpolar in the Northern Hemisphere; in North America predominantly near the coasts and in mountains, in the east from Newfoundland southward to the northern Appalachian Mountains; in the west from Alaska southward to California and inland to western Montana.
B.C. Distribution: Map 101, page 369.
Distinguishing Features: The strongly and regularly squarrose leaves of the frequently erect main stems serve as useful characters. In open sites, branching is very irregular. Often plants in lawns or on stabilized dunes form turf-like carpets of erect plants.
Similar Species: See notes under *R. loreus*.
Comments: This is one of the most frequent lawn pests in cities and towns near the coast; its growth is enhanced through late-season lawn mowing and fertilization, since it grows during the wet winter when the grass is dormant.

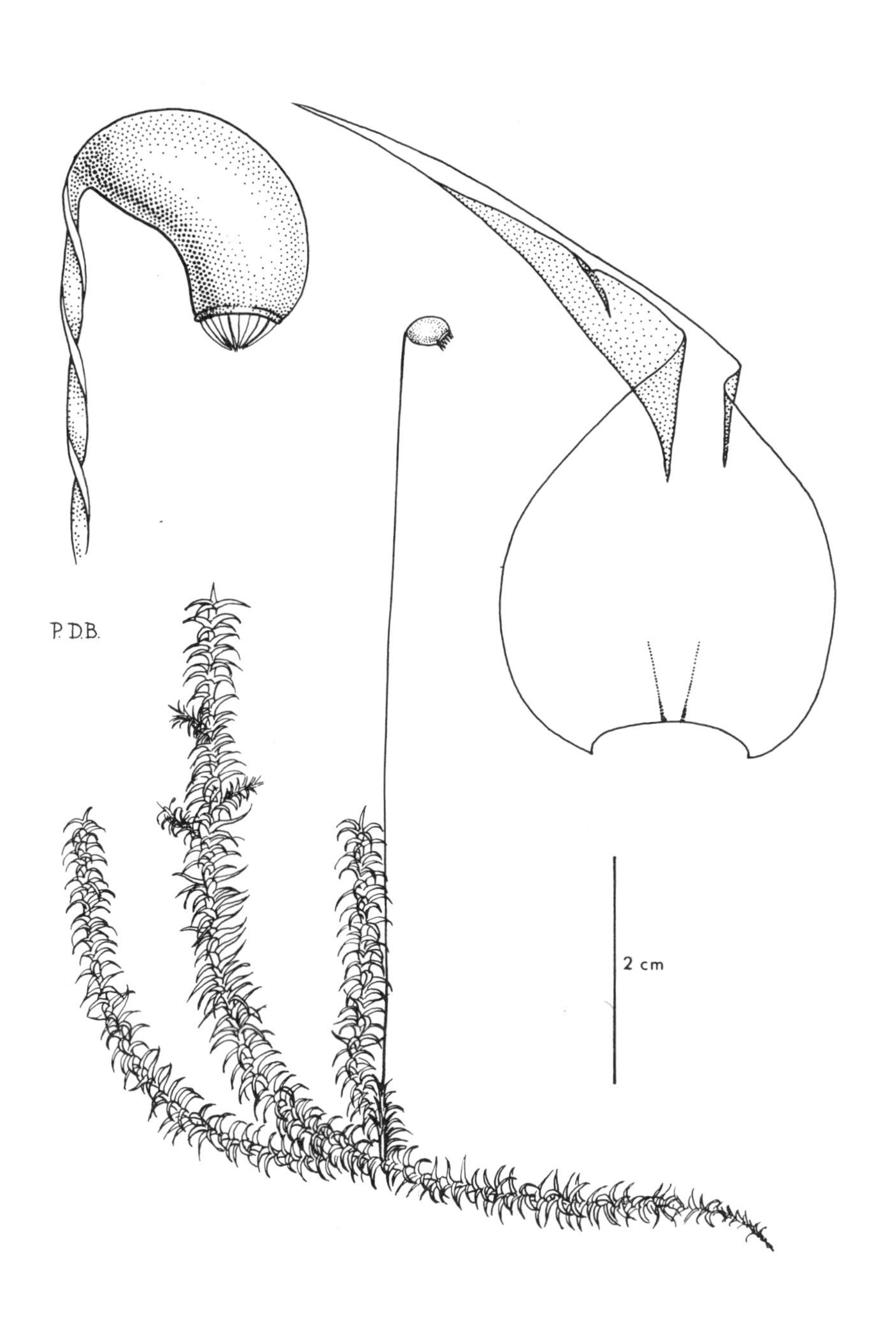

Rhytidiadelphus squarrosus

Rhytidiadelphus triquetrus (Hedw.) Warnst.

Name: Species name referring to the triangular leaves and the occasional three-rowed arrangement of the uppermost leaves of some stems.
Habit: Forming pale yellow-green, loose mats of coarse, interwoven, suberect, branched shoots in which the strongly divergent leaves form bristly, untidy shoot tips.
Habitat: Usually in well-drained sites in coniferous forests, on cliff shelves and over boulders and logs; occasionally epiphytic on tree trunks.
Reproduction: Sporophytes occasional, red-brown when mature in spring.
World Distribution: Circumpolar in the Northern Hemisphere; in North America across the boreal portion of Canada and Alaska, southward in the east in mountains to the southern Appalachians and Arkansas, in the west to California.
B.C. Distribution: Map 102, page 370.
Distinguishing Features: The very coarse, pale yellow-green plants with usually untidy divergent leaves of the main stem tip and upper branches, the strongly pleated, somewhat wrinkled leaves and the two strong midribs serve as useful characters. This species is more tolerant of drier climates than the other species of *Rhytidiadelphus*.
Similar Species: *Rhytidiopsis robusta* is of a similar size but the leaves are wrinkled, lack pleats and show a strong tendency to be falcate-secund. Plants tend to be golden yellow-green to brownish, rather than pale yellow-green. *Antitrichia curtipendula* is superficialy similar but the leaf shape differs, pleats are lacking and a radiating series of costae is usually apparent (see notes under *Antitrichia*).
Comments: Commonly called the rough neck moss or shaggy moss because of the untidy leaves at the shoot tips. A whimsical name, electrified cat tail moss, has gained some popularity in British Columbia.

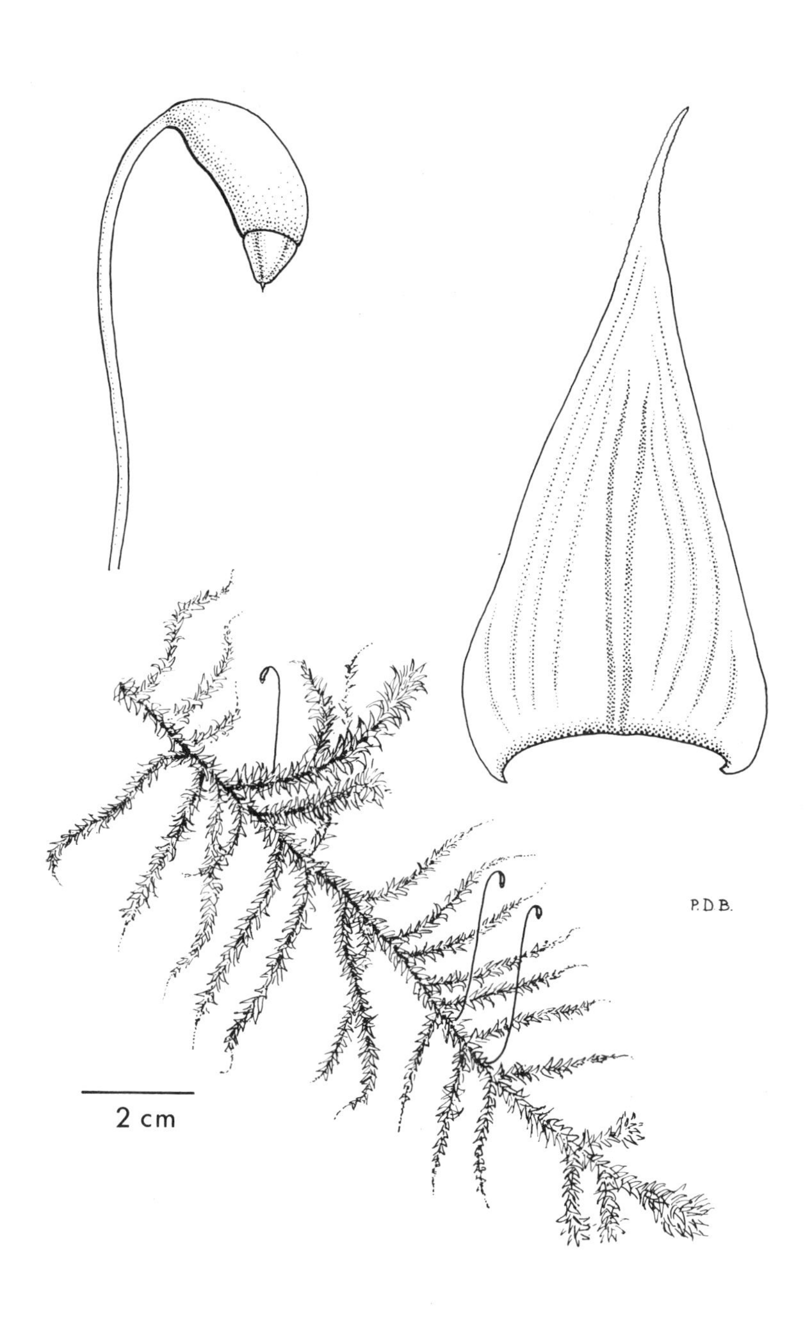

Rhytidiadelphus triquetrus

Rhytidiopsis robusta (Hedw.) Broth.

Name: Genus name meaning "the appearance of *Rhytidium*", which it resembles superficially. Species name denoting that it is very large.
Habit: Forming mats of irregular branching, pale brownish-green, intertangled plants that creep over the substratum.
Habitat: On humus of coniferous forest floors, especially in subalpine forests, also rarely on logs especially at lower elevations.
Reproduction: Sporophytes occasional, maturing in late autumn, red-brown when mature.
World Distribution: Confined to western North America, from southeastern Alaska southward to Oregon and eastward to southwestern Alberta and western Montana.
B.C. Distribution: Map 103, page 370.
Distinguishing Features: The coarse plants, usually of subalpine forest floors, plus the wrinkled surface of the falcate-secund leaves and the soft, not brittle, texture of the plants are useful features.
Similar Species: *Rhytidium rugosum* is superficially similar but the leaves have a single midrib rather than double as in *Rhytidiopsis*. Branching in *Rhytidium* is usually regular, at least in some plants, while in *Rhytidiopsis* branching is very irregular. *Rhytidium* usually occurs in dry, open sites while *Rhytidiopsis* is a forest species, reaching great abundance in humid, subalpine conifer forests. See also note under *Rhytidiadelphus triquetrus*.

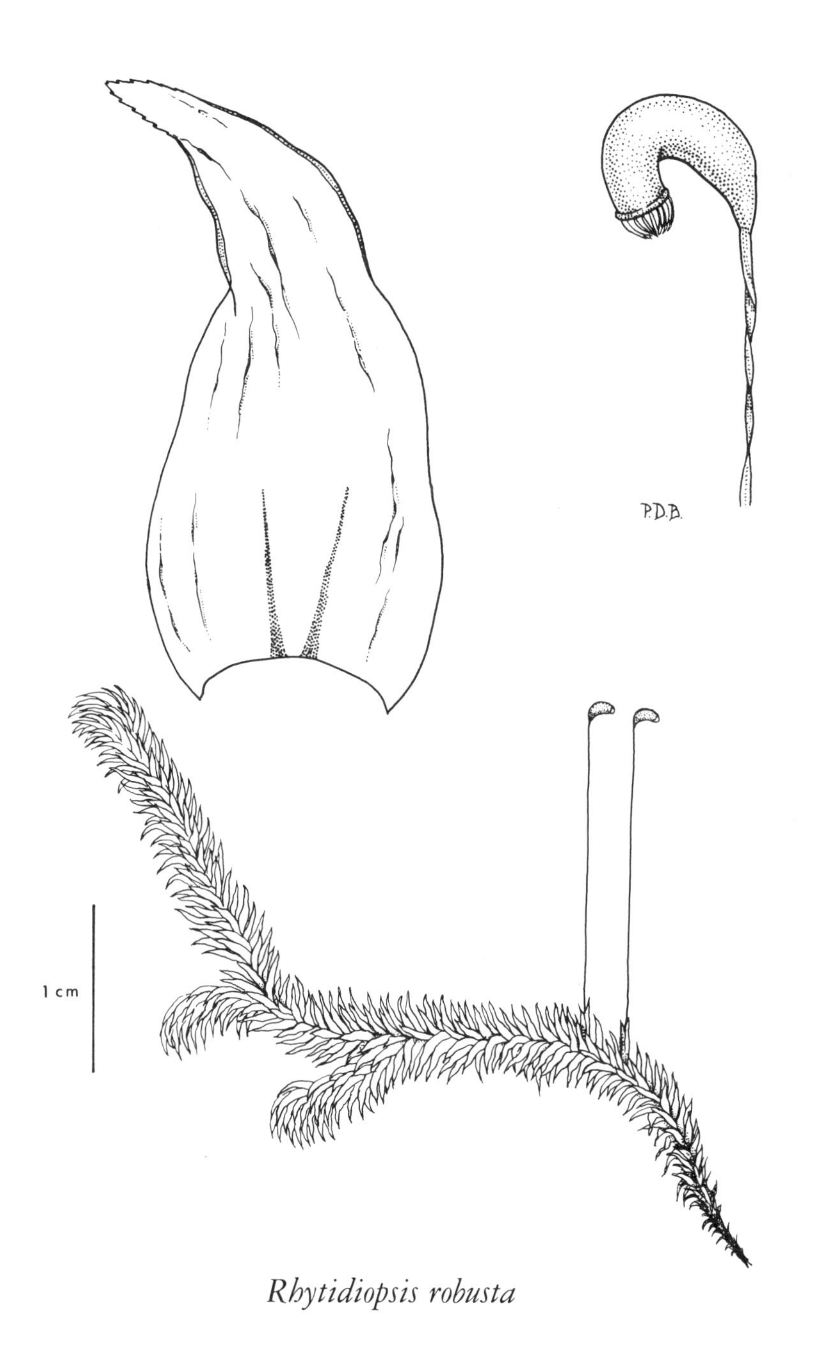

Rhytidiopsis robusta

Rhytidium rugosum (Hedw.) Kindb.

Name: Genus name denoting wrinkled, in reference to the leaves. The species name, meaning transversely wrinkled, further emphasizes the point.
Habit: Forming stiff, golden brownish-green to yellow-green, mats of reclining to suberect, interwoven shoots.
Habitat: On dry soil of usually open areas at higher elevations and latitudes and on calcareous substrata, also in open woodland and descending to lower elevations in the drier interior.
Reproduction: Sporophytes unknown in British Columbia. The plants are brittle when dry, thus the fragments undoubtedly serve in propagation.
World Distribution: Circumpolar in the Northern Hemisphere, extending southward to Bolivia in the Americas, and to north Africa; in North America widespread, extending southward in the east to the southern Appalachian Mountains, and in the west to northern Oregon and Colorado.
B.C. Distribution: Map 104, page 371.
Distinguishing Features: The leaves with a single midrib and wrinkled surface, plus the regular side branches of many shoots, and the dry, usually open, habitat, usually in calcium-rich areas, are useful features.
Similar Species: See notes under *Rhytidiopsis robusta*.
Comments: It is rather surprising that this species shows such a wide distribution, especially when sporophytes are so rare. It remains a mystery, therefore, what reproductive devices this moss possesses that allow it to be dispersed from one locality to another.

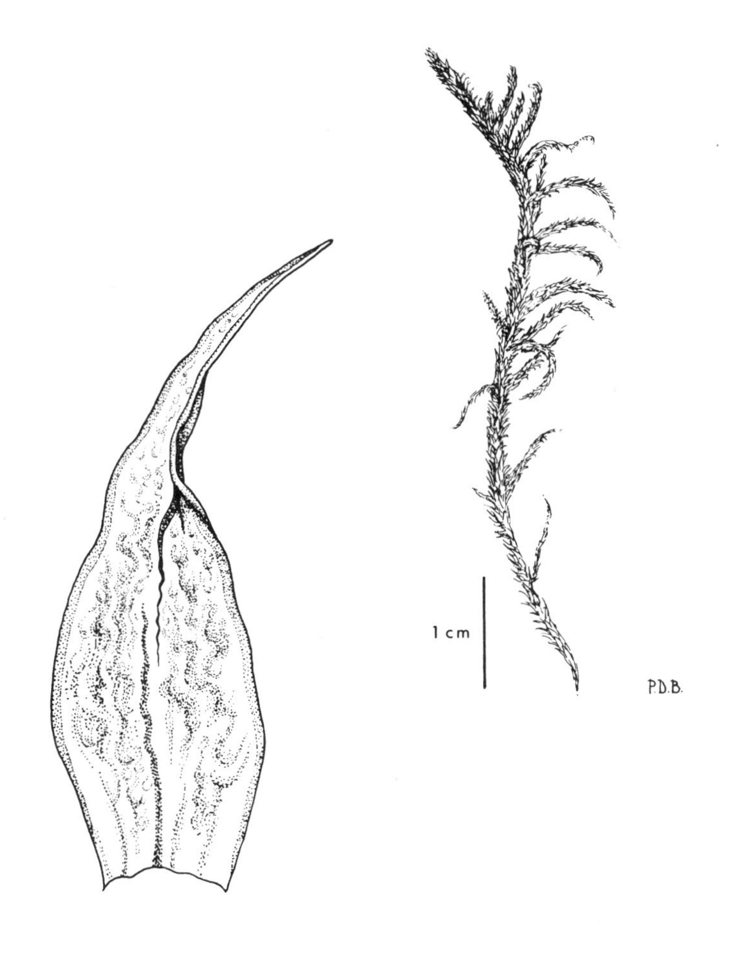

Rhytidium rugosum

Roellia roellii (Broth. *ex* Roell.) Andr. *ex* Crum

Name: Both genus and species named to honour the collector J. Roell, a German botanist (1846–1928), who collected mosses in western North America.
Habit: Loose turfs of erect, unbranched plants with rosettes of apical leaves, light to dark green and glossy, with reddish stems attached to the substratum by rhizoids at the base of the stem.
Habitat: Humid banks in subalpine forests, also on forest floor, especially on slopes.
Reproduction: Sporophytes common, red-brown and remarkably large; maturing in summer.
World Distribution: Confined to western North America from British Columbia to California and eastward to Colorado.
B.C. Distribution: Map 105, page 371.
Distinguishing Features: This is the largest *Bryum*-like moss; the apical rosettes of divergent, somewhat glossy leaves, combined with the very large nodding sporangia and the habitat on slopes in subalpine forests are distinctive characters.
Similar Species: Unlikely to be confused with any other moss except *Rhodobryum roseum*, an extremely rare moss in which the apical rosette is flat and flower-like while in *Roellia* the leaves diverge at an oblique angle from the apex. From species of *Mnium* it can be readily distinguished by the differentiated leaf margins of that genus.

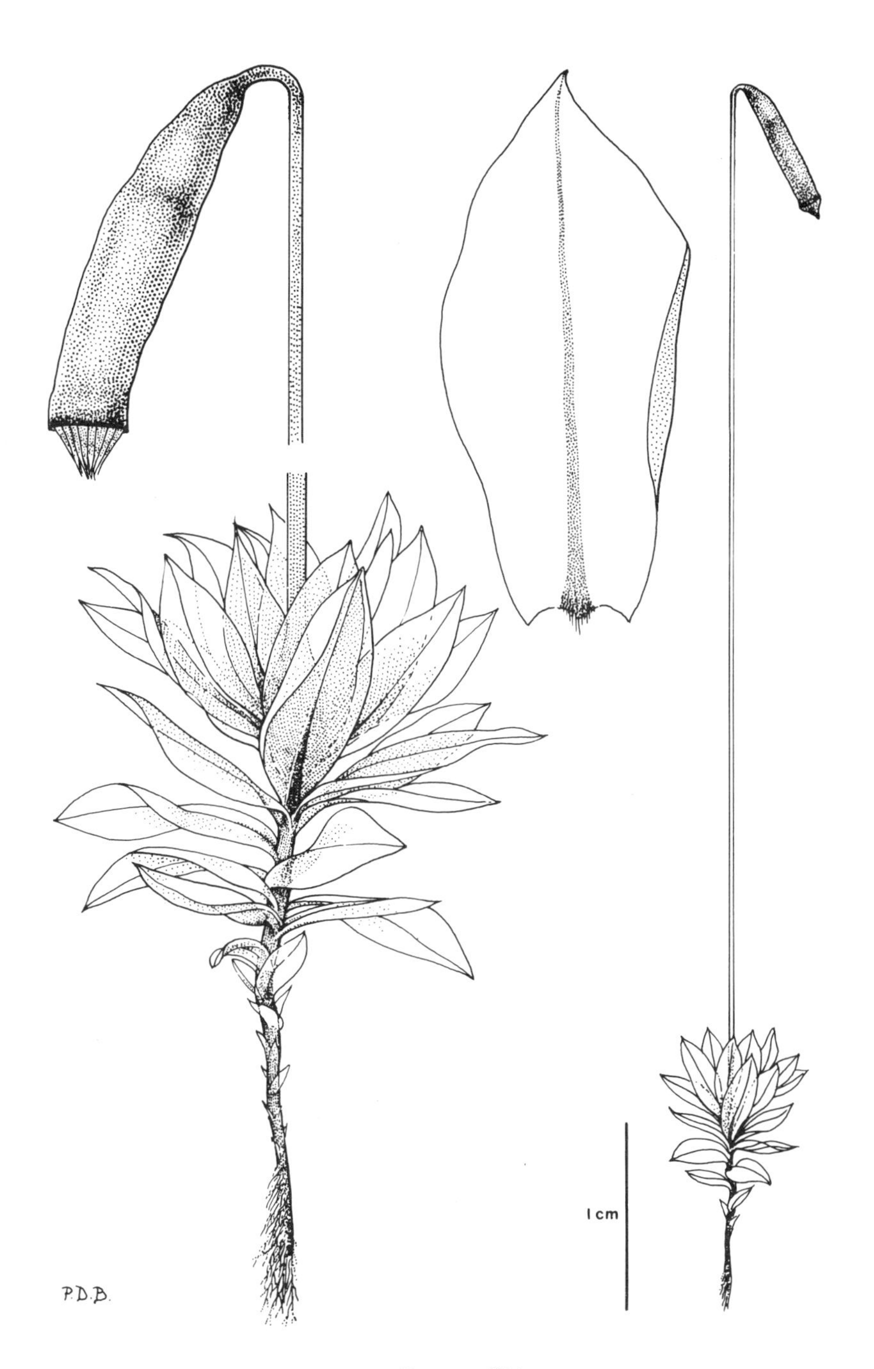

Roellia roellii

Schistidium apocarpum (Hedw.) B. & S. *in* B.S.G.

Name: Genus name denoting that the operculum breaks away, carrying the columella with it. Species meaning that the sporangia are free or separate, the meaning of which is obscure in this context.
Habit: Forming tufts or mats of dark green to red-brown, irregularly branched plants varying from 10–100 mm tall, usually firmly affixed to substratum by rhizoids at the base of the shoots.
Habitat: On sunny rock surfaces and crevices from sea level to alpine elevations.
Reproduction: Sporophytes frequent, red-brown when mature and with bright red peristome teeth that flare outward when dry; maturing in spring to summer.
World Distribution: Circumpolar in the Northern Hemisphere and widespread in the southern portion of the Southern Hemisphere: Australasia, subantarctic islands, southern South America and Tahiti; widespread in North America.
B.C. Distribution: Map 106, page 372.
Distinguishing Features: *S. apocarum* shows considerable variability, the var. *strictum* being the most frequent expression in the province. The plants of *strictum* tend to be irregularly branched, reddish-brown and form loose turfs over rock. Plants of crevices, however, are darker green and are densely tufted; on rock surfaces of the semi-arid interior, plants are often a rich red-brown. The immersed sporangia with brillant red teeth are striking. White hair points are usually frequent in the upper leaves of the shoots and in the leaves around the sporangia.
Similar Species: *S. rivulare* usually has dark green to nearly black leaves that are bluntish tipped, rather than sharply tipped as in *S. apocarpum*. *S. rivulare* is confined to irrigated or splashed rock surfaces while *S. apocarpum* is most frequent away from water bodies. *S. maritimum*, of seaside rock crevices, forms dense tufts and is dark green with leaves usually twice the length of those in *S. apocarpum*. The usual presence of immersed sporangia in *Schistidium* separates this genus from *Racomitrium* in which sporophytes have long setae.
Comments: Sometimes called *Grimmia apocarpa*, a synonym. Some authors divide *S. apocarpum* into many species which are distinguished from each other by microscopic features only.

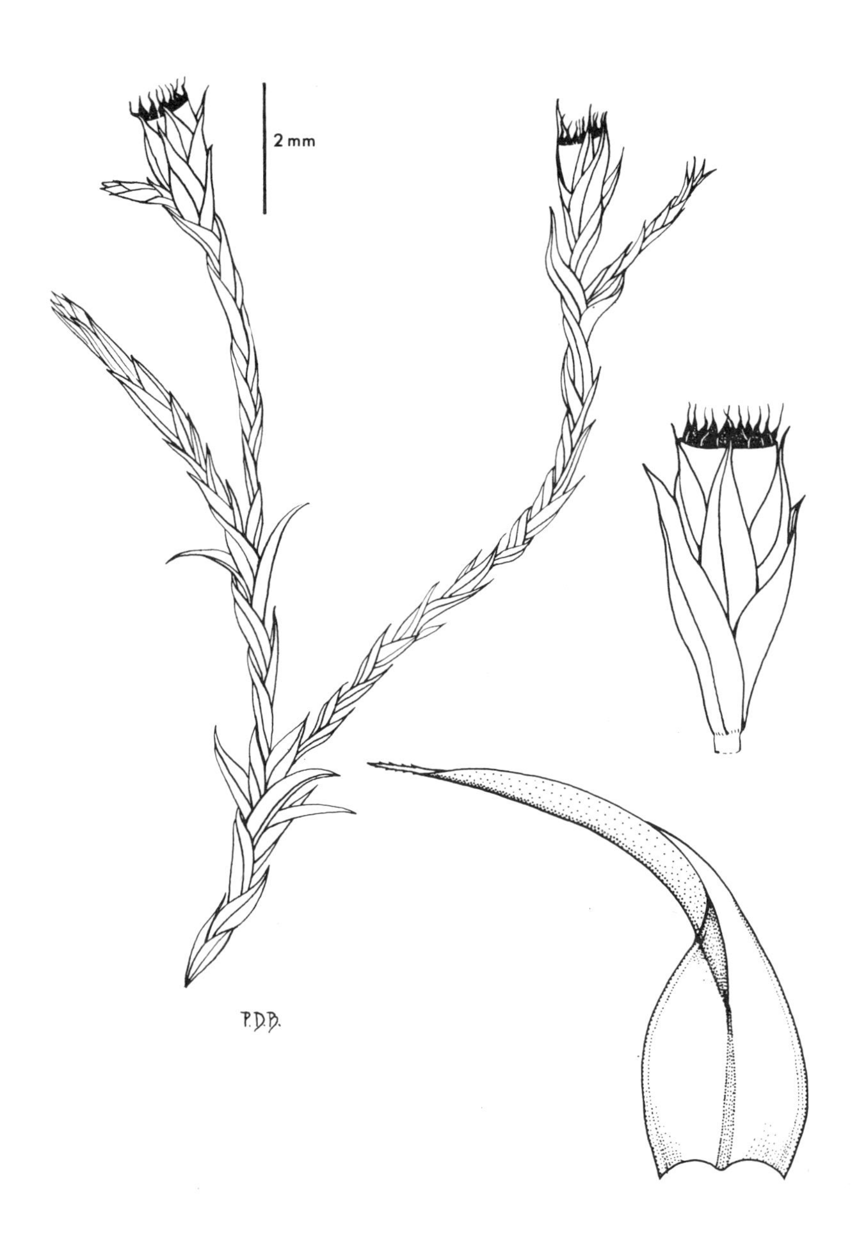

Schistidium apocarpum

Schistostega pennata (Hedw.) Web. & Mohr

Name: The genus name based on a wrong observation that the operculum splits. Species name describing the feather-like appearance of the vegetative leafy plant.
Habit: Short turfs, loosely affixed to substratum; the protonema glows with reflected light, producing a light yellow-green sheen on the substratum.
Habitat: Shaded cliff crevices or, more commonly, on shaded sterile mineral soil of overturned tree roots in swampy forests and near watercourses from near sea level to subalpine forests.
Reproduction: Sporophytes common, sporangia subspherical pale yellow-brown when mature in spring, lacking peristome teeth.
World Distribution: Circumtemperate but widely scattered in the Northern Hemisphere; in North America from Newfoundland to southern Alaska and, in the east, southward to New York and Ohio; in the west to Washington.
B.C. Distribution: Map 107, page 372.
Distinguishing Features: The very tiny plants with leaves clearly in two rows and laterally fused at the base, the seemingly leafless lower portion of the shoots, the tiny subspherical sporangia that lack peristome teeth, added to the very shaded habitat and the "luminous" protonema, when present, are distinctive features.
Similar Species: Small species of *Fissidens* can resemble *S. pennata* but the leaves of *Fissidens* show a clear sheathing flap and a midrib, and bright red peristome teeth are characteristic in the elongate sporangium.
Comments: Sometimes called luminous moss or fairy gold, based on the protonema that glows yellow-green from reflected light.

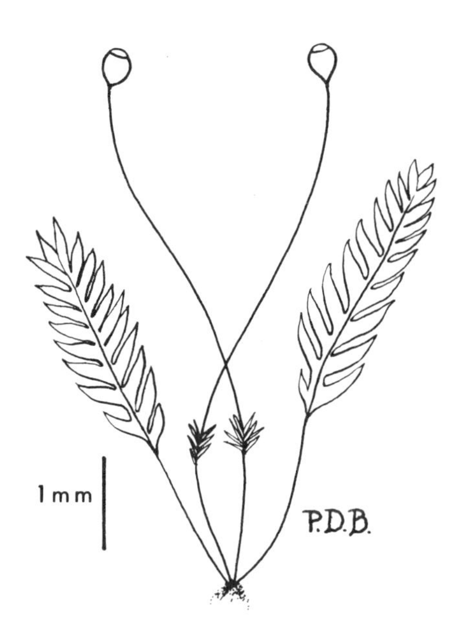

Schistostega pennata

Scleropodium obtusifolium (Jaeg. & Sauerb.) Kindb. *ex* Mac. & Kindb.

Name: Genus name meaning hard foot, possibly referring to the wiry stems. Species name describing the obtuse apex of the leaf.
Habit: Bright yellow or golden to pale green mats firmly affixed to the substratum by basal rhizoids.
Habitat: Irrigated rock surfaces; in seepage over sunny cliffs, and on boulders or outcrops in water courses from sea level to subalpine elevations.
Reproduction: Sporophytes occasional, red-brown when mature in spring.
World Distribution: Confined to western North America, from southeastern Alaska to California and Arizona eastward to Alberta and Colorado.
B.C. Distribution: Map 108, page 373.
Distinguishing Features: The swollen, worm-like glossy yellow to golden plants with broadly ovate leaves associated with the irrigated habitat are useful features.
Similar Species: *S. touretei* is similar, but smaller, and is usually on earth in shaded habitats rather than on irrigated rock of usually well-lit sites. Some specimens of *Isothecium stoloniferum* of rock surfaces, especially near watercourses, might be confused with *Scleropodium* but hand lens examination will reveal marginal teeth on the leaves of *Isothecium*; these are absent in the *Scleropodium*. *Rhynchostegium riparioides* grows in similar habitats to the *Scleropodium* but the leaves show marginal teeth and also tend to be widely spaced and divergent rather than strongly imbricate, thus the shoots are not strongly worm-like in appearance. Some species of *Hygrohypnum* may seem similar, but they generally have obscure or double midribs, and shoots tend not be be worm-like in appearance. *Cirriphyllum cirrosum* has abruptly apiculate leaves that separate it from the *Scleropodium*. *Cirriphyllum* occurs on cliff shelves, not in irrigated sites.

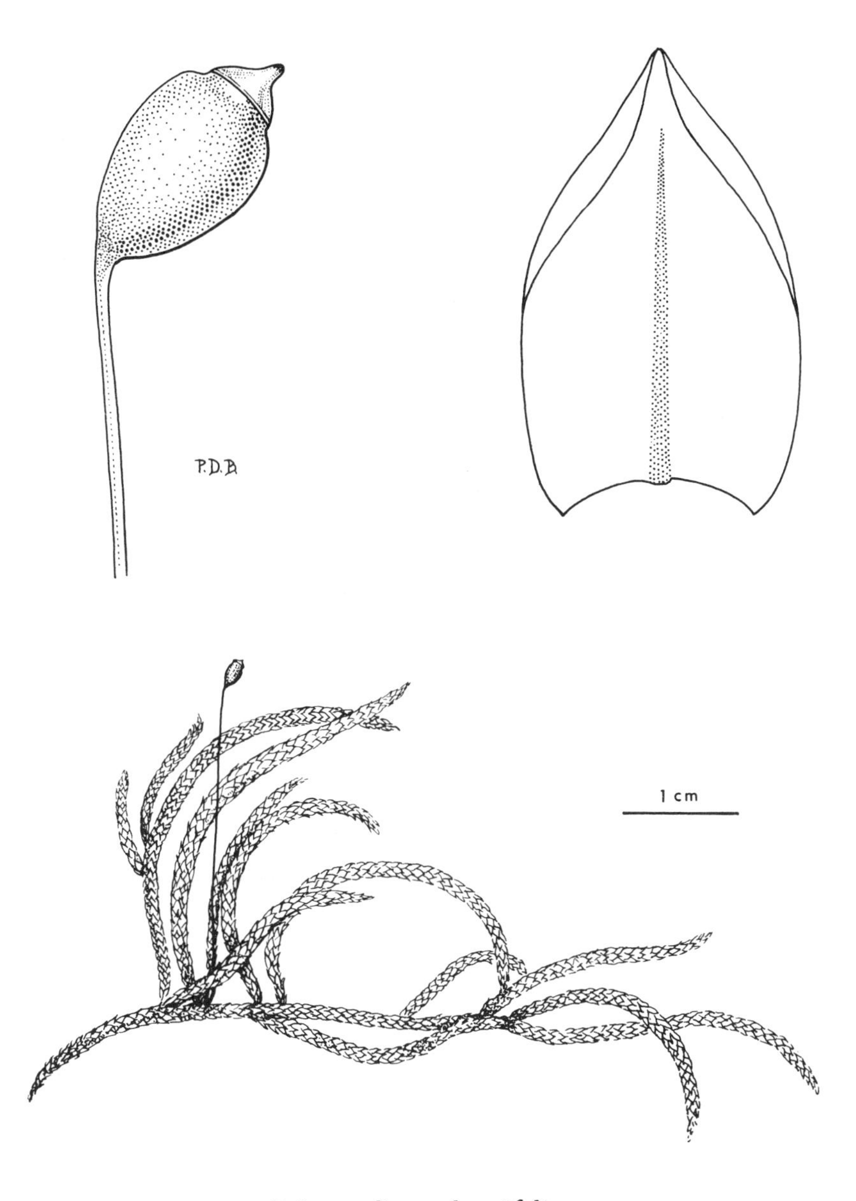

Scleropodium obtusifolium

Scouleria aquatica Hook.

Name: Genus named in honour of J. Scouler, a Scottish naturalist and surgeon who made collections in western North America in the 19th century. Species name denoting its aquatic habitat.
Habit: Dark green to nearly black turfs firmly attached to substratum by black rhizoids; leaves strongly divergent to squarrose when moist, imbricate when dry.
Habitat: Intermittently flooded, splashed or irrigated rock surfaces in or flanking watercourses from near sea level to subalpine elevations.
Reproduction: Sporophytes occasional, compressed-subspherical, immersed, glossy dark brown to black when mature in summer.
World Distribution: Eastern U.S.S.R. and northwestern North America; in North America from Alaska and Yukon to California and eastward to the Rocky Mountains of southern Colorado.
B.C. Distribution: Map 109, page 373.
Distinguishing Features: Plants of this species form remarkable dark green to blackish masses on the rocks that flank rivers; especially luxuriant in the Fraser River Canyon. The toothed leaf margins, frequent presence of rhizoids on the lower midrib, very strongly divergent leaves, when moist, and habitat of splashed or irrigated rock surfaces are useful features; the very short seta and compressed-subspherical, shiny black sporangia that retain the operculum on the columella long after spores are shed, are also useful features.
Similar Species: See notes under *Racomitrium aciculare*.

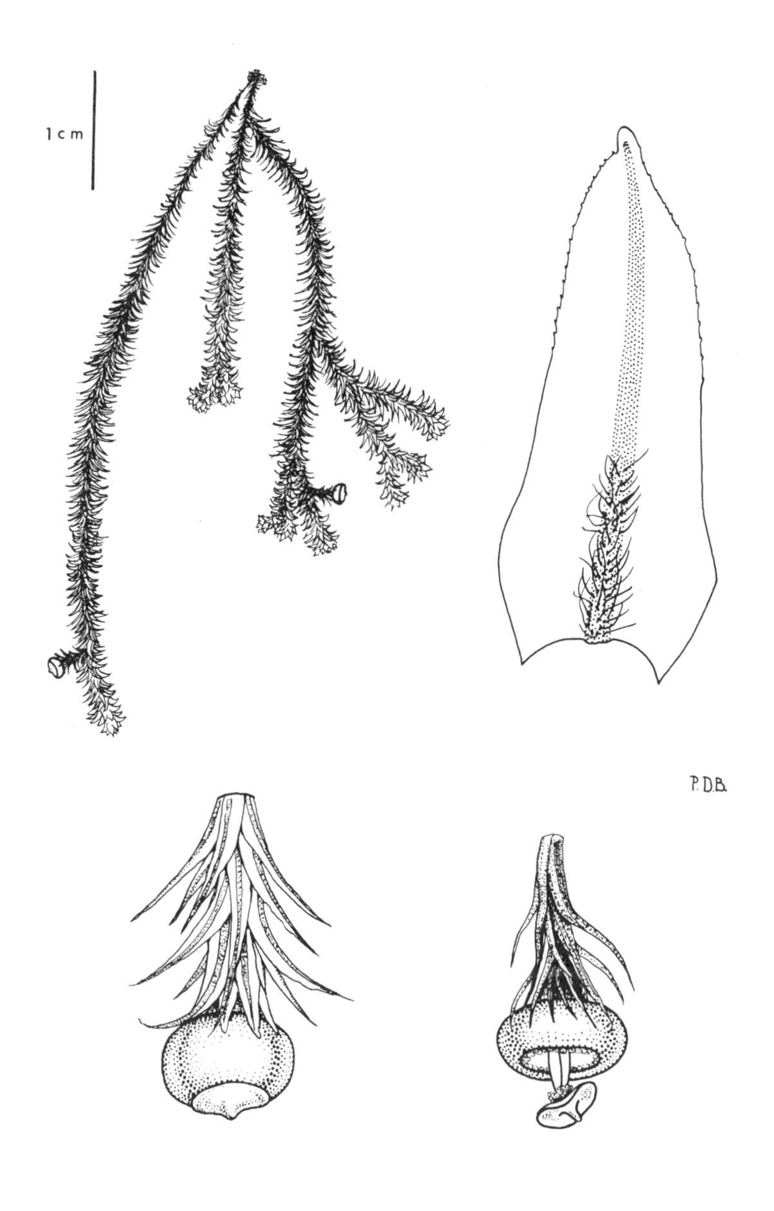

Scouleria aquatica

Sphagnum capillifolium (Ehrh.) Hedw.

Name: Genus name Greek, originally applied to a plant of unknown identity. Species name meaning hair-leafed, presumably referring to the narrower leaves, compared to many species.
Habit: Tall, pale green to pink to deep red turfs or rounded tufts of erect shoots.
Habitat: Frequent in bogs, usually forming cushions or mats that flourish in better-drained sites but some specimens (those that tend to be deep red, especially in autumn) occur in wetter depressions but never submerged for extended periods.
Reproduction: Sporophytes occasional, borne on a short stalk, nearly immersed, dark brown; sporangia globose, peristome teeth lacking, maturing in late spring and summer.
World Distribution: Circumpolar in the Northern Hemisphere; widespread in North America, south in the east to Georgia and in the west to California.
B.C. Distribution: Map 110, page 374.
Distinguishing Features: The rounded, condensed head of branches, the narrow, acute (actually flat-topped but appearing acute at 10X) branch leaves, and the frequent deep pink to red pigmentation of leaves and stems, serve as useful features although field characters are difficult to give because many species resemble *S. capillifolium*. Detailed microscopic features are needed to be confident of determination. When dry, the plants are very brittle and pulverize to light dusty fragments.
Similar Species: *S. rubellum* is often separated from *S. capillifolium*; deep red plants of *Sphagnum* at low elevations are likely to be *S. rubellum* rather than *S. capillifolium*, which may show only slight hints of pink. *S. warnstorfii* of high elevations and more northern latitudes is also very similar and is readily distinguished only on microscopic features. *S. fuscum* resembles *S. capillifolium* in general appearance but, in *S. fuscum*, the stems are dark brown and the whole plant is often rusty in appearance. Specimens of *S. girgensohnii* of higher elevations may resemble *S. capillifolium* but the leaves of the main stem of *S. girgensohnii* show a broad tattered apex when viewed with a hand lens; those of *S. capillifolium* are pointed and appear to have small apical teeth.
Comments: This is one of the "peat mosses" as the species of *Sphagnum* are commonly called. Some authors divide *S. capillifolium* into several species distinguished from each other using rather controversial characters.

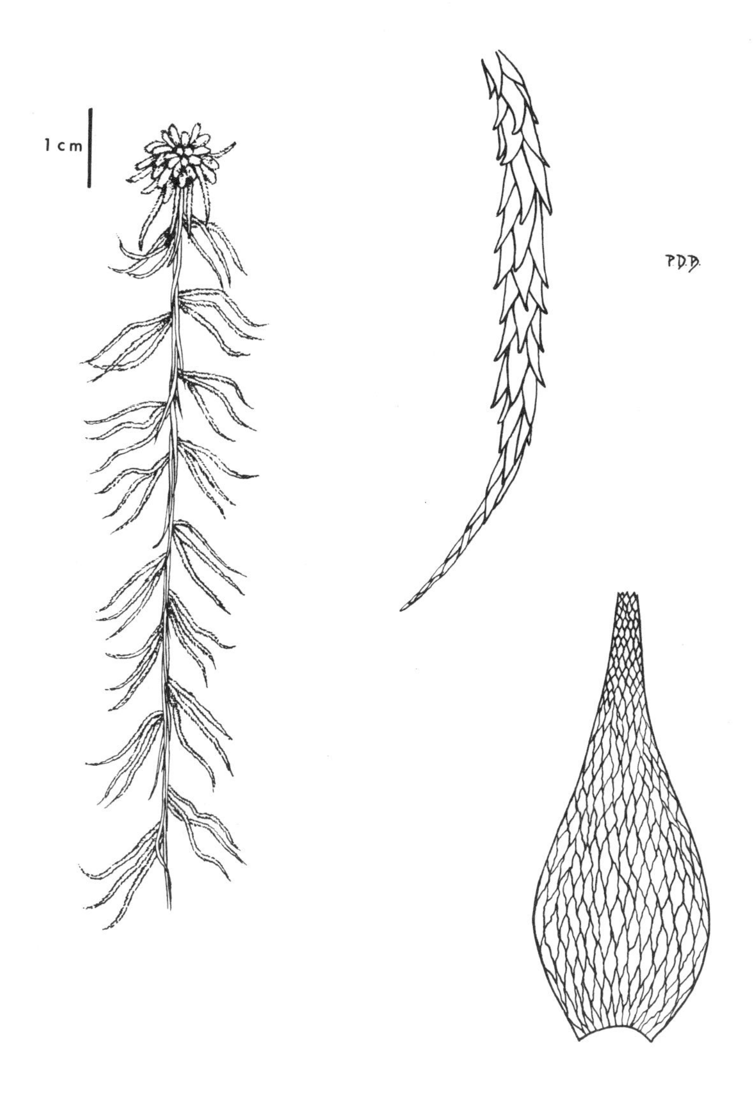

Sphagnum capillifolium

Sphagnum palustre L.

Name: Species name indicating the swampy habitat.
Habit: Tall, pale green to brownish turfs of closely to loosely packed plants with conspicuously swollen, divergent branches and heads of branches bearing broadly ovate leaves.
Habitat: Bog margins, forming rounded, well-drained mounds, sometimes also on splashed or damp cliff ledges and near watercourses from sea level to subalpine elevations.
Reproduction: Sporophytes infrequent, maturing in summer.
World Distribution: Extremely widespread in all continents except Antarctica.
B.C. Distribution: Map 111, page 374.
Distinguishing Features: The swollen divergent branches with incurved inflated leaves overlapping each other, leaf tips rounded, plus light green to brownish-green colour and dark green stems, are usually enough to separate this species, but several others very closely resemble it.
Similar Species: *S. henryense* is virtually impossible to distinguish from *S. palustre* except on technical microscopic features. *S. papillosum* tends to show rather dull green, leafy shoots but can be distinguished convincingly only on microscopic characters (papillae on the inner faces of cell walls). *S. austinii* forms orange-brown, relatively condensed tufts (microscopically the comb-like ornamentation of the walls of the elongate cells is distinctive). *S. magellanicum* is usually pale to wine-red, a colour absent in *S. palustre*. *S. compactum*, a species of subalpine cliffs and lowland peatland is frequently orange but can be troublesome to distinguish from *S. palustre* without microscopic examination (*S. compactum* lacks fibril thickenings in the outer cells of the stem).
Comments: *S. palustre* and its close relatives form the most valuable horticulturally used peat. In the province *S. henryense* is probably more common than *S. palustre*.

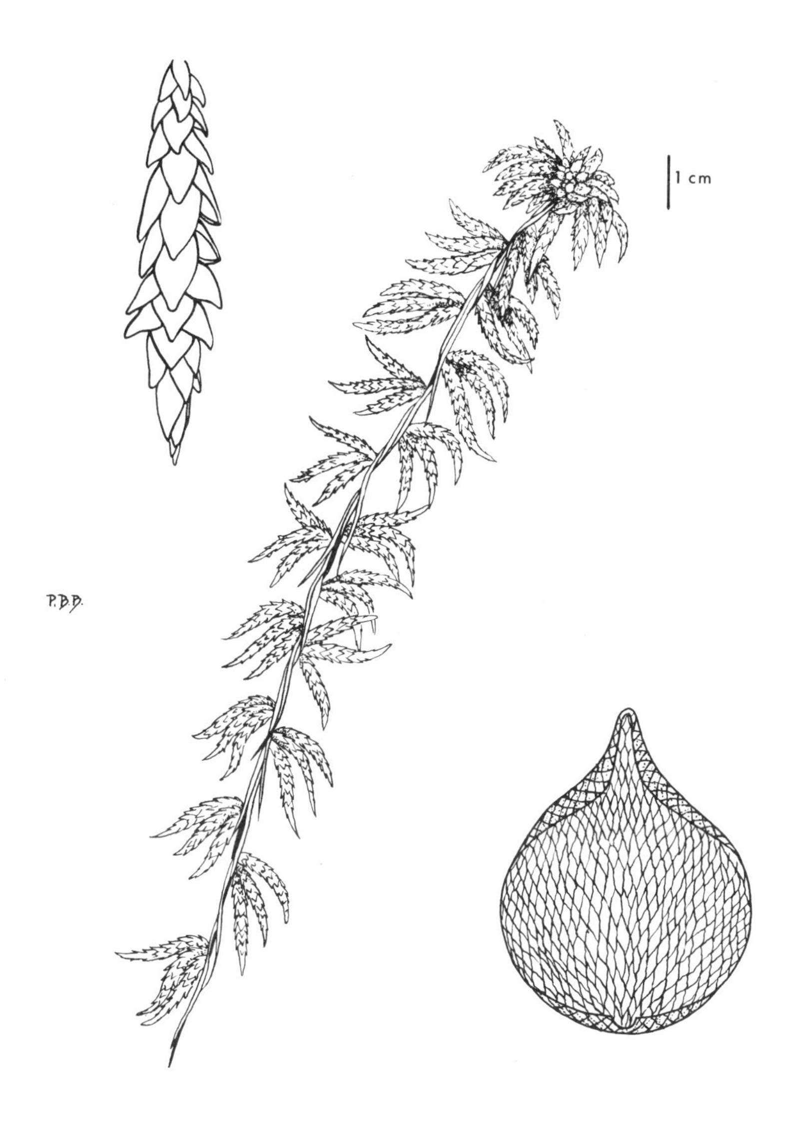

Sphagnum palustre

Sphagnum squarrosum Crome

Name: Species name referring to the leaf orientation in which the points bend abruptly outward, giving the leafy branches a prickly appearance.
Habit: Loose, pale, dull green turfs of sprawling or suberect to erect intertangled shoots, sometimes partially submerged.
Habitat: Predominantly a woodland species tolerant of some shade, in swampy or seepage sites or near waterfalls or watercourses from sea level to subalpine elevations.
Reproduction: Sporophytes occasional, maturing in spring to summer.
World Distribution: Widespread in the temperate portion of the Northern Hemisphere, also in New Zealand; in North America southward in the east to North Carolina and to California in the west.
B.C. Distribution: Map 112, page 375.
Distinguishing Features: The coarse plants with bristly squarrose leaves on the divergent branches are characteristic; the species is one of forest and cliff habitats, not in bogs.
Similar Species: *S. squarrosum* might be confused with *S. palustre* but the strongly squarrose leaves and the non-bog habitat of the former should separate them. Microscopically *S. squarrosum* lacks stem cells that have fibril thickenings.
Comments: Although called a peat moss, *S. squarrosum* is not an important peat-former.

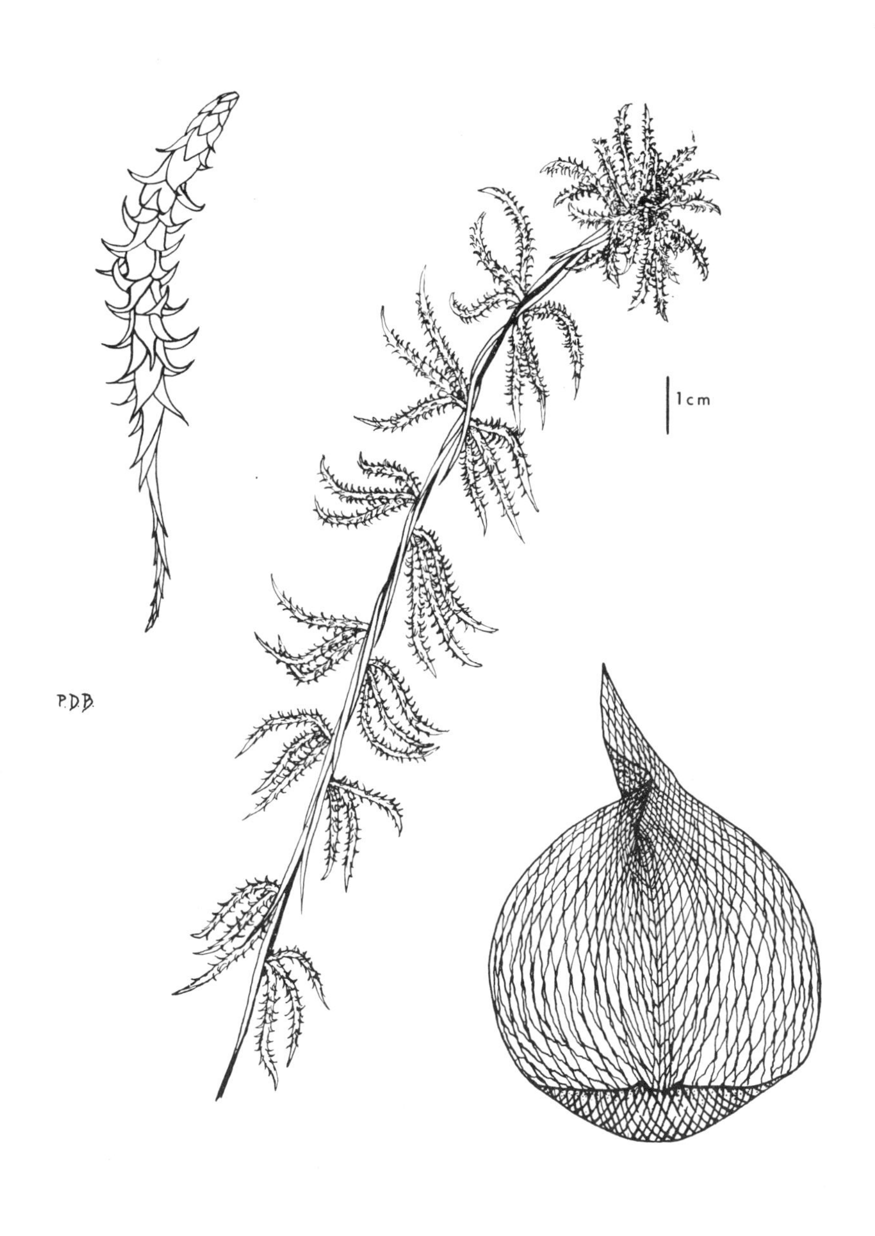

Sphagnum squarrosum

Splachnum rubrum Hedw.

Name: Name from Greek and applied originally to a foliose lichen. Species name referring to the brilliant opalescent red-purple sporangium.
Habit: Short, shiny, light to dark green turfs usually not obvious without sporophytes attached.
Habitat: On decaying animal waste in damp, often boggy or swampy areas in spruce forests, mainly at lower elevations.
Reproduction: Sporophytes frequent, opalescent red-purple when mature and with a slender soft seta.
World Distribution: Widely scattered in cool temperate portions of the Northern Hemisphere; in North America extending northward to Labrador and southward in the east to Nova Scotia and the Great Lakes area and scattered across the continent to Alaska and British Columbia.
B.C. Distribution: Map 113, page 375.
Distinguishing Features: The opalescent red-purple, umbrella-like sporangia are diagnostic.
Similar Species: *S. luteum* is of similar form but with pale yellow sporangia; it is found in open forests rather than boggy or swampy habitats. *S. ampullaceum* is also purplish to pink and occurs in habitats like those of *S. rubrum* but, in *S. ampullaceum*, the sporangium is much smaller with an inflated lower portion of the sporangium narrower than the skirt-like one of *S. rubrum*.
Comments: This is a strikingly beautiful moss for which few collections are known in the province.

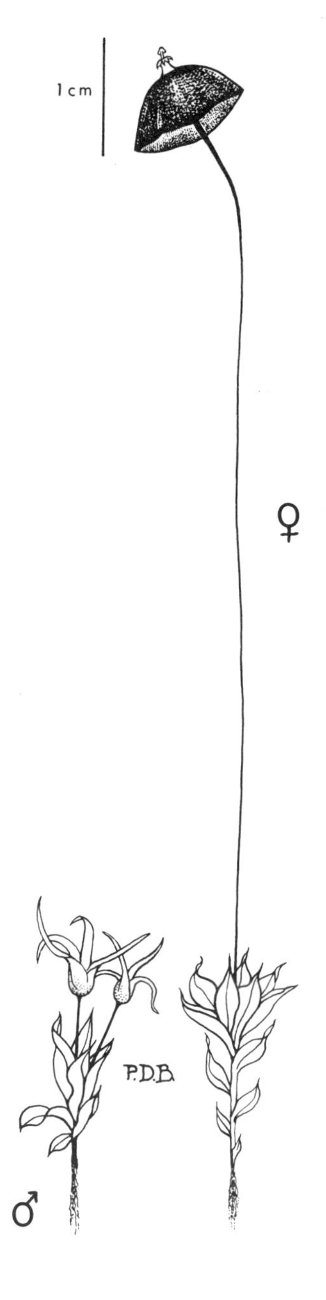

Splachnum rubrum

aphis pellucida Hedw.

..ame: Genus name refers to the four teeth of the peristome. Species name refers to the transparent leaves (when wet).
Habit: Forming short, dense to loose, dark to pale green turfs of erect, unbranched plants.
Habitat: Usually on decomposing stumps and logs of coniferous trees within coniferous forests also on peaty banks; from sea level to subalpine elevations.
Reproduction: Sporophytes common, reddish-brown when mature, maturing in late spring to summer. Gemma-producing shoots also abundant, especially conspicuous in late summer, producing cup-like terminal masses of leaves surrounding "nests" of slender-stalked, disc-shaped gemmae that are dispersed by splashing raindrops.
World Distribution: Circumpolar in the Northern Hemisphere; in North America widespread across the continent, extending southward in the east to South Carolina and in the west to California, Arizona and Colorado.
B.C. Distribution: Map 114, page 376.
Distinguishing Features: The gemma-cup bearing shoots, the four unjointed peristome teeth and the usual habitat on well decomposed, but friable wood are useful characters.
Similar Species: *T. geniculata* is very similar but the seta has a sharp, angular bend in the middle. *Aulacomnium androgynum* grows in similar, but usually drier, sites and produces gemma-bearing shoots which are terminated by a sphere of gemmae that are not enclosed in a "cup" of leaves; sporophytes of *Aulacomnium* are grooved and have many jointed peristome teeth.
Comments: With diligence, it is possible to find tiny, rounded, leaf-like flaps growing on the shaded, rotten-wood habitat of *Tetraphis*; these structures precede the appearance of the leafy shoots. They are clearly visible with a 10X hand lens but can be found with the unaided eye.

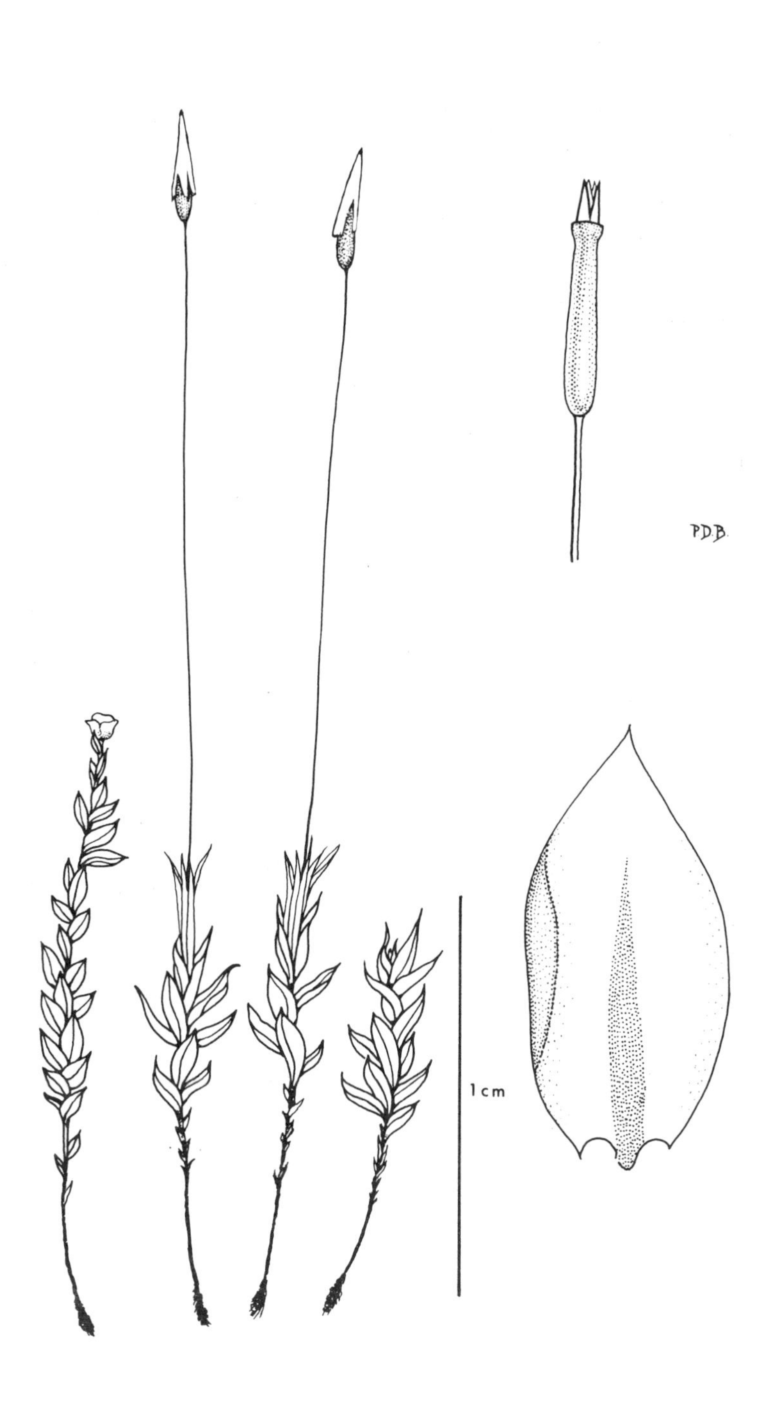

Tetraphis pellucida

Tetraplodon mnioides (Hedw.) B.S.G.

Name: Genus name describing the peristome teeth that are often fused in fours when young. Species name meaning *Mnium*-like, although this similarity is difficult to discern.
Habit: Forming hard, rounded to conical, dark green to light green tufts.
Habitat: Animal waste in bogs, open forest, and trail margins, from sea level to subalpine elevations.
Reproduction: Sporophytes frequent, with brilliant red setae and dark brown to red-brown sporangia, maturing in summer. Sometimes dark brown to nearly black when dried.
World Distribution: Circumpolar in the Northern Hemisphere, also in southern South America and New Guinea; in North America extending southward in the east to New York and the Great Lakes area, and in the west to southern Orgeon.
B.C. Distribution: Map 115, page 376.
Distinguishing Features: The habitat on decomposing animal waste, the conic habit, the sporangia that show a somewhat swollen lower portion and a cylindric portion above it, and the usually wine-red seta are usually reliable features.
Similar Species: *T. angustatus* is very similar but the leaves are toothed (clearly visible with hand lens); those of *T. mnioides* lack teeth. *T. pallidus* is similar but has yellow setae and pale sporangia; *T. urceolatus* is also similar but the seta is very short and the tufts are often very dense; the sporangia are also short and stout. *Tayloria* species usually do not form tufts and the sporangia lack a swollen lower portion. *Splachnum sphaericum* and *S. vasculosum* usually have very long, slender and readily collapsing setae; those of *Tetraplodon* are rigid.

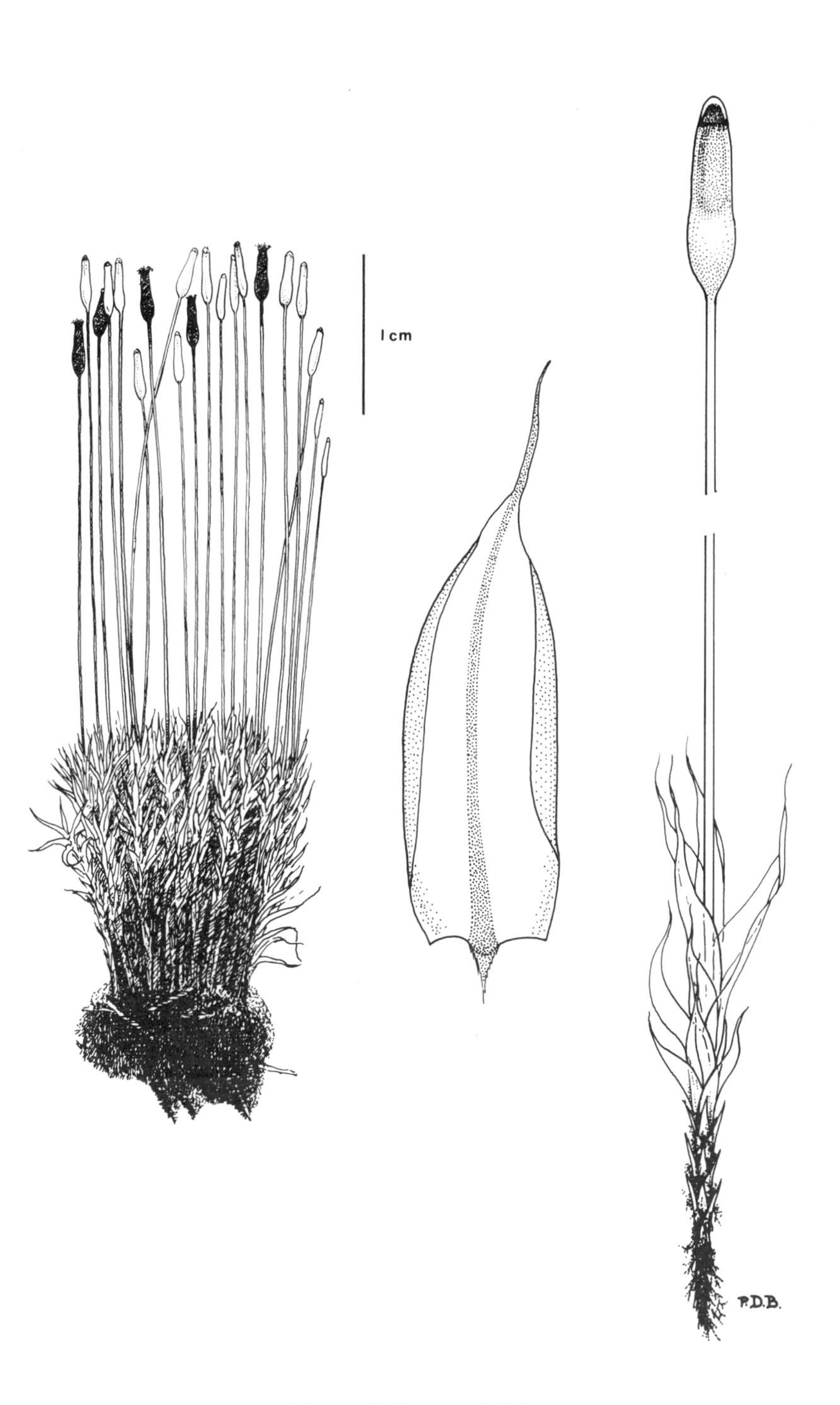

Tetraplodon mnioides

Thuidium recognitum (Hedw.) Lindb.

Name: Genus name referring to the fancied resemblance of the leafy plants to *Thuja* (arbor vitae or cedar). Species name derived from the fact that another species, differing from the widespread *T. delicatulum* (Hedw.) B.S.G., was recognizable.
Habit: Yellow-green to dark green, loosely interwoven, much-branched, somewhat arching plants not firmly attached to substratum.
Habitat: Forest floor and open slopes.
Reproduction: Sporophytes rare, red-brown, sporangia cylindric, curved, seta elongate, maturing in summer.
World Distribution: Circumpolar, but interruptedly, in the Northern Hemisphere; in North America from Greenland southward to Arkansas and Georgia in the east and from Alaska and Yukon southward to Oregon in the west.
B.C. Distribution: Map 116, page 377.
Distinguishing Features: The pale yellowish-green, non-glossy, complicatedly branched, lacy plants are highly distinctive. Green paraphyllia on the stem are usually visible with a hand lens.
Similar Species: *T. philibertii* is impossible to distinguish from *T. recognitum* with conviction on hand lens characters although the leaves on the main stem tend to have revolute margins (those of *T. recognitum* are not revolute) and the leaves tend to be compressed against the stem (those of *T. recognitum* have the tips wide spreading and recurved). From *Hylocomium splendens*, the shiny plants, red stems, and generally golden to brownish-green colour distinguish it from the yellow-green non-glossy plants of *Thuidium*. *Abietinella abietina* plants are once-pinnate; *Thuidium* always has the lateral branches bearing branchlets.

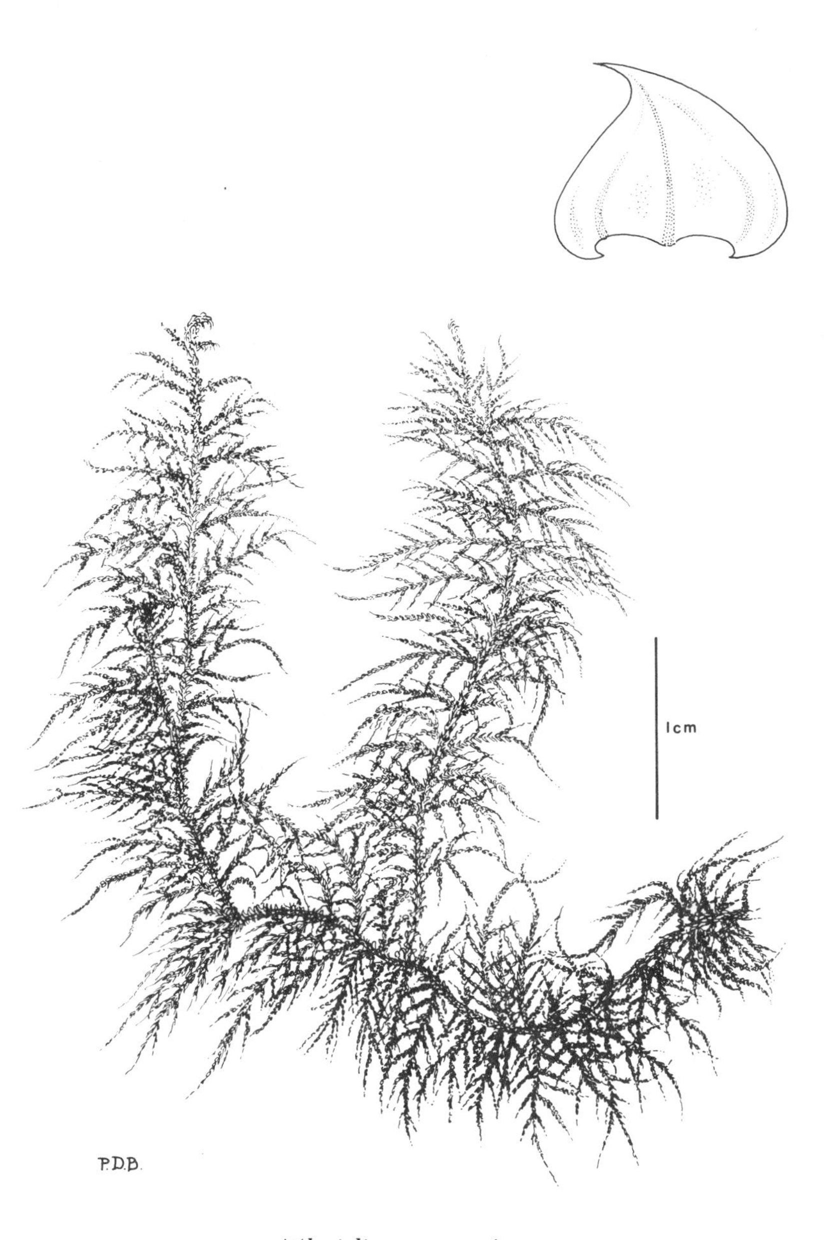

Thuidium recognitum

Timmia austriaca Hedw.

Name: Genus named in honour of J.C. Timm, an 18th-century, German botanist. Species named for Austria, where the species was first collected.
Habit: Tall, dull, greyish-green to dark green turfs, often suffused with the red-brown of the coloured sheathing leaf base and the unbranched stems; affixed to the substratum by brownish rhizoids.
Habitat: Forest floor, cliff shelves and crevices, usually in somewhat shaded sites from near sea level to alpine elevations.
Reproduction: Sporophytes occasional, the setae and sporangia light brown when mature in summer.
World Distribution: Circumpolar in the Northern Hemisphere, also in New Zealand; in North America southward in the east to New Brunswick and in the west to California and Colorado.
B.C. Distribution: Map 117, page 377.
Distinguishing Features: The somewhat sheathing often reddish-pigmented leaf bases, the light to dark green plants, the strongly divergent but stiff leaves, and the nodding sporangia are useful features for determination.
Similar Species: *T. austriaca* is sometimes confused with *Polytrichum* but the similarity is superficial since the leaves lack lamellae found in *Polytrichum* and the sporophyte is obviously different. The leaves that are subsheathing will immediately distinguish it from *Bryum*, *Mnium* or *Pohlia* in which the sporophyte is superficially similar.

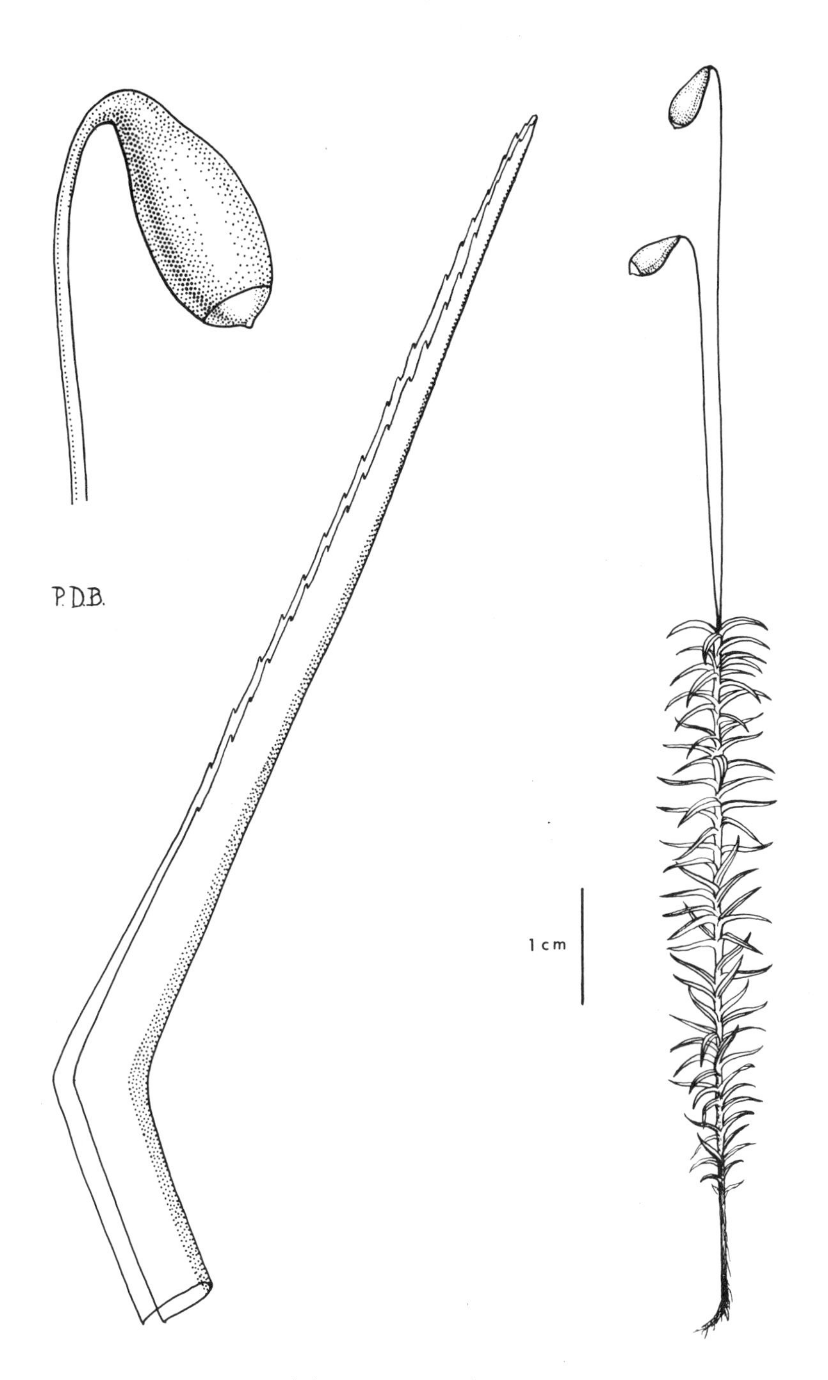

Timmia austriaca

Tomentypnum nitens (Hedw.) Loeske

Name: Genus name denoting a hairy moss, in reference to the abundance of rhizoids on the stem of most specimens of this moss. Species name meaning shining or polished, in reference to the leaves.
Habit: Forming turf-like mats of erect, regularly branched, glossy bright yellow-green to golden plants in which the stem is often matted with red-brown rhizoids.
Habitat: Usually calcareous swamps and fens but sometimes on bog margins from near sea level to subalpine elevations.
Reproduction: Sporophytes rare, red-brown, sporangia cylindric, curved, seta elongate, maturing in summer.
World Distribution: Circumpolar in the Northern Hemisphere; in North America extending southward in the east to Connecticut and the Great Lakes and in the west to Oregon.
B.C. Distribution: Map 118, page 378.
Distinguishing Features: The wetland, usually calcareous habitat, the erect, somewhat pinnately branched plants, the regularly pleated leaves, the glossy yellow-green colour, the usual felt of dark red-brown rhizoids on the stems and lower part of the leaf midrib are all useful features.
Similar Species: *T. falcifolium* has falcate-secund leaves, otherwise it is similar to *T. nitens*. *Orthothecium chryseum*, although it has pleated leaves, is usually irregularly branched or unbranched, lacks abundant rhizoids on the stem, and is reddish to golden-green. *Homalothecium* species do not occur in marshy habitats but show colour, pleating and form similar to *T. nitens*. *Brachythecium* species in the same habitat as *T. nitens* lack pleated leaves and abundant stem rhizoids. Unfortunately, some populations of *T. nitens* lack abundant stem rhizoids; however, in such cases the habitat and strongly tapered leaves are usually diagnostic, although confusion with *Brachythecium* or *Orthothecium* is possible.

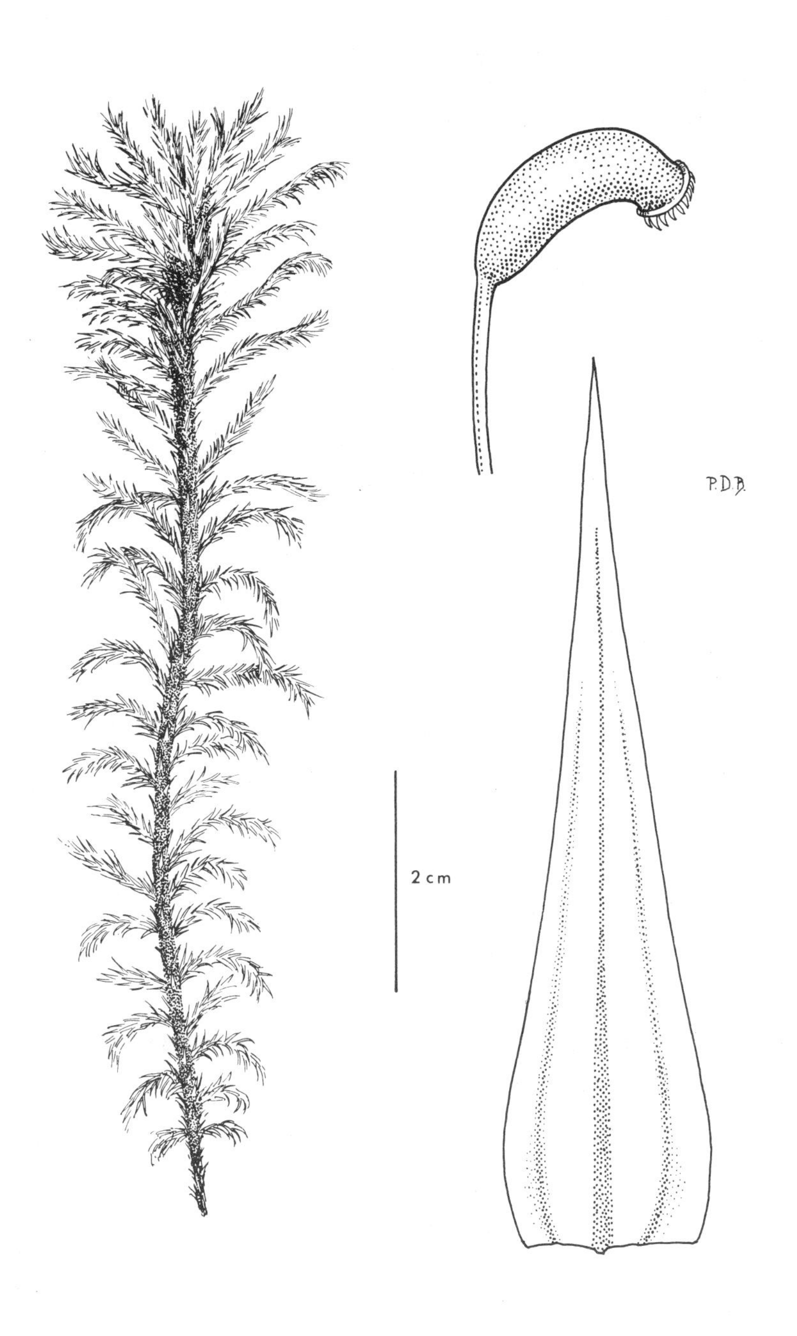

Tomentypnum nitens

Tortella tortuosa (Hedw.) Limpr.

Name: Genus name meaning "little twisted" in reference to the corkscrew twisted peristome teeth. Species name referring to the contorted leaves, when dry.
Habit: Forming dense turfs of bright yellow-green plants in which the shoot apex is often formed of pinwheel-like radiating leaves. Leaves strongly contorted when dry.
Habitat: Cliffs and cliff crevices, especially in calcareous areas, usually somewhat shaded and humid, but also on tree trunks (especially yellow cedar) and in tundra vegetation; from near sea level to alpine elevations.
Reproduction: Sporophytes occasional, maturing in spring to summer, red-brown and with markedly corkscrew twisted red-brown peristome teeth.
World Distribution: Circumpolar in the Northern Hemisphere, also in southern South America; in North America southward in the east to the southern Appalachian Mountains and in the west to Utah, Nevada and California.
B.C. Distribution: Map 119, page 378.
Distinguishing Features: The pinwheel-like arrangement of the leaves at the apex of the stem (best viewed when plants are damp), the pale yellow-green leaves that have undulate margins, the remarkably contorted leaves when dry, and the frequent restriction to calcium-rich sites are useful characters. The leaf, when removed shows a clear basal area of elongate cells that form an M-shaped area as it merges with the opaque elongate point of the leaf. This characterizes the genus *Tortella*.
Similar Species: *T. fragilis* has rigid leaves in which the tips break off readily; *T. inclinata* has bluntish leaf apices; *T. humilis* is a small plant and the leaves lack undulating margins. *Oxystegus tenuirostris* is superficially similar but lacks the M-shaped area of clear basal cells; these clear cells gradually merge with the opaque cells. *Geheebia gigantea* usually produces reddish-brown plants in which the leaves are markedly recurved and lack the hyaline basal area. *Leptodontium recurvifolium* has markedly toothed recurved leaves and lacks the basal hyaline area. *Ulota obtusiuscula* is similar in colour and in the contorted leaves when dry but plants are epiphytic and tufted, and the sporophytes are usually present and bear a hairy calyptra. *Trichostomum arcticum* can resemble *Tortella*, especially in the pinwheel-like arrangement of the leaves at the shoot apices. In *Trichostomum arcticum*, however, the leaf base has a hyaline area that grades gradually into the opaque point, rather than forming an M-shaped area.

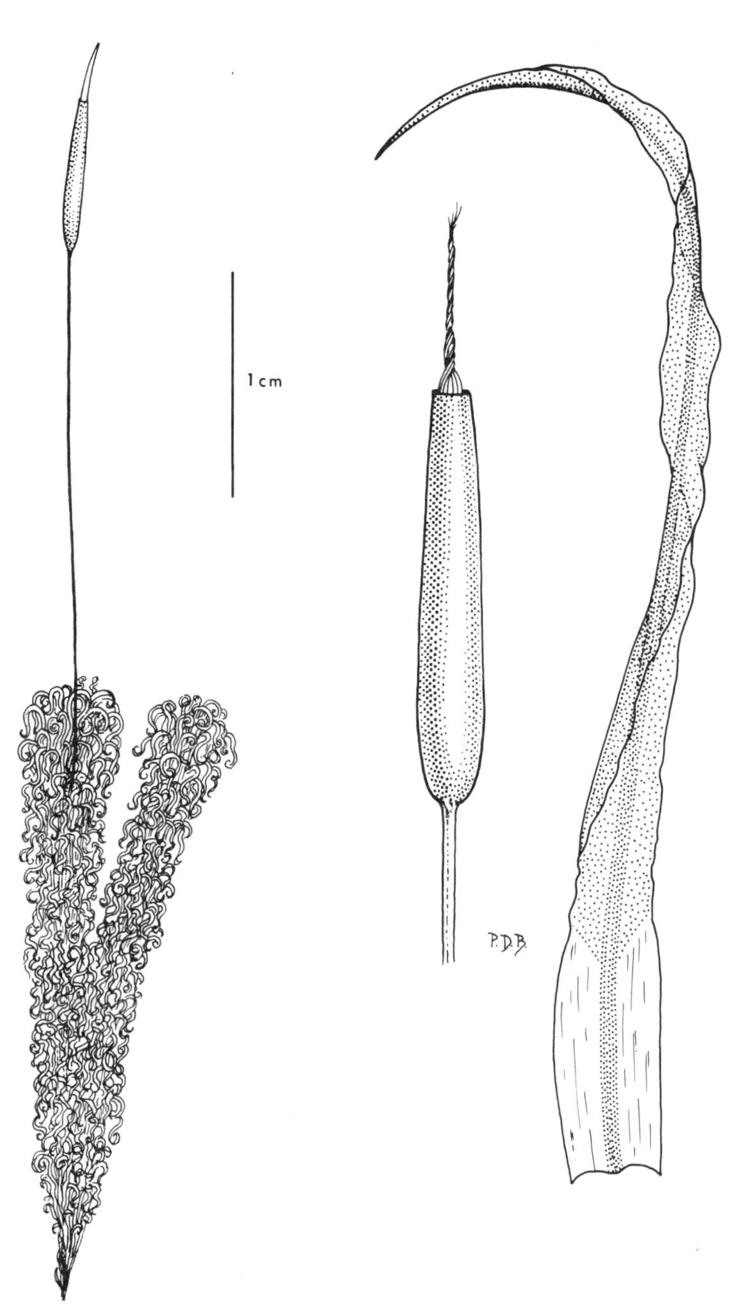

Tortella tortuosa

Tortula mucronifolia Schwaegr.

Name: Genus name referring to the twisted peristome teeth. Species name describing the leaf terminated by a short, abruptly tapering point (apiculus).
Habit: Forming loose, short turfs or scattered, dark green plants closely affixed to the substratum.
Habitat: Somewhat shaded calcareous soil of cliff shelves or tree bases.
Reproduction: Sporophytes common, red-brown and with pale red-brown to whitish, corkscrew-twisted peristome teeth.
World Distribution: Circumpolar in the Northern Hemisphere; in North America south in the east to Virginia and the Great Lakes and in the west to Baja California.
B.C. Distribution: Map 120, page 379.
Distinguishing Features: The dark green rosette of leaves, the cylindric sporangium with whitish, spirally twisted peristome teeth make this a distinctive species of calcareous somewhat shaded soil.
Similar Species: *T. subulata* is similar but the sporangium is twice to three times as long as that in *T. mucronifolia*. *T. amplexa* and *T. bolanderi* have red-brown rather than whitish peristome teeth, and sporangia mature in spring rather than summer. *T. muralis* and *T. brevipes* have leaves with white hair points. *T. latifolia* usually has round gemmae on the leaf surfaces, leaf apices are blunt, and is frequent on concrete walls or is epiphytic. *Crumia latifolia* is a species of wet calcareous cliffs near the coast and the peristome teeth are straight and short. *Desmatodon latifolius* is a terrestrial species of higher elevations and has red peristome teeth. *Aloina* has similar sporophytes to *Tortula* but the leaves are very short, shiny and the margins are folded in over filaments on the leaf surface. *Crossidium* also has hair pointed leaves but the peristome teeth are red and spirally twisted as in many species of *Tortula*.

Tortula mucronifolia

a muralis Hedw.

: Species name describing the usual habitat of walls.
Habit: Forming short, reddish-brown (when dry), bluish-green (when humid) to whitish turfs, with the white hair points often dominating the leafy plants.
Habitat: Frequent on concrete of walls, between stones and bricks and on sandstone cliffs, rarely epiphytic on the trunk bases of Garry oak.
Reproduction: Sporophytes abundant, red-brown, with red, cork-screw twisted peristome teeth; maturing in spring.
World Distribution: Interruptedly circumpolar in temperate portions of the Northern and Southern Hemispheres; in North America infrequent in the east from Newfoundland to North Carolina and Tennessee and in the west from British Columbia to California, where often abundant.
B.C. Distribution: Map 121, page 379.
Distinguishing Features: The usual habitat on mortar or concrete in urban areas, combined with the white hair points of the leaves and the erect-cylindric sporangia with red, spirally twisted peristome teeth are useful characters.
Similar Species: *T. brevipes* is similar to *T. muralis* but the peristome teeth have a basal cylinder from which the teeth emerge; the teeth in *T. muralis* lack this tube (shown in the figure of *T. princeps*). See note under *Grimmia pulvinata*.
Comments: This is an extremely common species in urban areas and infrequent in natural environments. The abundance of mortar surface and persistence of seasonal moisture no doubt favour it.

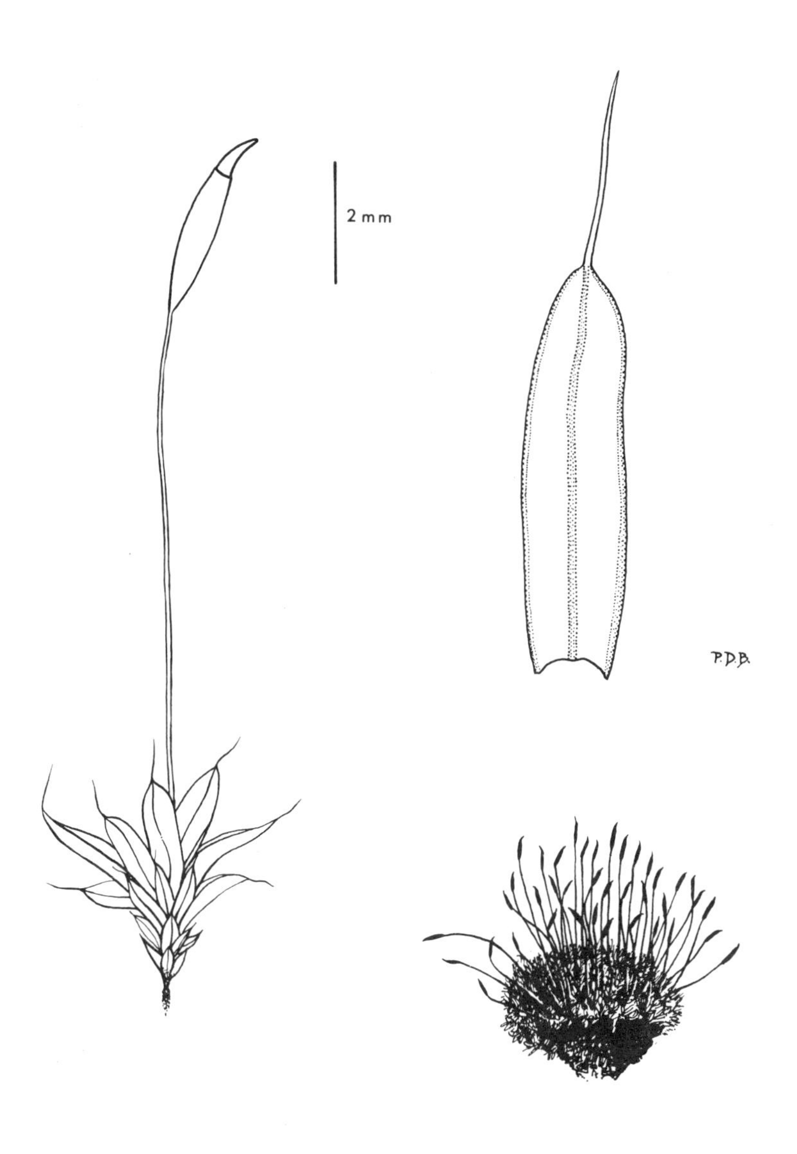

Tortula muralis

Tortula princeps De Not.

Name: Species name meaning foremost, perhaps in reference to the elegance of the moss.
Habit: Forming loose turfs or tufts of yellowish-green to dark green, erect, unbranched plants showing annual rosettes of enlarged leaves alternating with smaller leaves below. Leaves appearing somewhat twisted and brownish when dry.
Habitat: Concrete walls, between bricks and stones, and on cliff shelves and crevices, sometimes on sand, mainly near sea level and in urban areas.
Reproduction: Sporophytes frequent, red-brown with white, corkscrew-twisted peristome teeth, maturing in spring.
World Distribution: Europe, northern Africa, western Asia, Australasia; in North America confined to the west from Bristish Columbia to Arizona.
B.C. Distribution: Map 122, page 380.
Distinguishing Features: The usually yellowish-green to dark green plants that produce abundant sporophytes with white, twisted peristome teeth that arise from a basal tube, and the whorls of enlarged leaves marking the annual growth increments of the shoot are useful features to separate this species.
Similar Species: *T. ruralis* is similar but plants are usually reddish-green, lack the whorls of leaves that show annual growth increments and the leaves are usually strongly squarrose or recurved when moist; they are merely divergent in *T. princeps*.

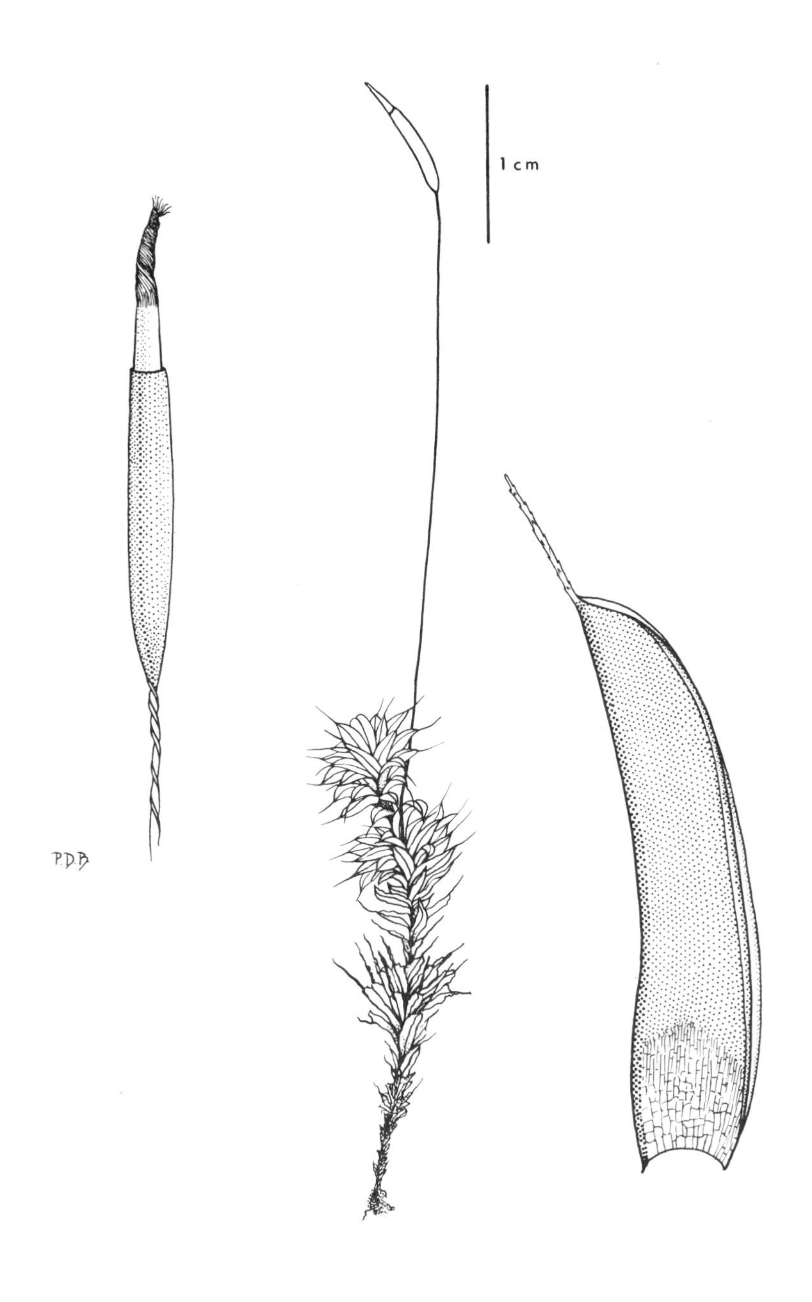

Tortula princeps

la ruralis (Hedw.) Gaertn., Meyer & Scherb.

..ne: Species name meaning belonging to the country; *T. ruralis* is frequent in much of the countryside in Europe, from where it was first described.

Habit: Usually reddish-brown, tall to short turfs, loosely affixed to the substratum, with leaves markedly divergent to squarrose when humid, imbricate when dry; white hair points very apparent.

Habitat: Terrestrial, on concrete and mortar, epiphytic on tree trunks (broad-leafed maple, poplar, Garry oak), from sea level (where on stabilized sand) to alpine elevations; frequent in the semi-arid interior, especially on calcium-rich mineral soil.

Reproduction: Sporophytes occasional to locally abundant, red-brown with whitish, corkscrew-twisted peristome teeth, maturing in spring to summer.

World Distribution: Almost cosmopolitan; in North America extending southward in the east to the Great Lakes area, and widely distributed in the west.

B.C. Distribution: Map 123, page 380.

Distinguishing Features: The strongly squarrose leaves, when moist, the abrupt colourless hair point, and the frequent reddish-brown tint to the dark green plants are useful features.

Similar Species: See note under *T. princeps*. *T. norvegica* is extremely similar, the reddish hair points and subalpine to alpine restriction are the best characters to separate it from *T. ruralis*; *T. norvegica* occurs most often in herb fields. *T. caninervis* also strongly resembles *T. ruralis* but the moist leaves are not squarrose and plants tend to be smaller than most specimens of *T. ruralis*. *T. laevipila*, a rare species on oak trees, has gemmae in the axils of leaves; otherwise it resembles a miniature *T. ruralis*.

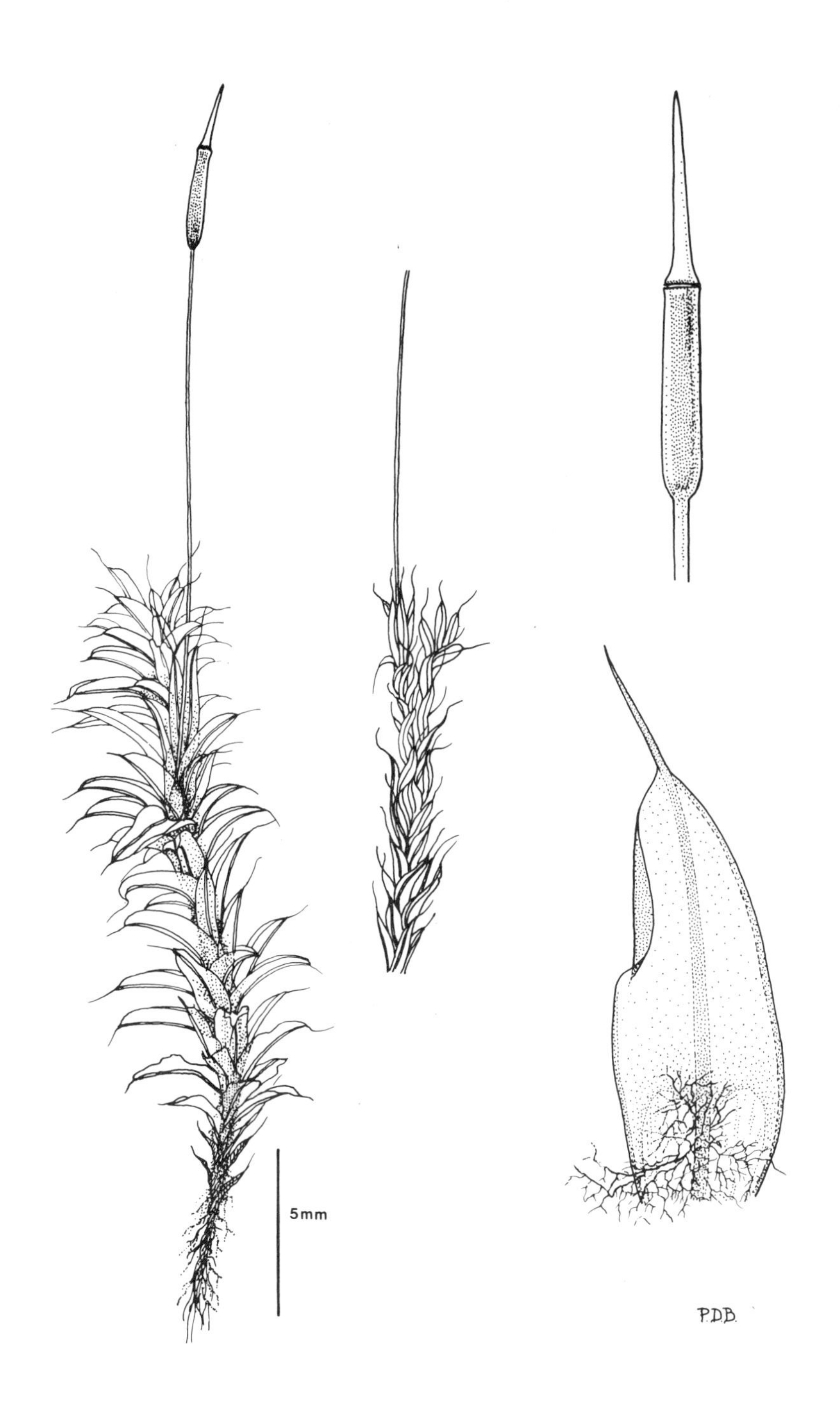

Tortula ruralis

Ulota obtusiuscula C. Muell. & Kindb. *ex* Mac. & Kindb.

Name: Genus name referring to the curled leaves of many species. Species name meaning somewhat obtuse, in reference to the leaf apex.
Habit: Forming yellowish-green tufts or rounded cushions.
Habitat: Predominantly epiphytic on deciduous trees near the coast (broad-leafed maple, alder, Garry oak) but also on coniferous trees (Sitka spruce, yew, coastal lodgepole pine). Mainly near sea level but extending to subalpine elevations on the immediate coast.
Reproduction: Sporophytes abundant, light brown when mature, with sporangia longitudinally grooved, maturing in spring; calyptra with erect hairs.
World Distribution: Confined to western North America from southeastern Alaska southward to California.
B.C. Distribution: Map 124, page 381.
Distinguishing Features: The rounded tufts on trees, the hairy calyptra, and the leaves much contorted when dry are useful characters for determination.
Similar Species: *Orthotrichum consimile* is like a miniature version of *Ulota* but the leaves are not strongly contorted when dry, the calyptra has very few (or no) hairs, and the plants are less than ¼ the size of *Ulota*. Of the other epiphytic species of *Ulota,* both *U. drummondii* and *U. megalospora* produce creeping shoots from the more flattened tufts. *U. phyllantha* is occasionally epiphytic and most populations bear clusters of dark green to brownish gemmae at the leaf tips; *U. phyllantha* is a maritime species, usually on rock. See also notes under *Tortella tortuosa,* an occasionally epiphytic moss.

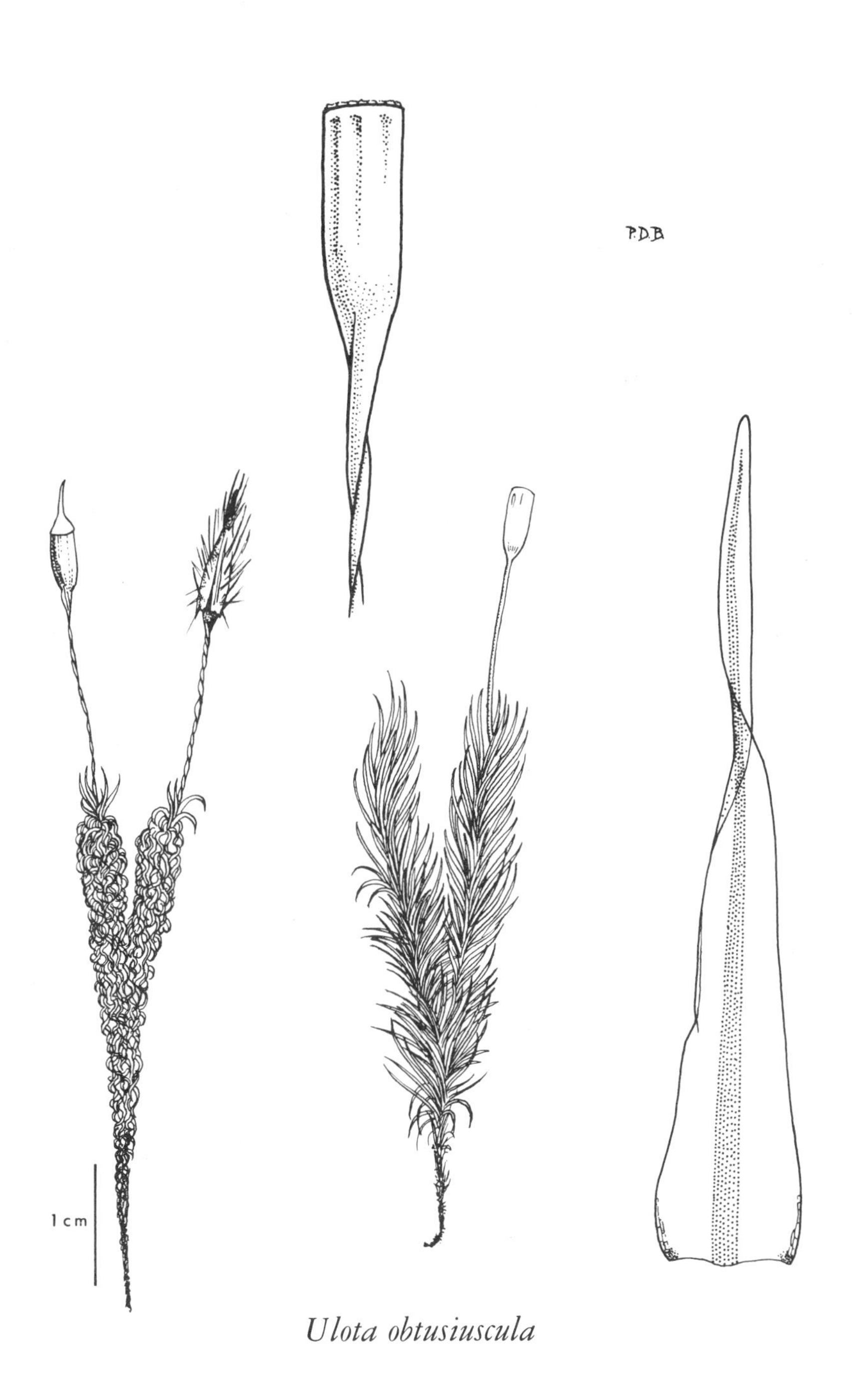

Ulota obtusiuscula

CHECKLIST OF THE MOSSES OF BRITISH COLUMBIA

The following list records all species known in the moss flora of the province for which there is an authenticated specimen. With further exploration and collection, the number increases annually, although the number of species that remain to be discovered in the province is probably not great. Some species, however, have been collected infrequently; a diligent amateur, with judicious collecting, can make extremely valuable contributions to the understanding of the flora.

Abietinella abietina (Hedw.) Fleisch.
Aloina bifrons (De Not.) Delg.
A. brevirostris (Hook. & Grev.) Kindb.
A. rigida (Hedw.) Limpr.
Alsia californica (Hook. & Arnott) Sull.
Amblyodon dealbatus (Hedw.) B.S.G.
Amblystegium compactum (C. Muell.) Aust.
A. fluviatile (Hedw.) B.S.G.
A. jungermannoides (Brid.) A.J.E. Sm.
A. noterophilum (Sull. & Lesq. *ex* Sull.) Holz.
A. riparium (Hedw.) B.S.G.
A. serpens (Hedw.) B.S.G.
A. tenax (Hedw.) C. Jens.
A. varium (Hedw.) Lindb.
Amphidium californicum (Hampe *ex* C. Muell.) Broth.
A. lapponicum (Hedw.) Schimp.
A. mougeotii (B.S.G.) Schimp.
Anacolia menziesii (Turn.) Paris
Andreaea alpestris (Thed.) Schimp.
A. blyttii Schimp.
A. megistospora B. Murray
A. mutabilis Hook. f. & Wils.
A. nivalis Hook.
A. obovata Thed.
A. rothii Web. & Mohr.
A. rupestris Hedw.
A. schofieldiana B. Murray
A. sinuosa B. Murray

Andreaeobryum macrosporum Steere & B. Murray
Anoectangium aestivum (Hedw.) Mitt.
A. sendtnerianum B.S.G.
A. tenuinerve (Limpr.) Par.
Anomobryum julaceum (Gaertn., Meyer & Scherb.) Schimp.
Antitrichia californica Sull. *ex* Lesq.
A. curtipendula (Hedw.) Brid.
Aongstroemia longipes (Somm.) B.S.G.
Arctoa fulvella (Dicks.) B.S.G.
Atrichum hausknechtii Jur. & Milde
A. selwynii Aust.
A. tenellum (Roehl.) B. & S.
A. undulatum (Brid.) P. Beauv.
Aulacomnium acuminatum (Lindb. & H. Arnell) Kindb.
A. androgynum (Hedw.) Schwaegr.
A. palustre (Hedw.) Schwaegr.
A. turgidum (Wahlenb.) Schwaegr.
Barbula amplexifolia (Mitt.) Jaeg.
B. convoluta Hedw.
B. unguiculata Hedw.
Bartramia halleriana Hedw.
B. ithyphylla Brid.
B. pomiformis Hedw.
B. stricta Brid.
Bartramiopsis lescurii (James)Kindb.
Blindia acuta (Hedw.) B.S.G.
Brachydontium olympicum (Britt. *ex* Frye) McIntosh & Spence
Brachythecium acutum (Mitt.) Sull.
B. albicans (Hedw.) B.S.G.
B. asperrimum (Mitt.) Sull.
B. calcareum Kindb.
B. campestre (C. Muell.) B.S.G.
B. erythrorrhizon B.S.G.
B. fendleri (Sull.) Jaeg.
B. frigidum (C. Muell.) Besch.
B. groenlandicum (C. Jens.) Schljak.
B. holzingeri (Grout) Grout
B. hylotapetum B. Hig. & N. Hig.
B. latifolium Kindb.
B. leibergii Grout
B. oedipodium (Mitt.) Jaeg. & Sauerb.
B. oxycladon (Brid.) Jaeg. & Sauerb.
B. plumosum (Hedw.) B.S.G.
B. populem (Hedw.) B.S.G.
B. rivulare B.S.G.
B. rutabulum (Hedw.) B.S.G.
B. salebrosum (Web. & Mohr) B.S.G.
B. starkei (Brid.) B.S.G.
B. trachypodium (Brid.)B.S.G.
B. turgidum (C.J. Hartm.) Kindb.
B. velutinum (Hedw.) B.S.G.
Brotherella roellii Ren. & Card. *ex* Roell.
Bryhnia hultenii Bartr. *ex* Grout
Bryobrittonia longipes (Mitt.) Horton
Bryoerythrophyllum alpigenum (Vent.) Chen
B. columbianum (Herm. & Lawt.) Zand.
B. ferruginascens (Stirt.) Giac.
B. recurvirostrum (Hedw.) Chen
Bryum algovicum Sendtn. *ex* C. Muell.
B. alpinum With.
B. arcticum (R. Br.) B.S.G.
B. argenteum Hedw.
B. bicolor Dicks.

B. blindii B.S.G.
B. caespiticium Hedw.
B. calobryoides Spence
B. calophyllum R. Br.
B. canariense Brid.
B. capillare Hedw.
B. cyclophyllum (Schwaegr.) B.S.G.
B. flaccidum Brid.
B. gemmiparum De Not.
B. lisae De Not.
B. meesioides Kindb. *in* Macoun
B. miniatum Lesq.
B. muehlenbeckii B.S.G.
B. pallens (Brid.) Sw. *ex* C. Roehl.
B. pallescens Schleich. *ex* Schwaegr.
B. pseudotriquetrum (Hedw.) Gaertn., Meyer & Scherb.
B. subapiculatum Hampe
B. turbinatum (Hedw.) Turn.
B. uliginosum (Brid.) B.S.G.
B. weigelii Spreng.
Buxbaumia aphylla Hedw.
B. piperi Best.
B. viridis (DC.) Moug. & Nestl.
Callicladium haldanianum (Grev.) Crum
Calliergon cordifolium (Hedw.) Kindb.
C. giganteum (Schimp.) Kindb.
C. richardsonii (Mitt.) Kindb. *ex* Warnst.
C. sarmentosum (Wahlenb.) Kindb.
C. stramineum (Brid.) Kindb.
C. trifarium (Web. & Mohr.) Kindb.
Calliergonella cuspidata (Hedw.) Loeske
Campylium calcareum Crundw. & Nyh.
C. chrysophyllum (Brid.) J. Lange
C. halleri (Hedw.) Lindb.
C. hispidulum (Brid.) Mitt.
C. polygamum (B.S.G.) C. Jens.
C. stellatum (Hedw.) J. Lange & C. Jens.
Campylopus atrovirens De Not.
C. flexuosus (Hedw.) Brid.
C. fragilis (Brid.) B.S.G.
C. schimperi Milde
C. schwarzii Schimp.
Catoscopium nigritum (Hedw.) Brid.
Ceratodon purpureus (Hedw.) Brid.
Cinclidium arcticum (B.S.G.) Schimp.
C. stygium Sw.
C. subrotundum Lindb.
Cirriphyllum cirrosum (Schwaegr. *ex* Schultes) Grout
Claopodium bolanderi Best
C. crispifolium (Hook.) Ren. & Card.
C. pellucinerve (Mitt.) Best
C. whippleanum (Sull.) Ren. & Card.
Climacium dendroides (Hedw.) Web. & Mohr
Cnestrum alpestre (Hueb.) Nyh. *ex* Mogensen
C. glaucescens (Lindb. & H. Arnell) Holmen *ex* Mogensen & Steere
C. schistii (Web. & Mohr) Hag.
Conostomum tetragonum (Hedw.) Lindb.
Coscinodon calyptratus (Hook. *in* Drumm.) C. Jens *in* Kindb.
C. cribrosus (Hedw.) Spruce
Cratoneuron commutatum (Hedw.)

Roth
C. filicinum (Hedw.) Spruce
Crossidium aberrans Holz. & Bartr.
C. rosei Williams
C. seriatum Crum & Steere
Crumia latifolia (Kindb. *ex* Mac.) Schof.
Ctenidium schofieldii Nishimura
Cynodontium jenneri (Schimp. *ex* Howie) Stirt.
C. strumiferum (Hedw.) Lindb.
C. tenellum (B.S.G.) Limpr.
Cyrtomnium hymenophylloides (Hueb.) Nyh. *ex* Kop.
C. hymenophyllum (B.S.G.) Holmen
Daltonia splachnoides (Sm. *ex* Sm. & Sowerby) Hook. & Tayl.
Dendroalsia abietina (Hook.) Britt
Desmatodon cernuus (Hueb.) B.S.G.
D. convolutus (Brid.) Grout
D. guepinii B.S.G.
D. heimii (Hedw.) Mitt.
D. latifolius (Hedw.) Brid.
D. leucostoma (R. Br.) Berggr.
D. obtusifolius (Schwaegr.) Schimp.
D. randii (Kenn.) Laz.
Dichelyma falcatum (Hedw.) Myr.
D. pallescens B.S.G.
D. uncinatum Mitt.
Dichodontium flavescens (With.) Lindb.
D. olympicum Ren. & Card.
D. pellucidum (Hedw.) Schimp.
Dicranella cerviculata (Hedw.) Schimp.
D. crispa (Hedw.) Schimp.
D. grevilleana (Brid.) Schimp.
D. heteromalla (Hedw.) Schimp.
D. pacifica Schof.
D. palustris (Dicks.) Crundw. *ex* Warb.
D. rufescens (With.) Schimp.
D. schreberiana (Hedw.) Hilf. *ex* Crum. & Anders.
D. subulata (Hedw.) Schimp.
D. varia (Hedw.) Schimp.
Dicranodontium asperulum (Mitt.) Broth.
D. denudatum (Brid.) Britt. *ex* Williams
D. subporodictyon Broth.
D. uncinatum (Harv. *ex* Hook.) Jaeg. & Sauerb.
Dicranoweisia cirrata (Hedw.) Lindb. *ex* Milde
D. crispula (Hedw.) Lindb. *ex* Milde
Dicranum acutifolium (Lindb. & H. Arnell) C. Jens *ex* Weinm.
D. affine Funck.
D. angustum Lindb.
D. brevifolium (Lindb.) Lindb.
D. elongatum Schleich. *ex* Schwaegr.
D. flagellare Hedw.
D. fragilifolium Lindb.
D. fuscescens Sm.
D. groenlandicum Brid.
D. majus Sm.
D. montanum Hedw.
D. muehlenbeckii B.S.G.
D. pallidisetum (Bail. *ex* Holz.) Irel.
D. polysetum Sw.
D. scoparium Hedw.
D. spadiceum Zett.
D. tauricum Sapeh.
Didymodon acutus (Brid.) Saito
D. asperifolius (Mitt.) Crum,

Steere & Anderson
D. *fallax* (Hedw.) Zand
D. *insulanus* (De Not.) M. Hill
D. *johansenii* (Williams) Crum
D. *luridus* Hornsch. *ex* Spreng.
D. *michiganensis* (Steere) K. Saito
D. *nicholsonii* Culm.
D. *nigrescens* (Mitt.) Saito
D. *occidentalis* (Mitt.) Zand.
D. *rigidulus* Hedw.
D. *subandreaeoides* (Kindb.) Zand.
D. *tophaceus* (Brid.) Lisa
D. *vinealis* (Brid.) Zand.
Diphyscium foliosum (Hedw.) Mohr
Discelium nudum (Dicks.) Brid.
Distichium capillaceum (Hedw.) B.S.G.
D. *inclinatum* (Hedw.) B.S.G.
Ditrichum ambiguum Best
D. *crispatissimum* (C. Muell.) Par.
D. *flexicaule* (Schwaegr.) Hampe
D. *heteromallum* (Hedw.) Britt.
D. *montanum* Leib.
D. *pusillum* (Hedw.) Hampe
D. *schimperi* (Lesq.) Kuntze
D. *zonatum* (Brid.) Kindb.
Drepanocladus aduncus (Hedw.) Warnst.
D. *capillifolius* (Warnst.) Warnst.
D. *crassicostatus* Janssens
D. *exannulatus* (B.S.G.) Warnst.
D. *fluitans* (Hedw.) Warnst.
D. *lapponicus* (Norrl.) Smirn.
D. *pseudostramineus* (C. Muell.) Roth
D. *revolvens* (Sw.) Warnst.
D. *sendtneri* (Schimp. *ex* H. Muell) Warnst.
D. *trichophyllus* (Warnst.) Podp.
D. *tundrae* (H. Arnell) Loeske
D. *uncinatus* (Hedw.) Warnst.
D. *vernicosus* (Mitt.) Warnst.
Dryptodon patens (Hedw.) Brid.
Encalypta affinis Hedw.f.
E. *alpina* Sm.
E. *brevicolla* (B.S.G.) Bruch *ex* Aongstr.
E. *brevipes* Schljak.
E. *ciliata* Hedw.
E. *intermedia* Mitt.
E. *longicolla* Bruch
E. *mutica* Hag.
E. *procera* Bruch
E. *rhaptocarpa* Schwaegr.
E. *spathulata* C. Muell.
E. *vulgaris* Hedw.
Entodon concinnus (De Not.) Par.
Entosthodon fascicularis (Hedw.) C. Muell.
E. *rubiginosus* (Williams) Grout
Epipterygium tozeri (Grev.) Lindb.
Eucladium verticellatum (Brid.) B.S.G.
Eurhynchium pulchellum (Hedw.) Jenn.
Fabronia pusilla Raddi
Fissidens adianthoides Hedw.
F. *aphelotaxifolius* Pursell
F. *bryoides* Hedw.
F. *fontanus* (B.-Pyl.) Stend.
F. *grandifrons* Brid.
F. *limbatus* Sull.
F. *osmundoides* Hedw.
F. *pauperculus* Howe
F. *ventricosus* Lesq.
Fontinalis antipyretica Hedw.
F. *howellii* Ren. & Card.
F. *hypnoides* C. J. Hartm.
F. *neomexicana* Sull. & Lesq.
Funaria hygrometrica Hedw.
F. *muhlenbergii* Hedw.f. *ex* Turn.
Geheebia gigantea (Funck.) Boul.

Gollania turgens (C. Muell.) Ando
Grimmia affinis Hoppe & Hornsch. *ex* Hornsch.
G. alpestris Nees
G. anodon B.S.G.
G. anomala Hampe *ex* Schimp.
G. brittoniae Williams
G. donniana Sm.
G. elatior Bruch. *ex* Bals. & De Not.
G. elongata Kaulf. *in* Sturm
G. heterophylla Kindb. *ex* Mac. & Kindb.
G. holzingeri Card. & Ther.
G. laevigata (Brid.) Brid.
G. montana B.S.G.
G. ovalis (Hedw.) Lindb.
G. plagiopodia Hedw.
G. pulvinata (Hedw.) Sm. *ex* Sm. & Sowerby
G. teretinervis Limpr.
G. torquata Hornsch. *ex* Grev.
G. trichophylla Grev.
G. unicolor Hook. *ex* Grev.
Gymnostomum aeruginosum Sm.
Hedwigia ciliata (Hedw.) P. Beauv.
Helodium blandowii (Web. & Mohr) Warnst.
Herzogiella adscendens (Lindb.) Iwats. & Schof.
H. seligeri (Brid.) Iwats.
H. striatella (Brid.) Iwats.
Heterocladium dimorphum (Brid.) B.S.G.
H. macounii Best.
H. papillosum (Lindb.) Lindb.
H. procurrens (Mitt.) Rau & Herv.
Homalia trichomanoides (Hedw.) B.S.G.
Homalothecium aeneum (Mitt.) Lawt.
H. arenarium (Lesq.) Lawt.
H. fulgescens (Mitt. *ex* C. Muell.) Lawt.
H. nevadense (Lesq.) Ren. & Card.
H. nuttallii (Wils.) Jaeg. & Sauerb.
H. pinnatifidum (Sull. & Lesq.) Lawt.
Hookeria acutifolia Hook. & Grev.
H. lucens (Hedw.) Sm.
Hydrogrimmia mollis (B.S.G.) Loeske
Hygrohypnum alpestre (Hedw.) Loeske
H. alpinum (Lindb.) Loeske
H. bestii (Ren. & Bryhn *ex* Ren.) Holz. *ex* Broth.
H. cochlearifolium (Vent. *ex* De Not.) Broth.
H. duriusculum (De Not.) Jamieson
H. luridum (Hedw.) Jenn.
H. molle (Hedw.) Loeske
H. norvegicum (B.S.G.) Amann
H. ochraceum (Turn. *ex* Wils.) Loeske
H. polare (Lindb.) Loeske
H. smithii (Sw. *ex* Lilj.) Broth.
H. styriacum (Limpr.) Broth.
Hylocomium pyrenaicum (Spruce) Lindb.
H. splendens (Hedw.) B.S.G.
H. umbratum (Hedw.) B.S.G.
Hymenostylium insigne (Dix.) Podp.
H. recurvirostrum (Hedw.) Dix.
Hypnum bambergeri Schimp.
H. callichroum Funck *ex* Brid.
H. circinale Hook.
H. cupressiforme Hedw.
H. dieckii Ren. & Card. *ex* Roell.

H. lindbergii Mitt.
H. pallescens (Hedw.) P. Beauv.
H. plicatulum (Lindb.) Jaeg. & Sauerb.
H. pratense W. Koch *ex* Spruce
H. procerrimum Mol.
H. recurvatum (Lindb. & H. Arnell) Kindb.
H. revolutum (Mitt.) Lindb.
H. subimponens Lesq.
H. vaucheri Lesq.
Hypopterygium fauriei Besch.
Isopterygiopsis muelleriana (Schimp.) Iwats.
I. pulchella (Hedw.) Iwats.
Isothecium cristatum (Hampe.) Robins.
I. stoloniferum Brid.
Iwatsukiella leucotricha (Mitt.) Buck & Crum
Kiaeria blyttii (Schimp.) Broth.
K. falcata (Hedw.) I. Hag.
K. glacialis (Berggr.) I. Hag.
K. starkei (Web. & Mohr.) I. Hag.
Kindbergia oregana (Sull.) Ochyra
K. praelonga (Hedw.) Ochyra
Leptobryum pyriforme (Hedw.) Wils.
Leptodontium recurvifolium (Tayl.) Linbb.
Lescuraea atricha (Kindb. *ex* Mac. & Kindb.) Lawt.
L. baileyi (Best & Grout *ex* Grout) Lawt.
L. incurvata (Hedw.) Lawt.
L. julacaea Besch. & Card *ex* Card.
L. patens (Lindb.) H. Arnell & C. Jens.
L. radicosa (Mitt.) Moenk.
L. saxicola (B.S.G.) Milde
L. stenophylla (Ren. & Card. *ex* Roell.) Kindb.
Leskea polycarpa Hedw.
Leskeella nervosa (Brid.) Loeske
Leucolepis acanthoneuron (Schwaegr.) Lindb.
Loeskypnum badium (C.J. Hartm.) Paul
L. wickesiae (Grout) Tuom.
Meesia longiseta Hedw.
M. triquetra (Richt.) Aongstr.
M. uliginosa Hedw.
Metaneckera menziesii (Hook. *ex* Drumm.) Steere
Micromitrium tenerum (B.S.G.) Crosby
Mielichhoferia macrocarpa (Hook. *ex* Drumm.) Bruch & Schimp. *ex* Jaeg. & Sauerb.
M. mielichhoferi (Funck *ex* Hook.) Wijk. & Marg.
Mnium ambiguum H. Muell.
M. arizonicum Amann
M. blyttii B.S.G.
M. marginatum (With.) Brid. *ex* P. Beauv.
M. spinulosum B.S.G.
M. thomsonii Schimp.
Myrinia pulvinata (Wahlenb.) Schimp.
Myurella julacea (Schwaegr.) B.S.G.
M. tenerrima (Brid.) Lindb.
Neckera douglasii Hook.
N. oligocarpa Bruch.
N. pennata Hedw.
Oedipodium griffithianum (Dicks.) Schwaegr.
Oligotrichum aligerum Mitt.
O. hercynicum (Hedw.) DC.
O. parallelum (Mitt.) Kindb.

Oncophorus virens (Hedw.) Brid.
O. wahlenbergii Brid.
Oreas martiana (Hoppe & Hornsch. *ex* Hornsch.) Brid.
Orthothecium chryseum (Schwaegr. *ex* Schultes) B.S.G
O. intricatum (C.J. Hartm.) B.S.G.
O. strictum Lor.
Orthotrichum affine Brid.
O. alpestre Hornsch. *ex* B.S.G.
O. anomalum Hedw.
O. consimile Mitt.
O. cupulatum Brid.
O. diaphanum Brid.
O. hallii Sull. & Lesq. *in* Sull.
O. laevigatum Zett.
O. lyellii Hook. & Tayl.
O. obtusifolium Brid.
O. pallens Bruch *ex* Brid.
O. pellucidum Lindb.
O. pulchellum Brunt. *ex* Winch. & Gateh.
O. pylaisii Brid.
O. rivulare Turn.
O. rupestre Schleich. *ex* Schwaegr.
O. speciosum Nees *ex* Sturm
O. striatum Hedw.
O. tenellum Bruch *ex* Brid.
Oxystegus tenuirostris (Hook. & Tayl.) A.J.E. Sm.
Paludella squarrosa (Hedw.) Brid.
Paraleucobryum enerve (Thed. *ex* C.J. Hartm.) Loeske
P. longifolium (Hedw.) Loeske
Phascum cuspidatum Hedw.
P. vlassovii Laz.
Philonotis capillaris Lindb. *ex* C.J. Hartm.
P. fontana (Hedw.) Brid.
P. marchica (Hedw.) Brid.
P. yezoana Besch. & Card. *in* Card.
Physcomitrella patens (Hedw.) B.S.G.
Physcomitrium immersum Sull.
P. megalocarpum Kindb. *ex* Mac.
Plagiobryum demissum (Hook.) Lindb.
P. zierii (Hedw.) Lindb.
Plagiomnium ciliare (C. Muell.) Kop.
P. cuspidatum (Hedw.) Kop.
P. drummondii (Bruch & Schimp.) Kop.
P. ellipticum (Brid.) Kop.
P. insigne (Mitt.) Kop.
P. medium (B.S.G.) Kop.
P. rostratum (Schrad.) Kop.
P. venustum (Mitt.) Kop.
Plagiopus oederi (Brid.) Limpr.
Plagiothecium cavifolium (Brid.) Iwats.
P. denticulatum (Hedw.) B.S.G.
P. laetum B.S.G.
P. piliferum (Sw. *ex* C.J. Hartm.) B.S.G.
P. platyphyllum Moenk.
P. undulatum (Hedw.) B.S.G.
Platygyrium repens (Brid.) B.S.G.
Pleuridium bolanderi C. Muell. *ex* Jaeg.
Pleuroziopsis ruthenica (Weinm.) Kindb. *ex* Britt.
Pleurozium schreberi (Brid.) Mitt.
Pogonatum contortum (Brid.) Lesq.
P. dentatum (Brid.) Brid.
P. urnigerum (Hedw.) P. Beauv.
Pohlia andalusica (Hoehn.) Broth.
P. annotina (Hedw.) Lindb.

P. atropurpurea (Wahlenb.) Lindb. f.
P. bolanderi (Sull.) Broth.
P. bulbifera (Warnst.) Warnst.
P. camptotrachela (Ren. & Card.) Broth.
P. cardotii (Ren. *ex* Ren. & Card.) Broth.
P. columbica (Kindb. *ex* Mac. & Kindb.) Andr.
P. cruda (Hedw.) Lindb.
P. drummondii (C. Muell.) Andr.
P. elongata Hedw.
P. erecta Lindb.
P. filum (Schimp.) Mart.
P. lescuriana (Sull.) Grout
P. longibracteata Broth. *ex* Roell.
P. longicollis (Hedw.) Lindb.
P. ludwigii (Spreng. *ex* Schwaegr.) Broth.
P. melanodon (Brid.) J. Shaw
P. nutans (Hedw.) Lindb.
P. obtusifolia (Brid.) L. Koch
P. pacifica J. Shaw
P. proligera (Kindb. *ex* Limpr.) Lindb. *ex* H. Arnell
P. sphagnicola (B.S.G.) Lindb. & H. Arnell
P. tundrae J. Shaw
P. vexans (Limpr.) Lindb. f.
P. wahlenbergii (Web. & Mohr) Andr.
Polytrichum alpinum Hedw.
P. commune Hedw.
P. formosum Hedw.
P. juniperinum Hedw.
P. longisetum Brid.
P. lyallii (Mitt.) Kindb.
P. piliferum Hedw.
P. sexangulare Brid.
P. sphaerothecium (Besch.) C. Muell.
P. strictum Brid.
Porotrichum bigelowii (Sull.) Kindb.
P. vancouveriensis (Kindb. *ex* Mac.) Crum
Pottia bryoides (Dicks.) Mitt.
P. nevadensis Card. & Ther.
P. truncata (Hedw.) Fuernr. *ex* B.S.G.
P. wilsonii (Hook.) B.S.G.
Pseudephemerum nitidum (Hedw.) Loeske
Pseudobraunia californica (Lesq.) Broth.
Pseudobryum cinclidioides (Hueb.) Kop.
Pseudocrossidium hornschuchianum (Schultz) Zand.
P. revolutum (Brid. *in* Schrad.) Zand.
Pseudoleskeella tectorum (Funck *ex* Brid.) Kindb. *ex* Broth.
Pseudoscleropodium purum (Hedw.) Fleisch.
Pseudotaxiphyllum elegans (Brid.) Iwats.
Psilopilum cavifolium (Wils.) Hag.
Pterigynandrum filiforme Hedw.
Pterogonium gracile (Hedw.) Sm.
Pterygoneurum kozlovii Laz.
P. lamellatum (Lindb.) Jur.
P. ovatum (Hedw.) Dix.
P. subsessile (Brid.) Jur.
Ptilium crista-castrensis (Hedw.)
Ptychomitrium gardneri Lesq.
Pylaisia intricata (Hedw.) B.S.G.
P. polyantha (Hedw.) Schimp.

Racomitrium aciculare (Hedw.) Brid.
R. affine (Schleich. *ex* Web. & Mohr) Lindb.
R. aquaticum (Brid. *ex* Schrad.) Brid.
R. brevipes Kindb. *in* Mac.
R. canescens (Hedw.) Brid.
R. elongatum Ehrh. *ex* Frisv.
R. ericoides (Web. *ex* Brid.) Brid.
R. fasciculare (Hedw.) Brid.
R. heterostichum (Hedw.) Brid.
R. lanuginosum (Hedw.) Brid.
R. lawtonae Irel.
R. macounii Kindb. *ex* Kindb. *in* Mac.
R. microcarpon (Hedw.) Brid.
R. muticum (Kindb. in Mac.) Frisv.
R. obesum Frisv.
R. occidentale (Ren. & Card.) Ren. & Card.
R. pacificum Irel. & Spence
R. pygmaeum Frisv.
R. sudeticum (Funck) B. & S. *in* B.S.G.
R. varium (Mitt.) Jaeg. & Sauerb.
Rhabdoweisia crispata (With.) Lindb.
Rhizomnium glabrescens (Kindb.) Kop.
R. gracile Kop.
R. magnifolium (Horik.) Kop.
R. nudum (Britt. & Williams) Kop.
R. pseudopunctatum (Bruch & Schimp.) Kop.
R. punctatum (Hedw.) Kop.
Rhodobryum roseum (Hedw.) Limpr.
Rhynchostegium serrulatum (Hedw.) Jaeg. & Sauerb.
R. riparioides (Hedw.) Card.
Rhytidiadelphus loreus (Hedw.) Warnst.
R. squarrosus (Hedw.) Warnst.
R. subpinnatus (Lindb.) Kop.
R. triquetrus (Hedw.) Warnst.
Rhytidiopsis robusta (Hedw.) Broth.
Rhytidium rugosum (Hedw.) Kindb.
Roellia roellii (Broth. *ex* Roell.) Andr. ex Crum
Saelania glaucescens (Hedw.) Bomanss. & Broth.
Schistidium agassizii Sull. & Lesq. *ex* Sull.
S. apocarpum (Hedw.) B. & S. *in* B.S.G.
S. maritimum (Turn.) B.S.G.
S. rivulare (Brid.) Podp.
S. tenerum (Zett.) Nyh.
S. trichodon (Brid.) Poelt
Schistostega pennata (Hedw.) Web. & Mohr.
Scleropodium cespitans (C. Muell.) L. Koch
S. obtusifolium (Jaeg. & Sauerb.) Kindb. *ex* Mac. & Kindb.
S. touretii (Brid.) L. Koch
Scorpidium scorpioides (Hedw.) Limpr.
S. turgescens (T. Jens.) Loeske
Scouleria aquatica Hook.
S. marginata Britt.
Seligeria acutifolia Lindb. *in* C. Hartm.
S. campylopoda Kindb. *ex* Mac. & Kindb.

S. careyana Vitt & Schof.
S. donniana (Sm.) C. Muell.
S. recurvata (Hedw.) B.S.G.
S. tristichoides Kindb.
Sematophyllum micans (Mitt.) Braithw.
Sphagnum angustifolium (C. Jens. *ex* Russ.) C. Jens. *in* Tolf.
S. aongstroemii C. Hartm.
S. austinii Sull.
S. balticum (Russ.) C. Jens.
S. capillifolium (Ehrh.) Hedw.
S. centrale C. Jens. *ex* H. Arnell & C. Jens.
S. compactum DC. *ex* Lam. & DC.
S. contortum Schultz
S. cuspidatum Ehrh. *ex* Hoffm.
S. fallax (Klinggr.) Klinggr.
S. fimbriatum Wils. *in* Wils. & Hook. f.
S. fuscum (Schimp.) Klinggr.
S. girgensohnii Russ.
S. henryense Warnst.
S. jensenii Lindb. f.
S. junghuhnianum Dozy & Molk.
S. lindbergii Schimp. *ex* Lindb.
S. magellanicum Brid.
S. majus (Russ.) C. Jens.
S. mendocinum Sull. & Lesq. *ex* Sull.
S. pacificum Flatberg
S. palustre L.
S. papillosum LIndb.
S. platyphyllum (Lindb. *ex* Braithw.) Sull. *ex* Warnst.
S. pulchrum (Lindb. *ex* Braithw.) WArnst.
S. quinquefarium (Lindb. *ex* Braithw.) Warnst.
S. riparium Aongstr.
S. rubellum Wils.
S. russowii Warnst.
S. schofieldii Crum
S. squarrosum Crome
S. subnitens Russ. & Warnst. *ex* Warnst.
S. subobesum Warnst.
S. subsecundum Nees *ex* Sturm
S. subtile (Russ.) Warnst.
S. tenellum (Brid.) Pers. *ex* Brid.
S. teres (Schimp.) Aongstr. *ex* C. Hartm.
S. warnstorfii Russ.
S. wilfii Crum
S. wulfianum Girg.
Splachnum ampullaceum Hedw.
S. luteum Hedw.
S. rubrum Hedw.
S. sphaericum Hedw.
S. vasculosum Hedw.
Stegonia latifolia (Schwaegr. *ex* Schultes) Vent. *ex* Broth.
Tayloria froelichiana (Hedw.) Mitt. *ex* Broth.
T. lingulata (Dicks.) Lindb.
T. serrata (Hedw.) B.S.G.
T. splachnoides (Schleich. *ex* Schwaegr.) Hook.
Tetraphis geniculata Girg. *ex* Milde
T. pellucida Hedw.
Tetraplodon angustatus (Hedw.) B.S.G.
T. blyttii Frisv.
T. mnioides (Hedw.) B.S.G.
T. pallidus Hag.
T. urceolatus B.S.G. *ex* Schimp.
Tetrodontium brownianum (Dicks.) Schwaegr.
T. repandum (Funck. *ex* Sturm) Schwaegr.

Thamnobryum neckeroides (Hook.) Lawt.
Thuidium philibertii Limpr.
T. recognitum (Hedw.) Lindb.
Timmia austriaca Hedw.
T. megapolitana Hedw.
T. norvegica Zett.
T. sibirica Lindb. & H. Arnell
Timmiella crassinervis (Hampe) L. Koch
Tomentypnum falcifolium (Ren. *ex* Nich.) Tuom.
T. nitens (Hedw.) Loeske
Tortella fragilis (Drumm.) Limpr.
T. humilis (Hedw.) Jenn.
T. inclinata (Hedw. f.) Limpr.
T. tortuosa (Hedw.) Limpr.
Tortula amplexa (Lesq.) Steere
T. atrovirens (Sm.) Lindb.
T. bolanderi (Lesq.) Howe
T. brevipes (Lesq.) Broth.
T. caninervis (Mitt.) Broth.
T. laevipila (Brid.) Schwaegr.
T. latifolia Bruch *ex* C.J. Hartm.
T. mucronifolia Schwaegr.
T. muralis Hedw.
T. norvegica (Web.) Wahlenb. *ex* Lindb.
T. princeps De Not.
T. ruralis (Hedw.) Gaertn., Meyer & Scherb.
T. subulata Hedw.
Trachybryum megaptilum (Sull.) Schof.
Trematodon ambiguus (Hedw.) Hornsch.
T. boasii Schof.
T. montanus Belland & Brass.
Trichodon cylindricus (Hedw.) Schimp.
Trichostomopsis australasiae (Grev. & Hook.) Robins.
Trichostomum arcticum Kaal.
Tripterocladium leucocladulum (C. Muell.) Jaeg. & Sauerb.
Ulota curvifolia (Wahlenb.) Lilj.
U. drummondii (Hook. & Brev.) *ex* Grev.) Brid.
U. japonica (Sull. & Lesq.) Mitt.
U. megalospora Vent. *ex* Roell.
U. obtusiuscula C. Muell. & Kindb. *ex* Mac. & Kindb.
U. phyllantha Brid.
Weissia brachycarpa (Nees & Hornsch.) Jur.
W. controversa Hedw.
Wijkia carlottae (Schof.) Crum
Zygodon gracilis (Hornsch. *ex* Reinw. & Hornsch.) A. Braun *ex* B.S.G.
Z. viridissimus (Dicks.) A. Br.

DISTRIBUTION MAPS

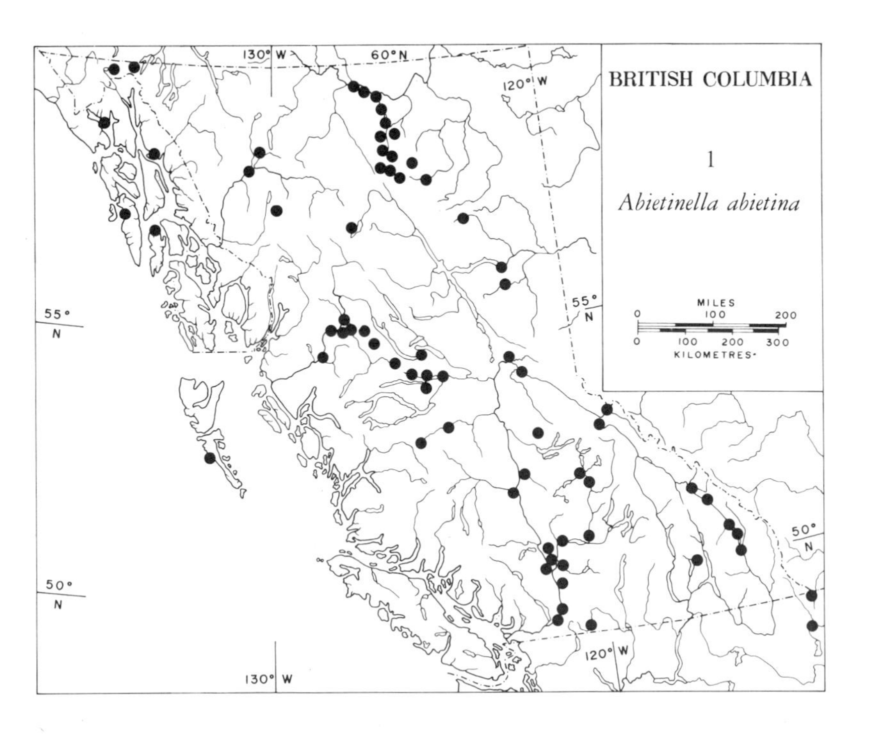

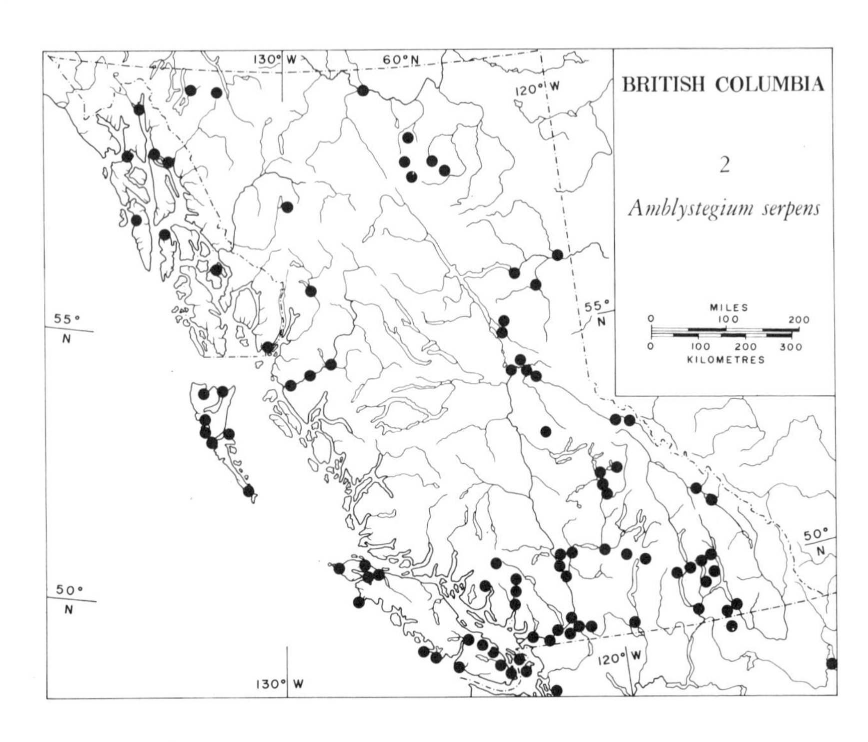
BRITISH COLUMBIA
2
Amblystegium serpens
MILES
0 100 200
0 100 200 300
KILOMETRES
130° W
60° N
120° W
55° N
50° N
120° W
130° W
55° N
50° N

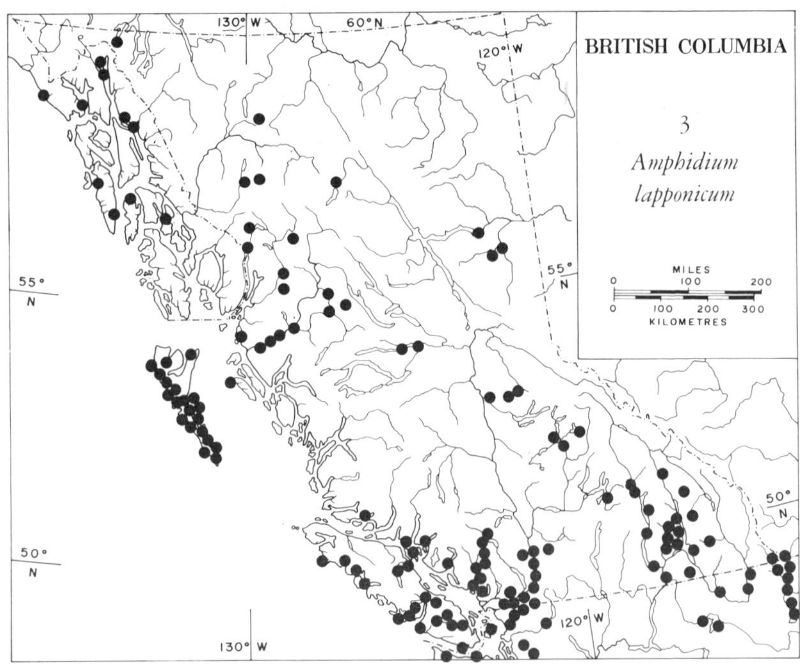
BRITISH COLUMBIA
3
Amphidium lapponicum
MILES
0 100 200
0 100 200 300
KILOMETRES
130° W
60° N
120° W
55° N
50° N
120° W
130° W
55° N
50° N

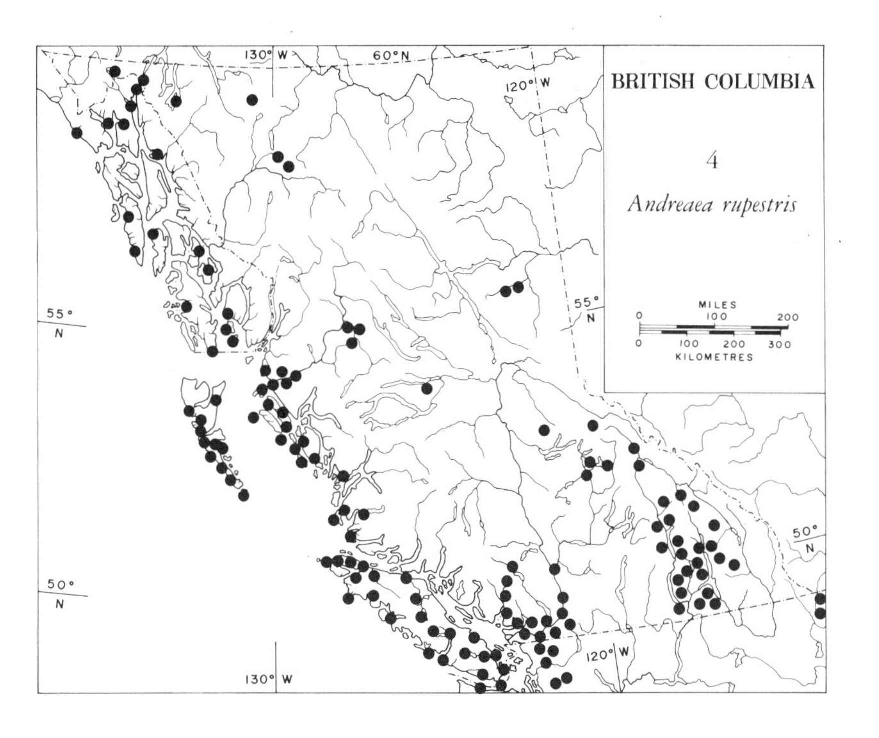
130° W
60° N
120° W
BRITISH COLUMBIA
4
Andreaea rupestris
55° N
MILES
0 100 200
0 100 200 300
KILOMETRES
50° N
130° W
120° W

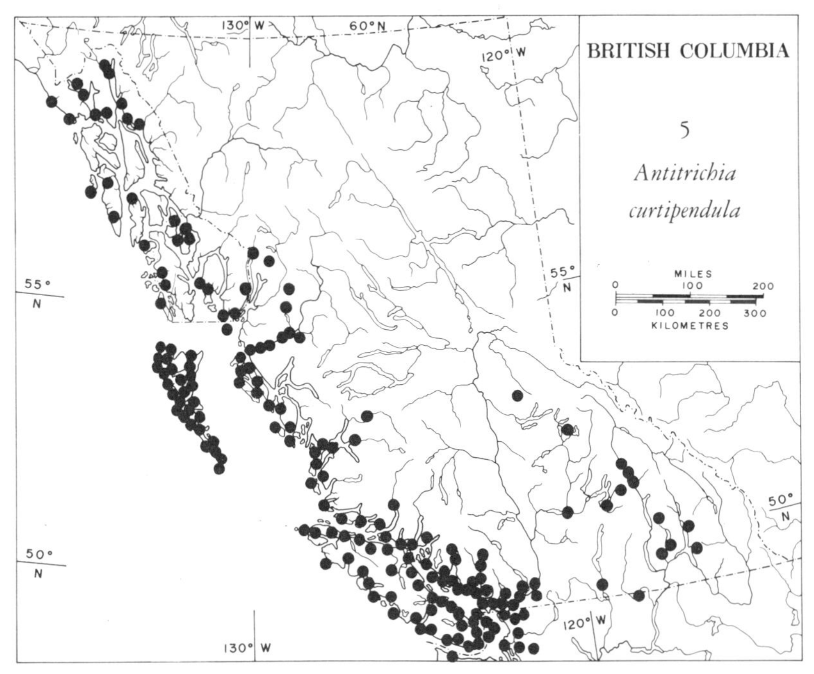
130° W
60° N
120° W
BRITISH COLUMBIA
5
Antitrichia curtipendula
55° N
MILES
0 100 200
0 100 200 300
KILOMETRES
50° N
130° W
120° W

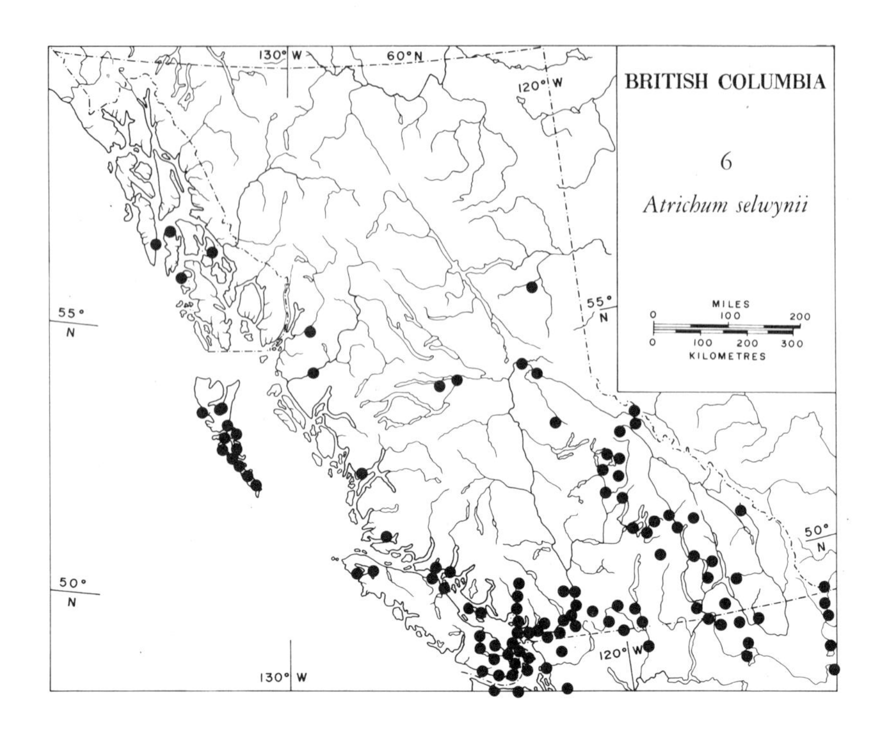
BRITISH COLUMBIA
6
Atrichum selwynii
130° W
60° N
120° W
55° N
50° N
MILES
0 100 200
0 100 200 300
KILOMETRES

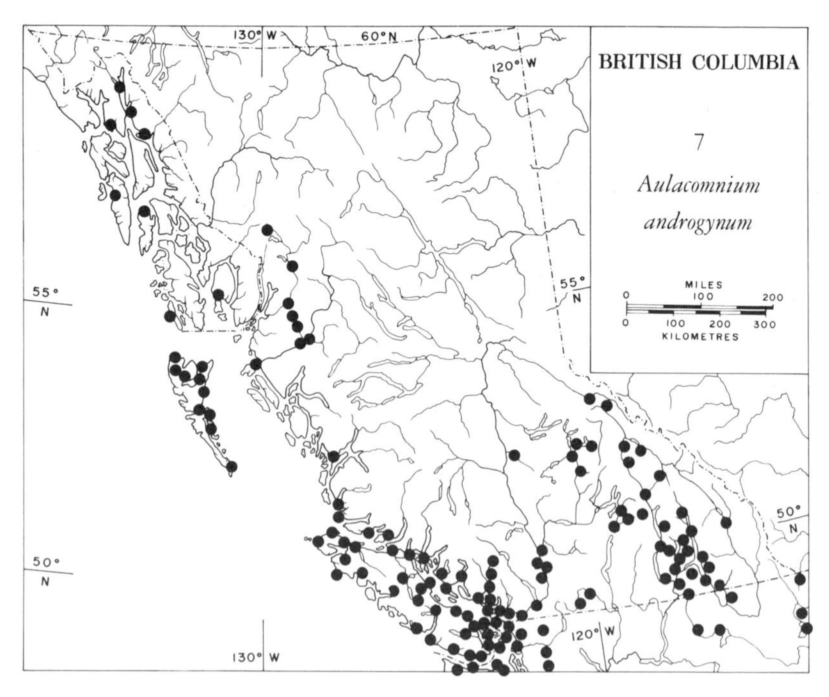
BRITISH COLUMBIA
7
Aulacomnium androgynum
130° W
60° N
120° W
55° N
50° N
MILES
0 100 200
0 100 200 300
KILOMETRES

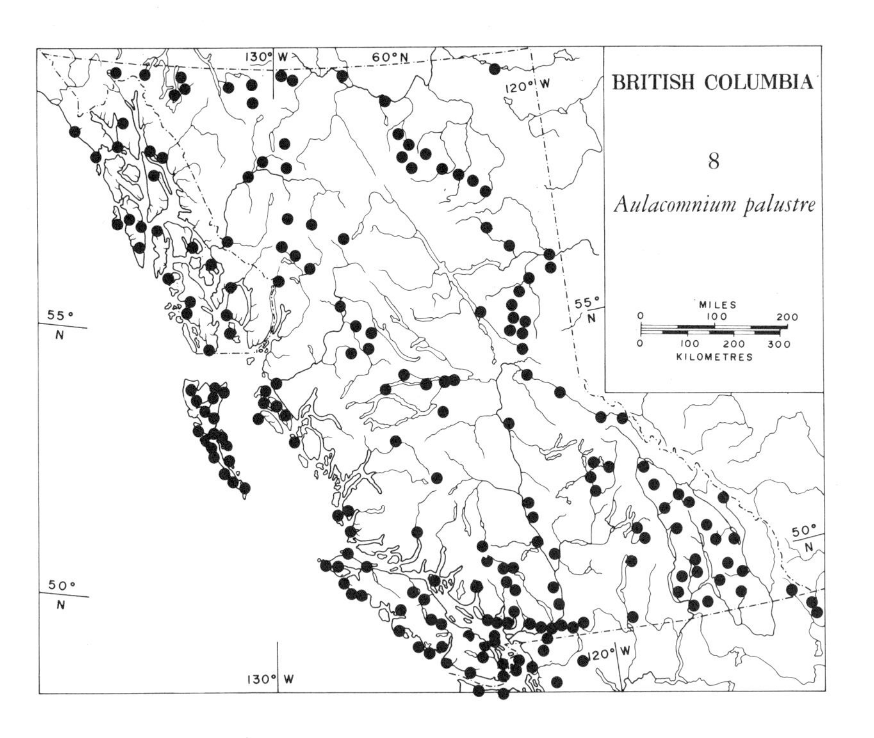
130° W
60° N
120° W
BRITISH COLUMBIA
8
Aulacomnium palustre
55° N
MILES
0 100 200
0 100 200 300
KILOMETRES
50° N
130° W
120° W

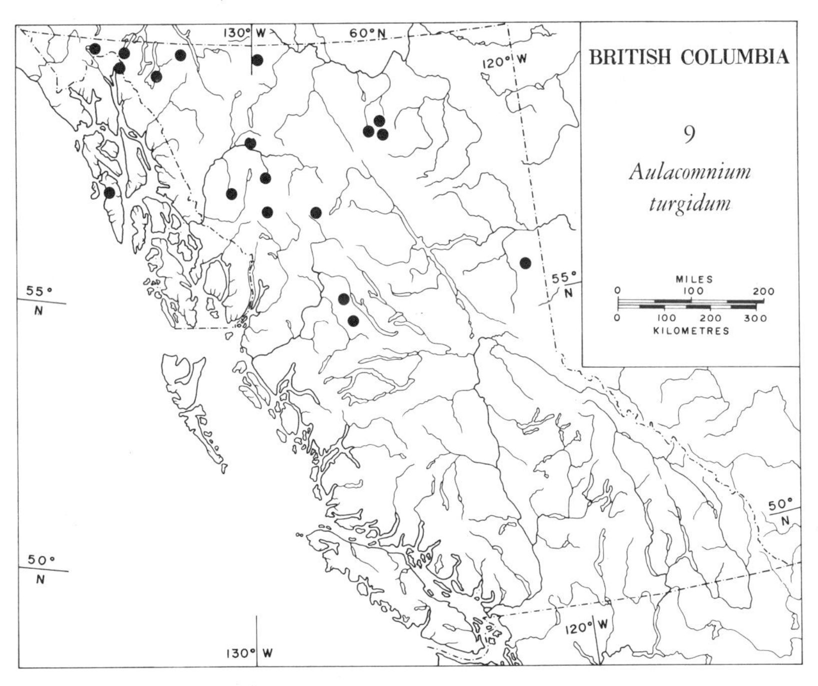
130° W
60° N
120° W
BRITISH COLUMBIA
9
Aulacomnium turgidum
55° N
MILES
0 100 200
0 100 200 300
KILOMETRES
50° N
130° W
120° W

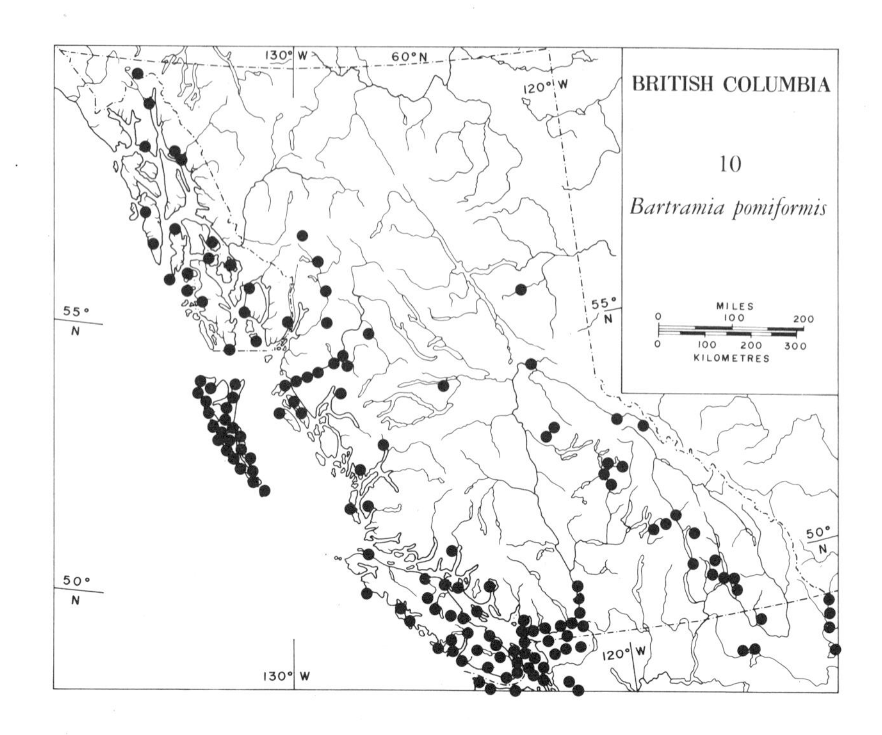
130° W
60° N
120° W
BRITISH COLUMBIA
10
Bartramia pomiformis
55° N
MILES
0 100 200
0 100 200 300
KILOMETRES
50° N
50° N
120° W
130° W

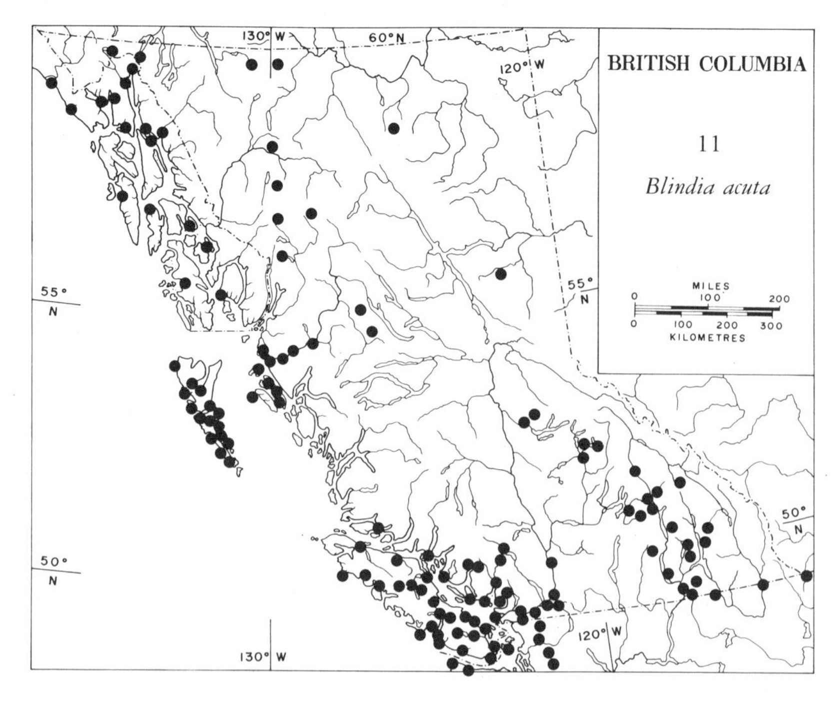
130° W
60° N
120° W
BRITISH COLUMBIA
11
Blindia acuta
55° N
MILES
0 100 200
0 100 200 300
KILOMETRES
50° N
50° N
120° W
130° W

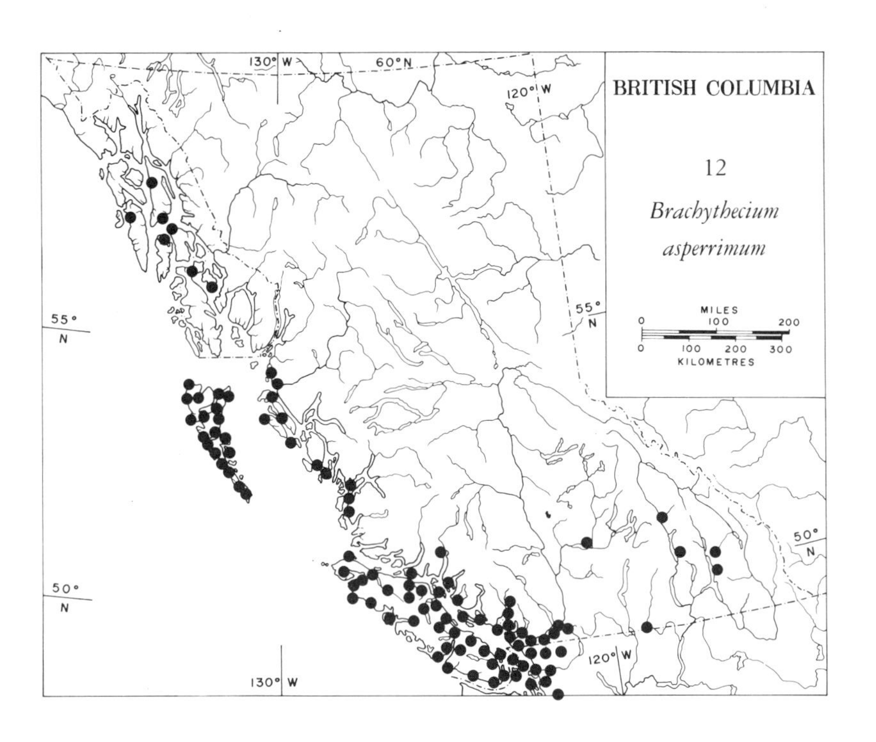
130° W
60° N
120° W
BRITISH COLUMBIA
12
Brachythecium asperrimum
55° N
MILES
0 100 200
0 100 200 300
KILOMETRES
50° N
130° W
120° W

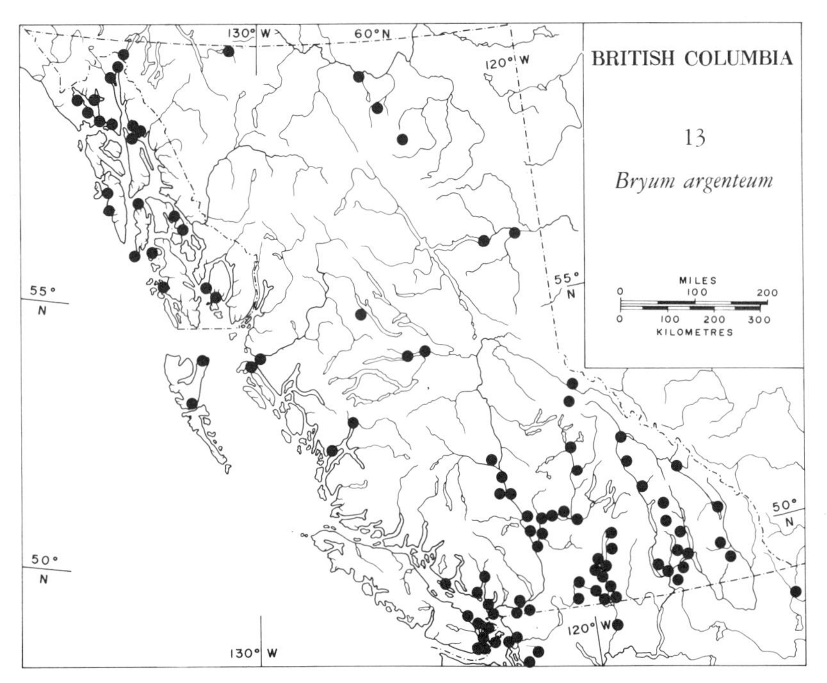
130° W
60° N
120° W
BRITISH COLUMBIA
13
Bryum argenteum
55° N
MILES
0 100 200
0 100 200 300
KILOMETRES
50° N
130° W
120° W

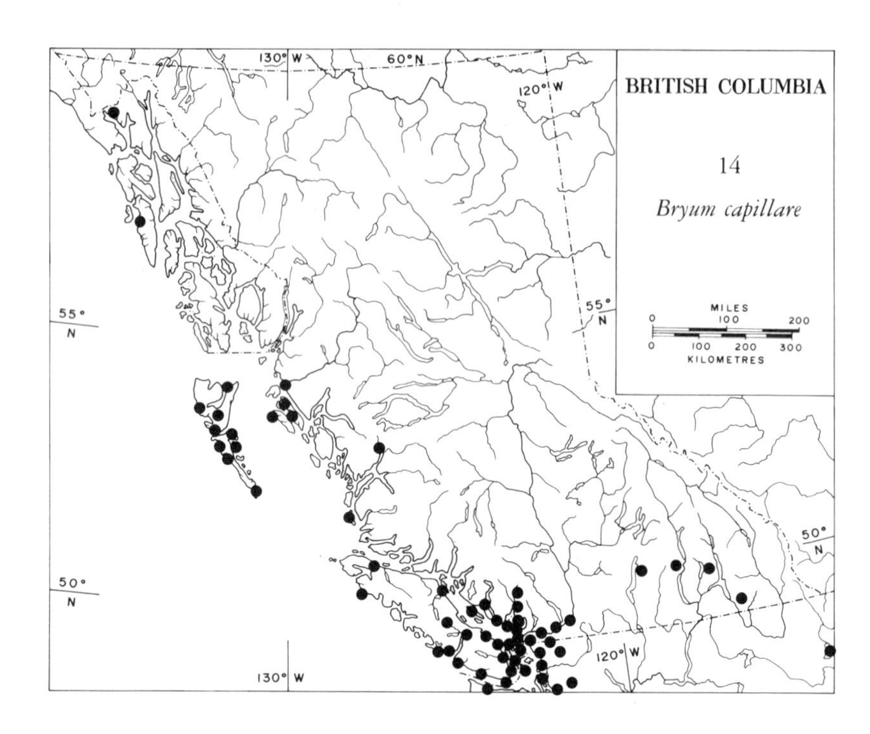
130° W
60° N
120° W
BRITISH COLUMBIA
14
Bryum capillare
55° N
MILES
0 100 200
0 100 200 300
KILOMETRES
50° N
120° W
130° W

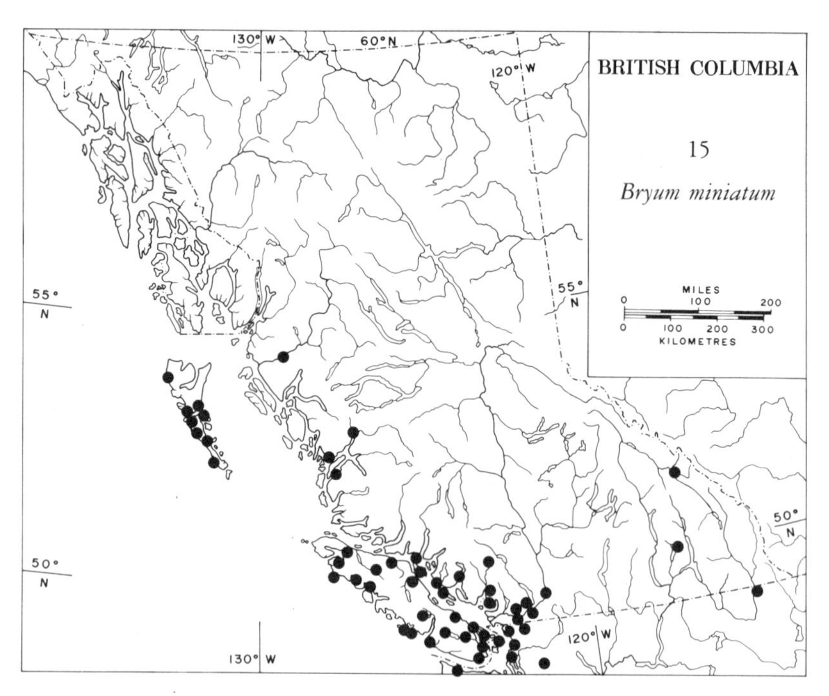
130° W
60° N
120° W
BRITISH COLUMBIA
15
Bryum miniatum
55° N
MILES
0 100 200
0 100 200 300
KILOMETRES
50° N
120° W
130° W

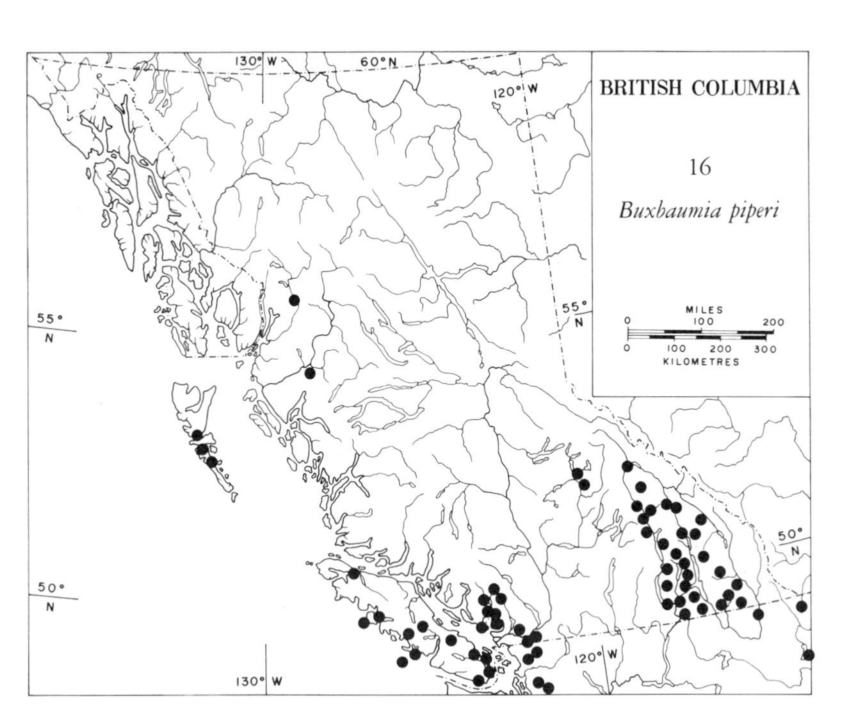
130° W
60° N
120° W
BRITISH COLUMBIA
16
Buxbaumia piperi
55° N
MILES
0 100 200
0 100 200 300
KILOMETRES
55° N
50° N
50° N
120° W
130° W

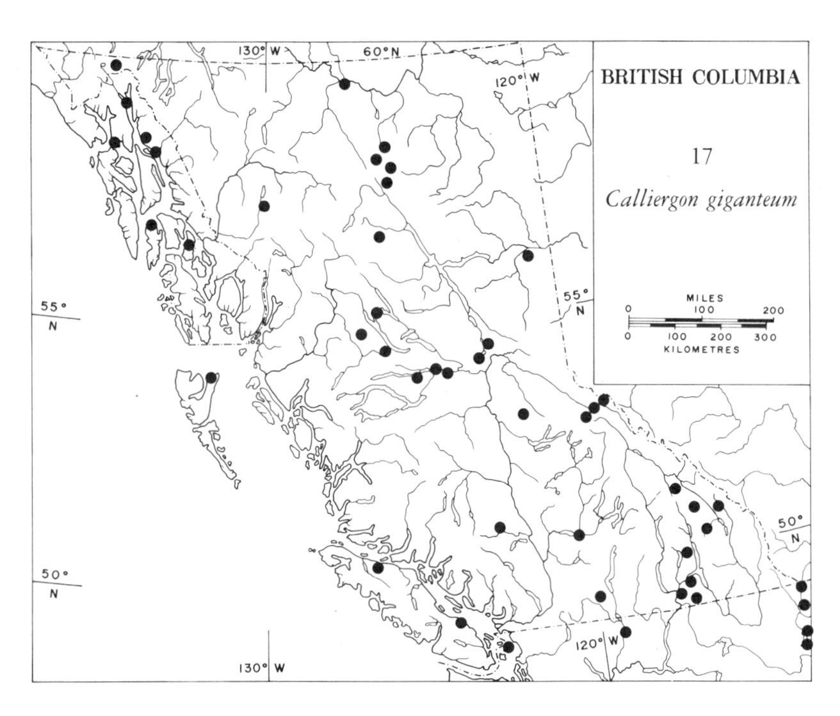
130° W
60° N
120° W
BRITISH COLUMBIA
17
Calliergon giganteum
55° N
MILES
0 100 200
0 100 200 300
KILOMETRES
55° N
50° N
50° N
120° W
130° W

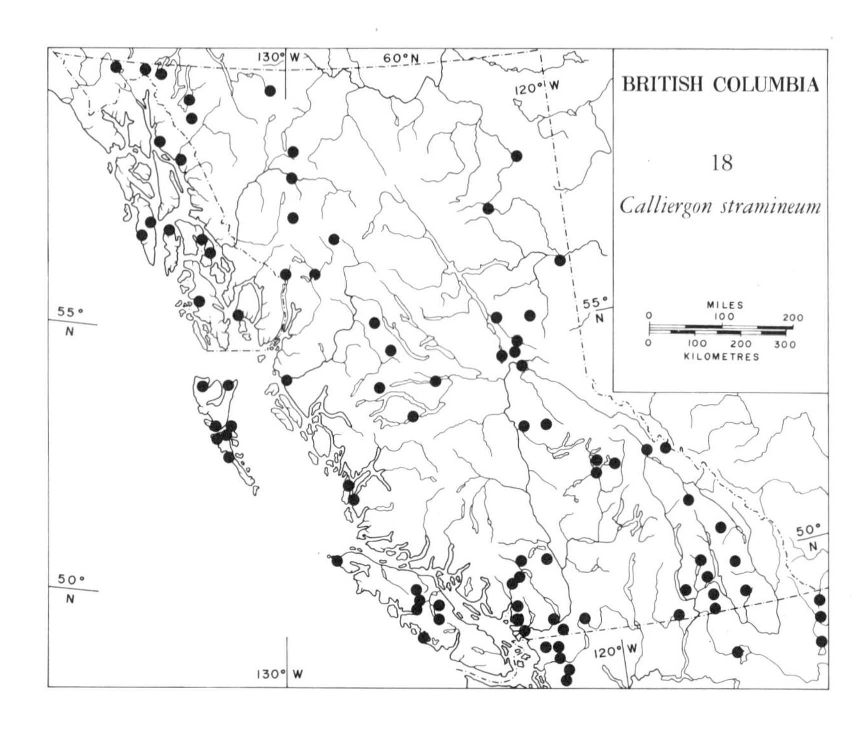
130° W
60° N
120° W
BRITISH COLUMBIA
18
Calliergon stramineum
MILES
0
100
200
0
100
200
300
KILOMETRES
55° N
55° N
50° N
50° N
130° W
120° W

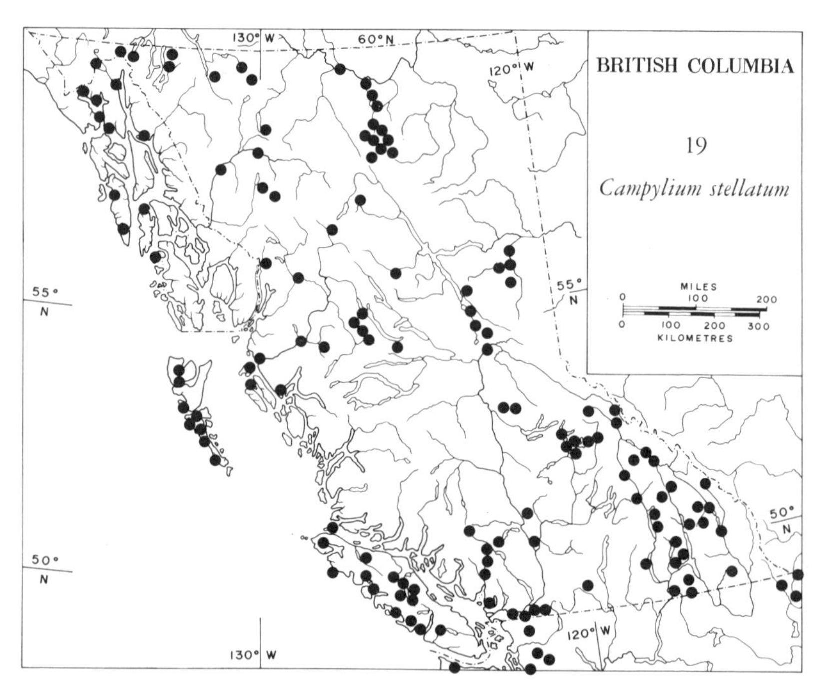
130° W
60° N
120° W
BRITISH COLUMBIA
19
Campylium stellatum
MILES
0
100
200
0
100
200
300
KILOMETRES
55° N
55° N
50° N
50° N
130° W
120° W

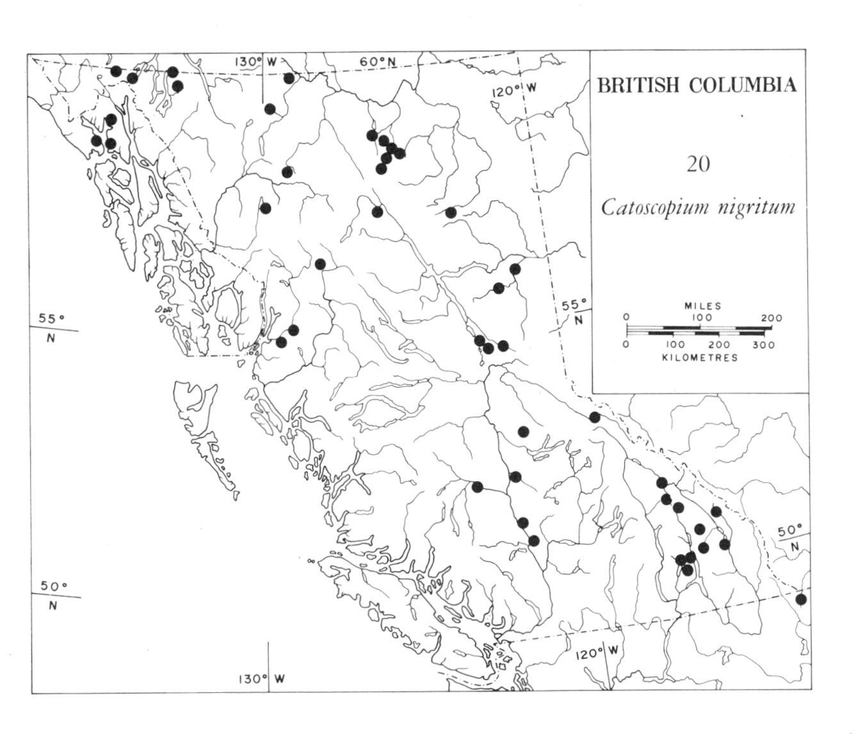
130° W
60° N
120° W
BRITISH COLUMBIA
20
Catoscopium nigritum
55° N
MILES
0 100 200
0 100 200 300
KILOMETRES
50° N
120° W
130° W

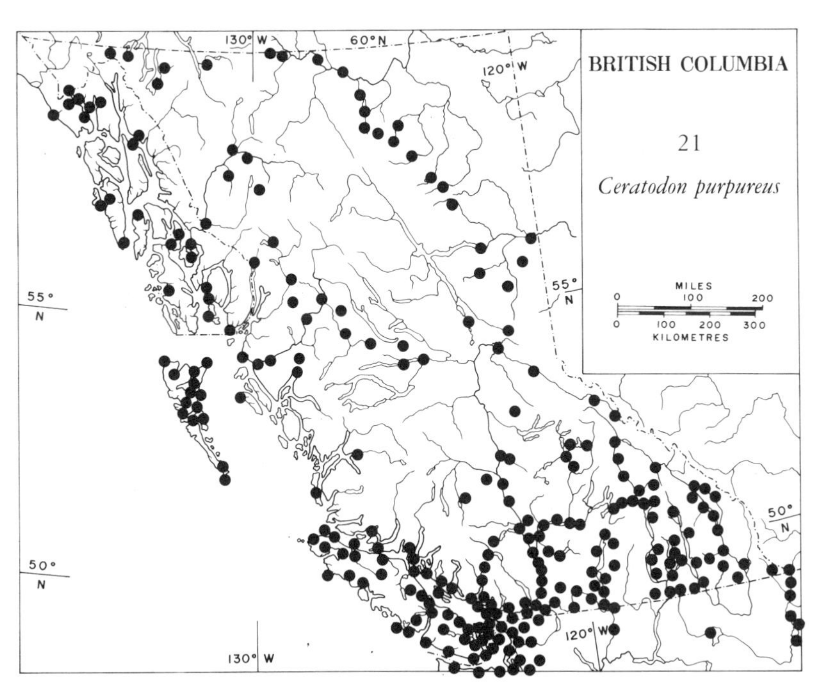
130° W
60° N
120° W
BRITISH COLUMBIA
21
Ceratodon purpureus
55° N
MILES
0 100 200
0 100 200 300
KILOMETRES
50° N
120° W
130° W

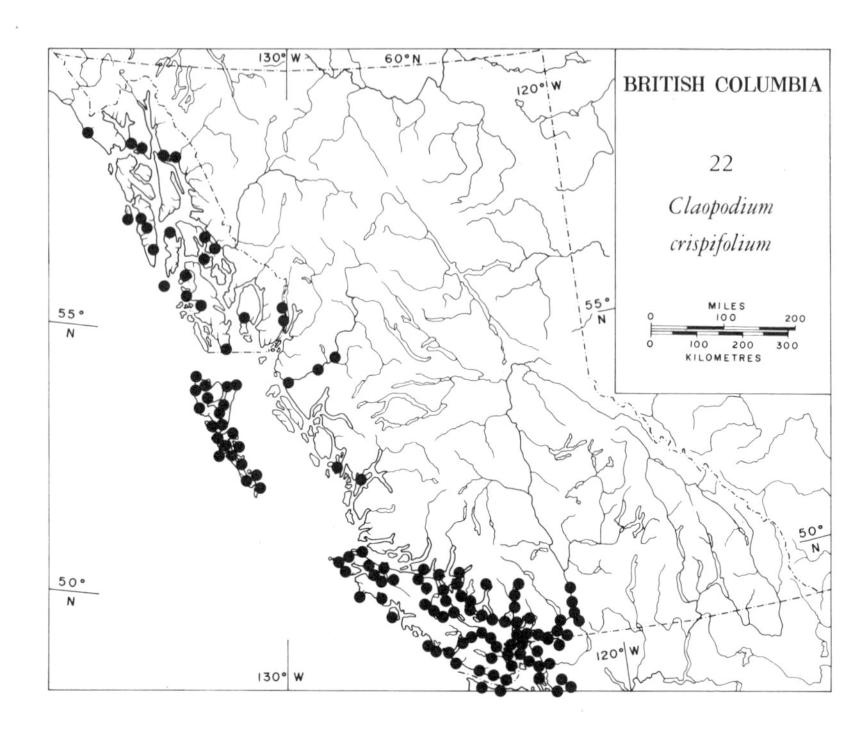
BRITISH COLUMBIA
22
Claopodium crispifolium
MILES
0 100 200
0 100 200 300
KILOMETRES
130° W
60° N
120° W
55° N
50° N

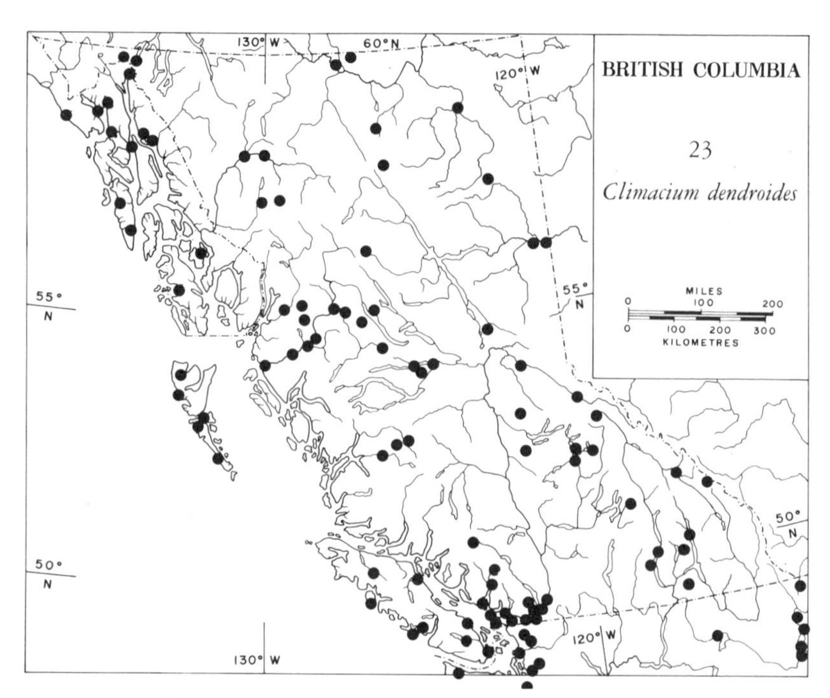
BRITISH COLUMBIA
23
Climacium dendroides
MILES
0 100 200
0 100 200 300
KILOMETRES
130° W
60° N
120° W
55° N
50° N

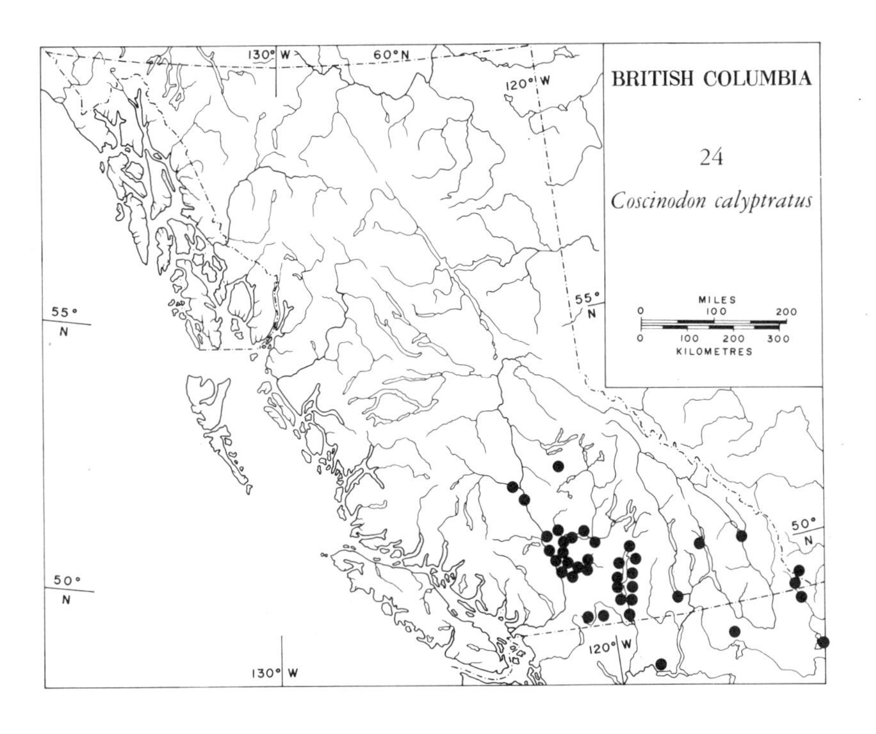
BRITISH COLUMBIA
24
Coscinodon calyptratus
MILES
0
100
200
0
100
200
300
KILOMETRES
130° W
60° N
120° W
55° N
55° N
50° N
50° N
120° W
130° W

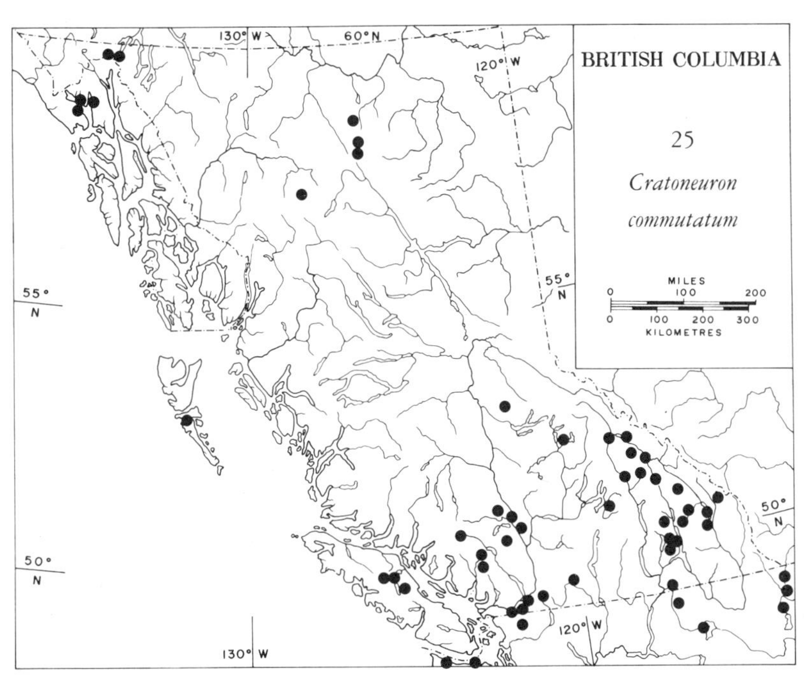
BRITISH COLUMBIA
25
Cratoneuron commutatum
MILES
0
100
200
0
100
200
300
KILOMETRES
130° W
60° N
120° W
55° N
55° N
50° N
50° N
120° W
130° W

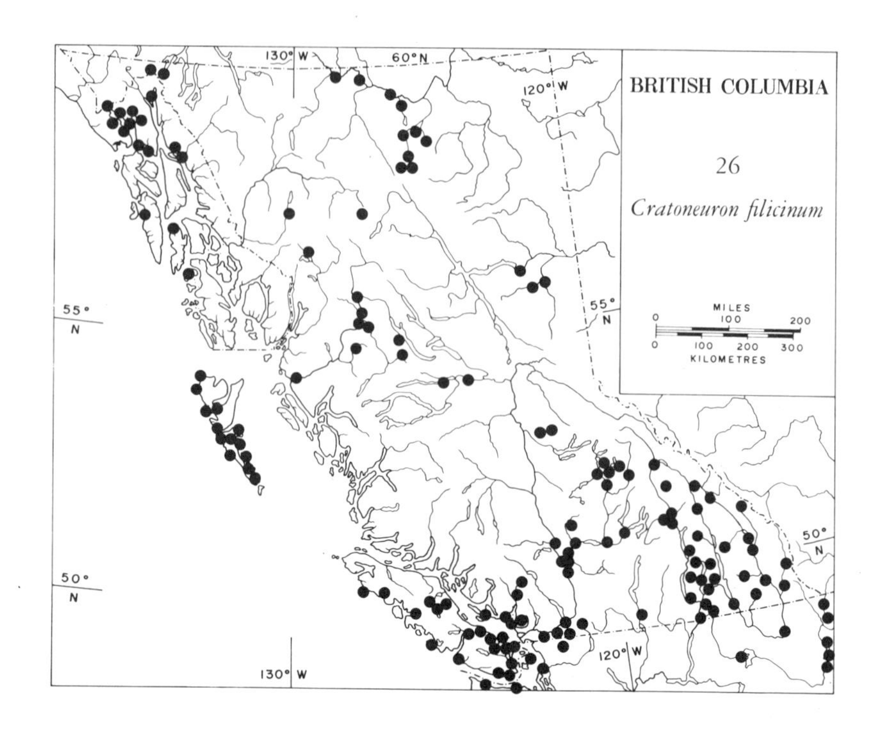
130° W
60° N
120° W
BRITISH COLUMBIA
26
Cratoneuron filicinum
55° N
55° N
MILES
0
100
200
0
100
200
300
KILOMETRES
50° N
50° N
120° W
130° W

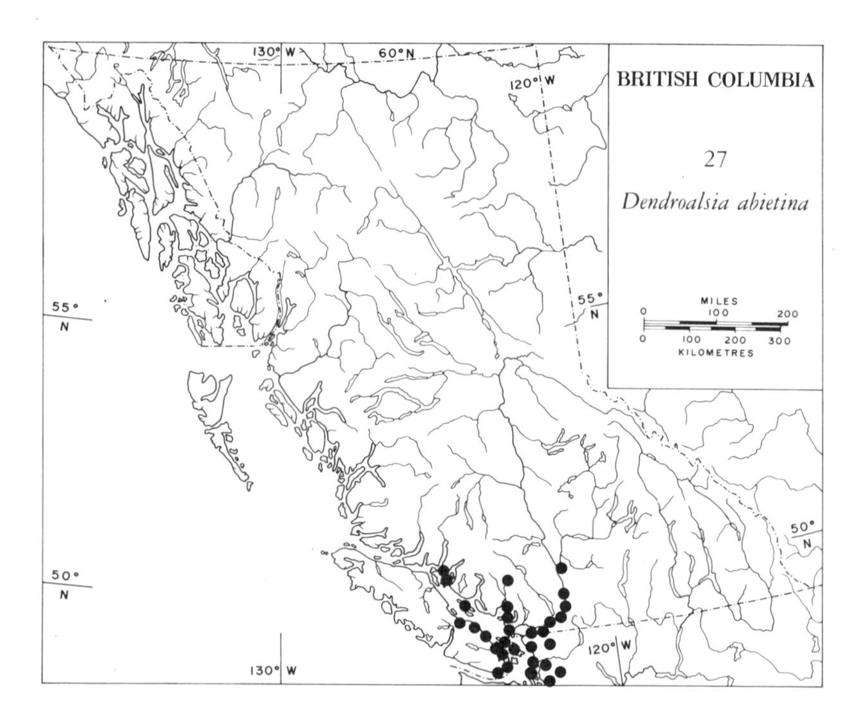
130° W
60° N
120° W
BRITISH COLUMBIA
27
Dendroalsia abietina
55° N
55° N
MILES
0
100
200
0
100
200
300
KILOMETRES
50° N
50° N
120° W
130° W

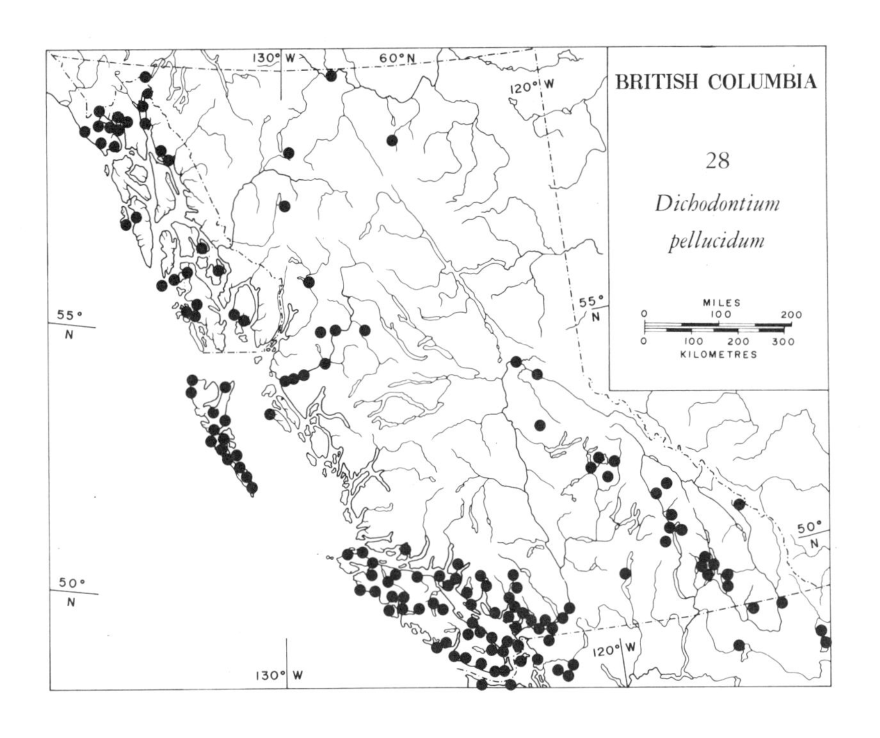
130° W
60° N
120° W
BRITISH COLUMBIA
28
Dichodontium pellucidum
MILES
0
100
200
0
100
200
300
KILOMETRES
55° N
55° N
50° N
50° N
120° W
130° W

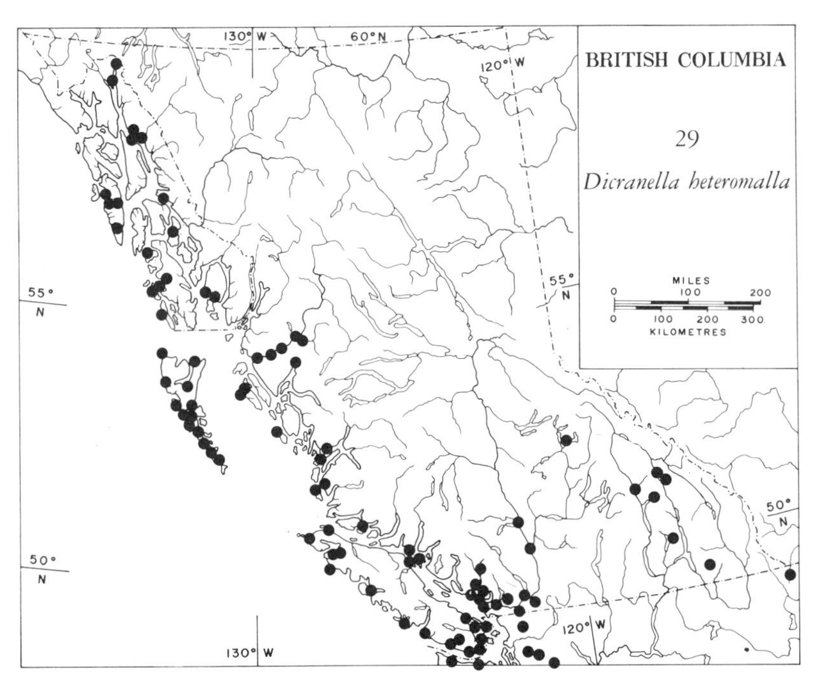
130° W
60° N
120° W
BRITISH COLUMBIA
29
Dicranella heteromalla
MILES
0
100
200
0
100
200
300
KILOMETRES
55° N
55° N
50° N
50° N
120° W
130° W

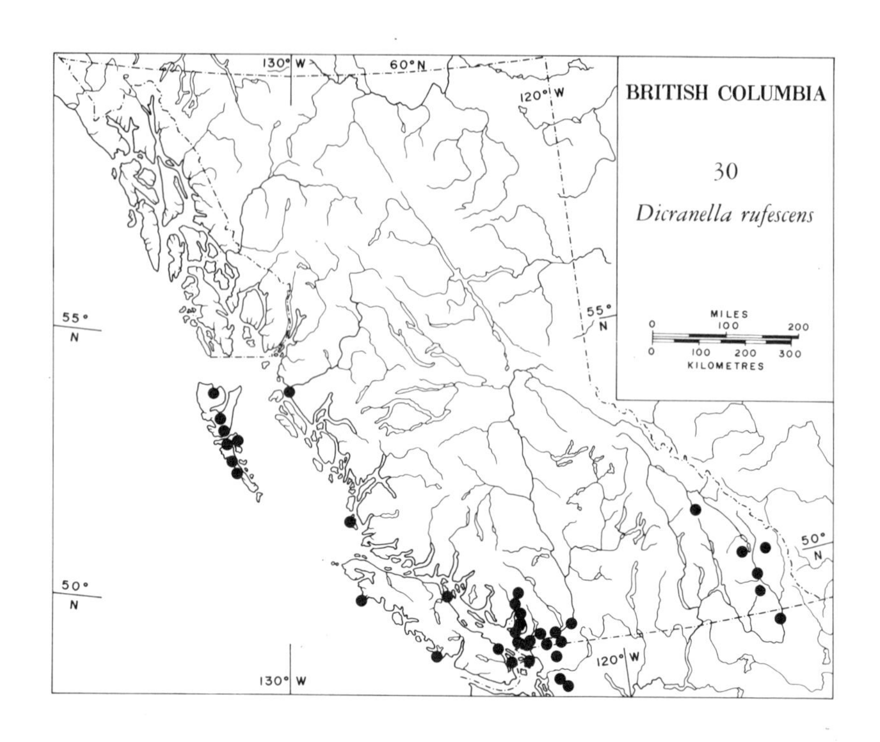
BRITISH COLUMBIA
30
Dicranella rufescens
MILES
0
100
200
0
100
200
300
KILOMETRES
130° W
60° N
120° W
55° N
50° N
120° W
130° W

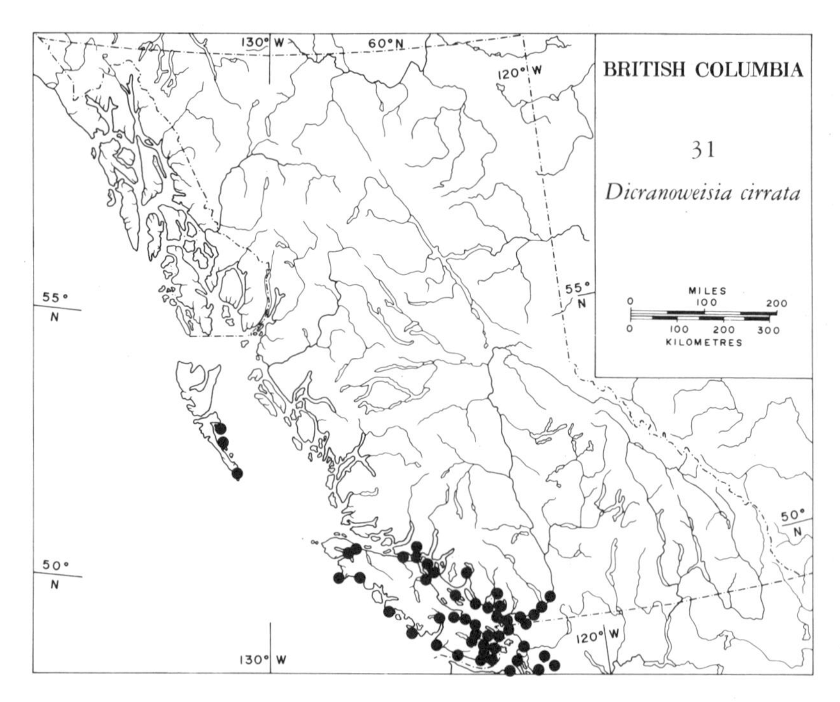
BRITISH COLUMBIA
31
Dicranoweisia cirrata
MILES
0
100
200
0
100
200
300
KILOMETRES
130° W
60° N
120° W
55° N
50° N
120° W
130° W

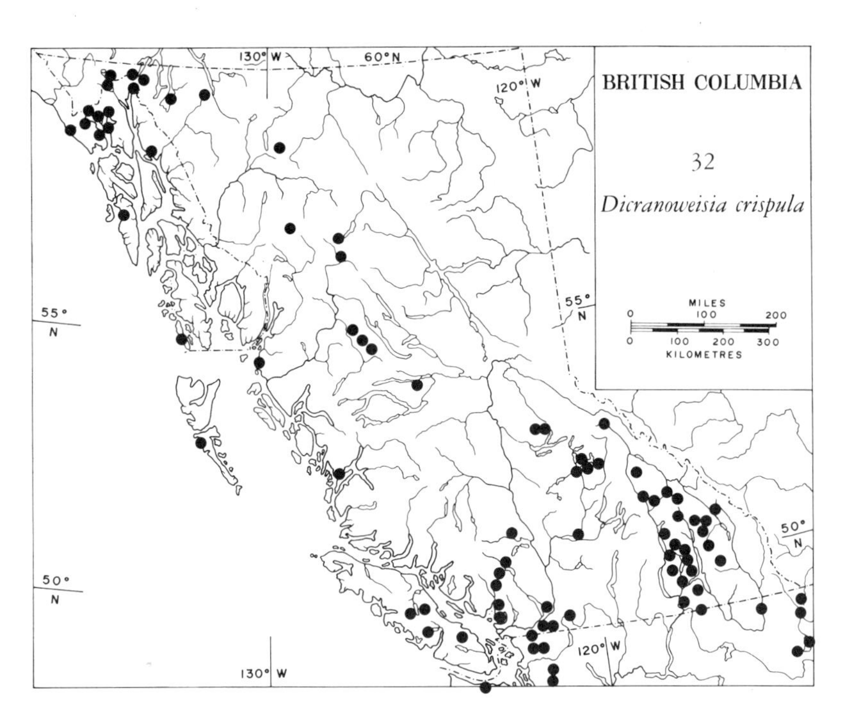
130° W
60° N
120° W
BRITISH COLUMBIA
32
Dicranoweisia crispula
55° N
MILES
0
100
200
0
100
200
300
KILOMETRES
50° N
50° N
120° W
130° W

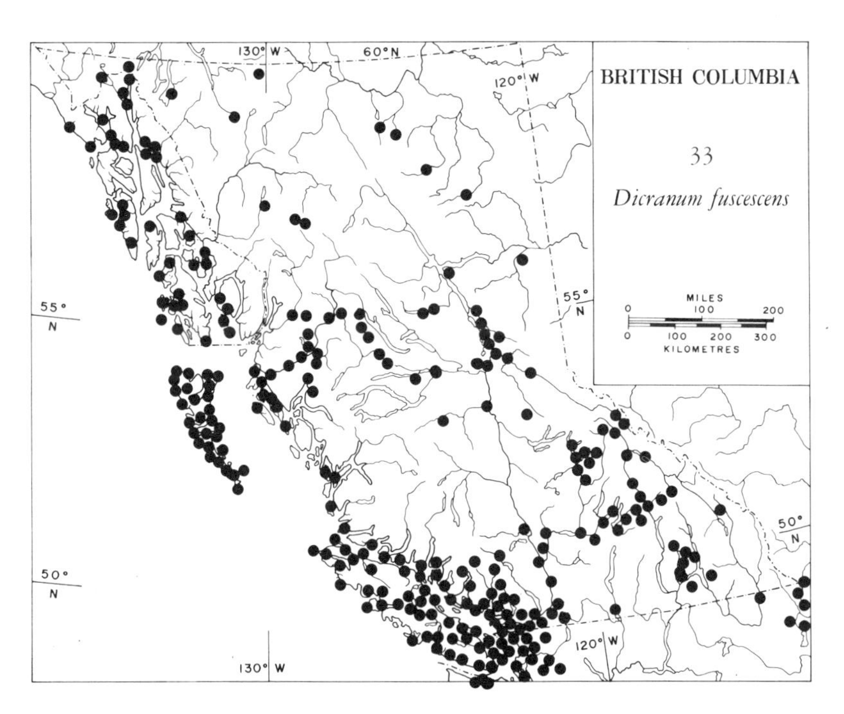
130° W
60° N
120° W
BRITISH COLUMBIA
33
Dicranum fuscescens
55° N
MILES
0
100
200
0
100
200
300
KILOMETRES
55° N
50° N
50° N
120° W
130° W

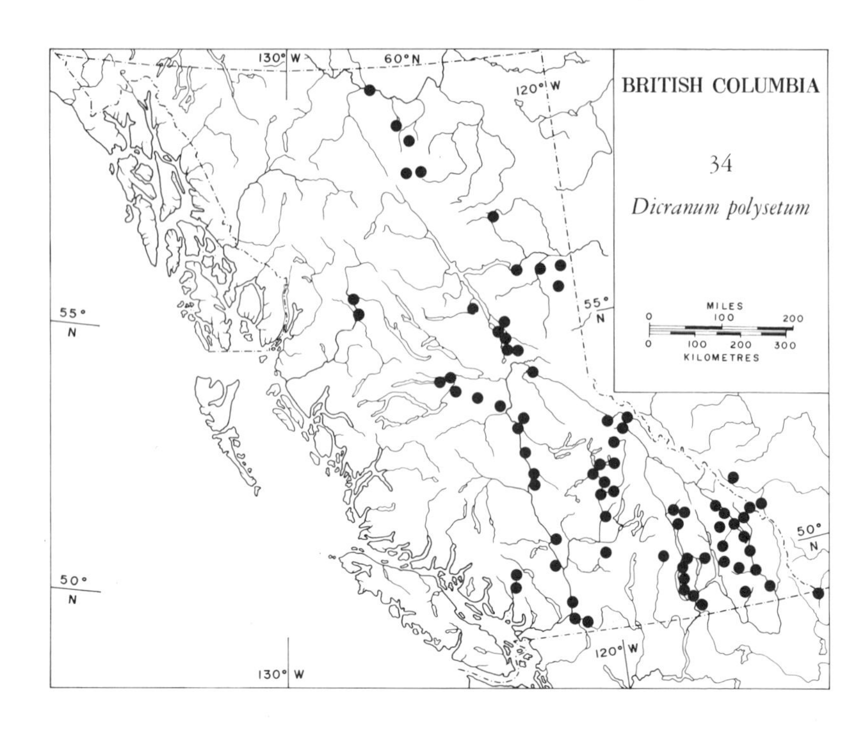
130° W
60° N
120° W
BRITISH COLUMBIA
34
Dicranum polysetum
55° N
MILES
0 100 200
0 100 200 300
KILOMETRES
50° N
50° N
120° W
130° W

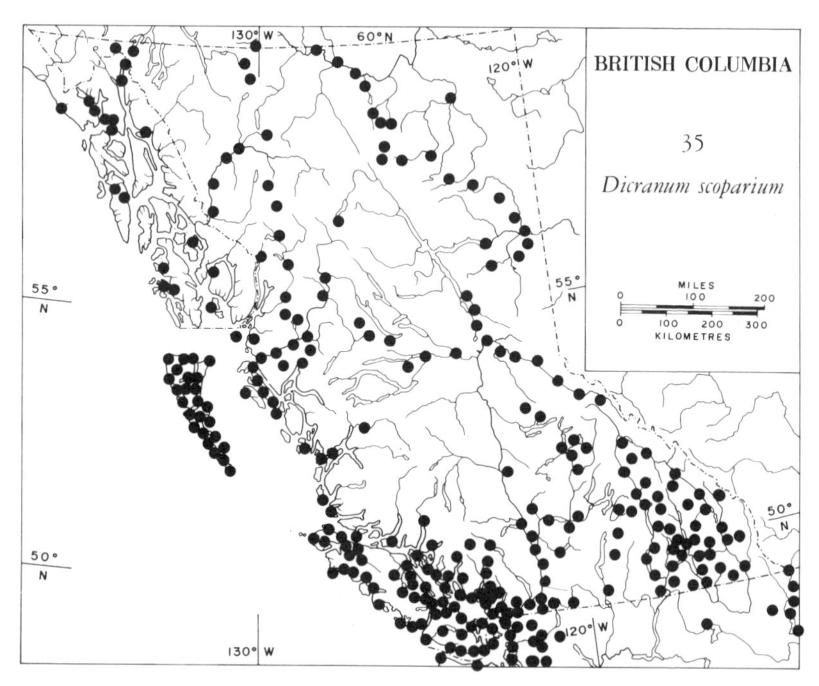
130° W
60° N
120° W
BRITISH COLUMBIA
35
Dicranum scoparium
55° N
MILES
0 100 200
0 100 200 300
KILOMETRES
50° N
50° N
120° W
130° W

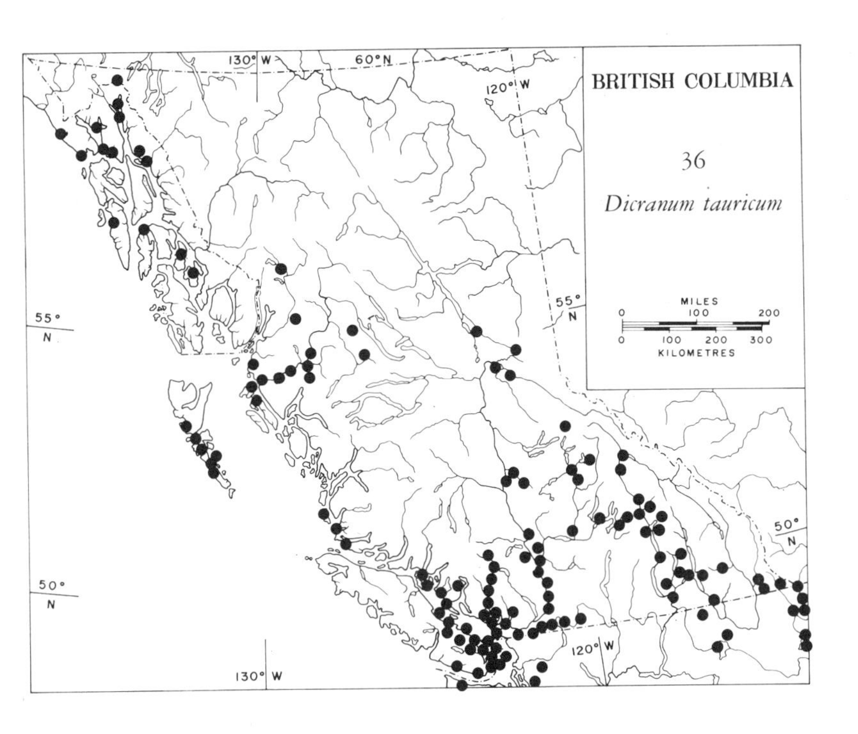

BRITISH COLUMBIA
36
Dicranum tauricum
MILES
0 100 200
0 100 200 300
KILOMETRES
130° W
60° N
120° W
55° N
50° N

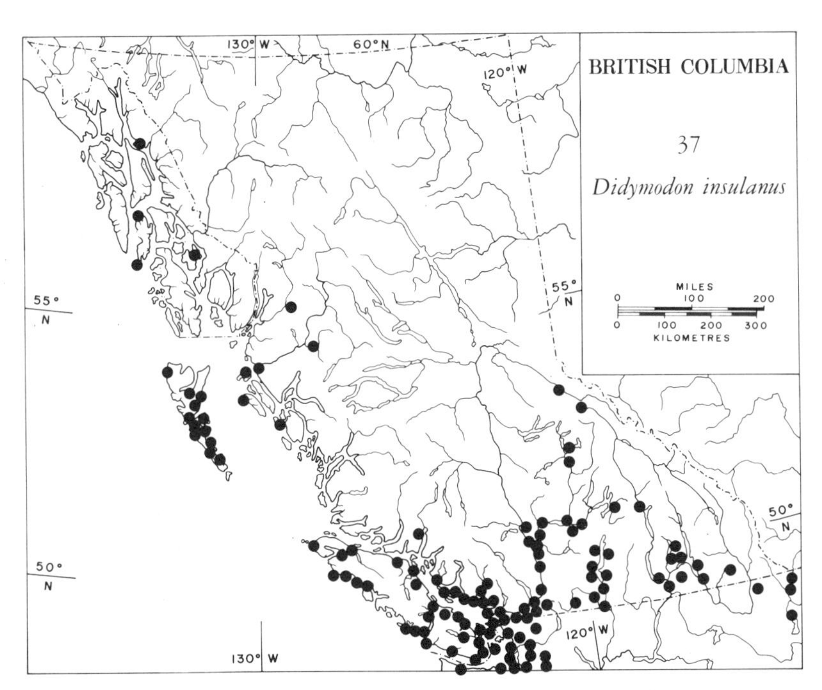

BRITISH COLUMBIA
37
Didymodon insulanus
MILES
0 100 200
0 100 200 300
KILOMETRES
130° W
60° N
120° W
55° N
50° N

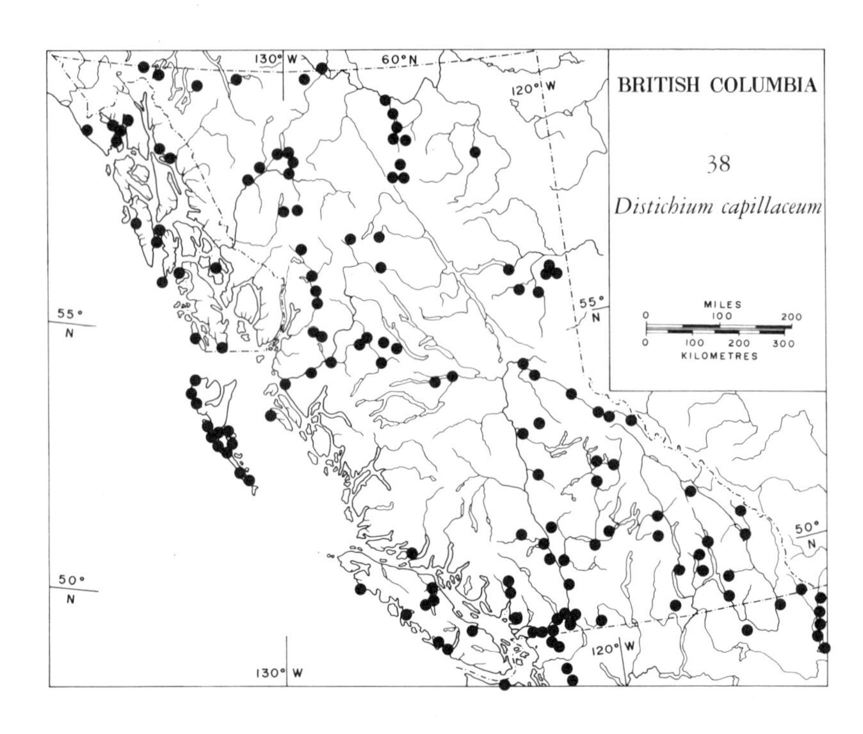
130° W
60° N
120° W
BRITISH COLUMBIA
38
Distichium capillaceum
55° N
MILES
0 100 200
0 100 200 300
KILOMETRES
50° N
120° W
130° W

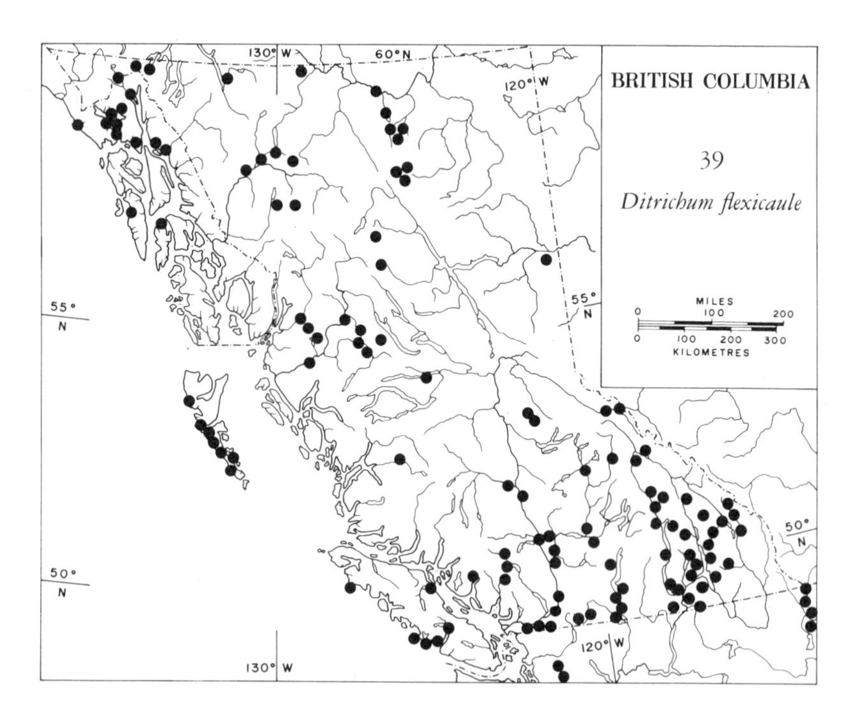
130° W
60° N
120° W
BRITISH COLUMBIA
39
Ditrichum flexicaule
55° N
MILES
0 100 200
0 100 200 300
KILOMETRES
50° N
120° W
130° W

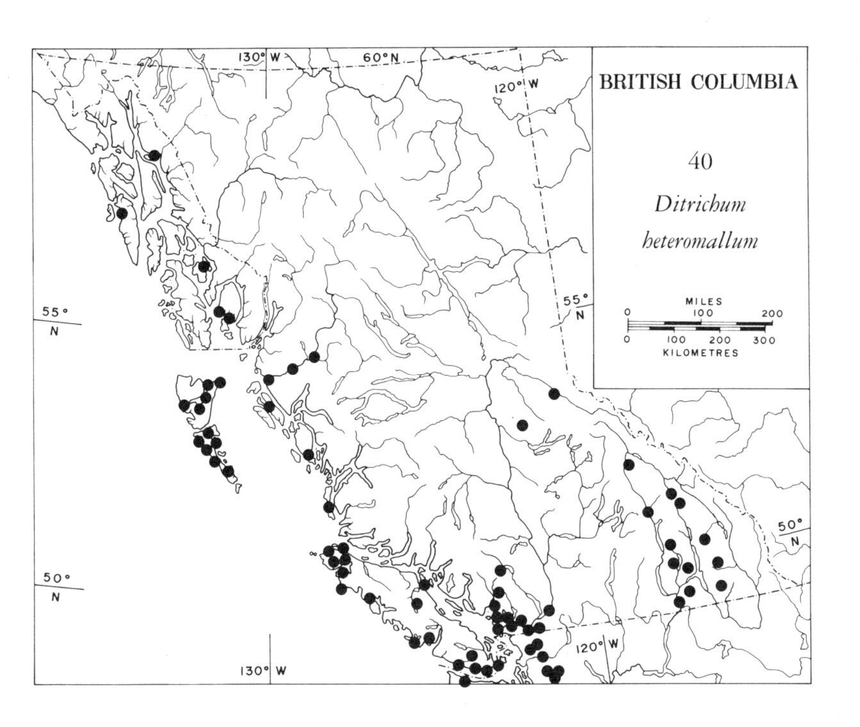
BRITISH COLUMBIA
40
Ditrichum heteromallum
MILES
0 100 200
0 100 200 300
KILOMETRES
130° W
60° N
120° W
55° N
50° N
130° W
120° W
55° N
50° N

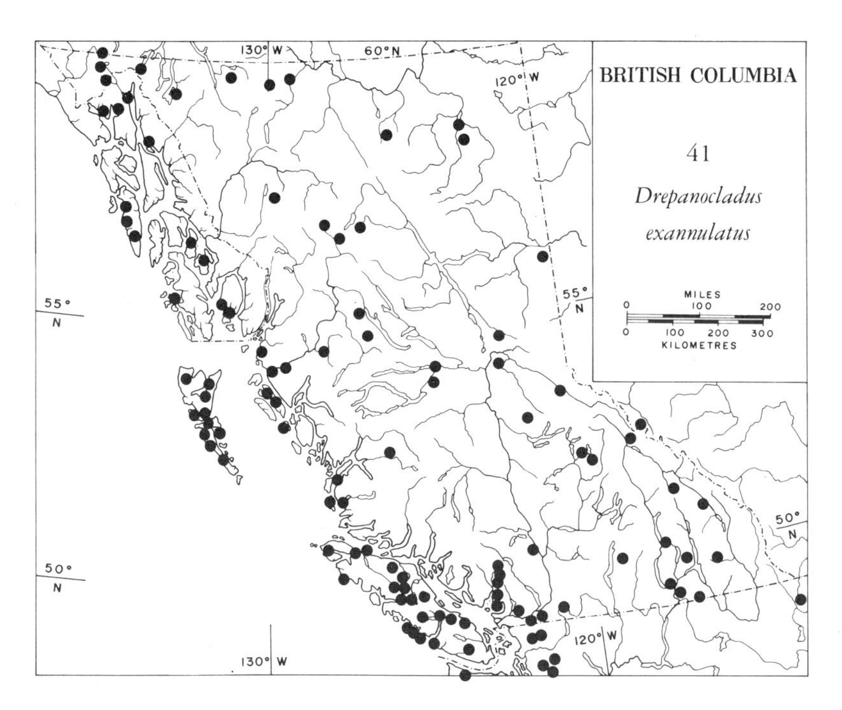
BRITISH COLUMBIA
41
Drepanocladus exannulatus
MILES
0 100 200
0 100 200 300
KILOMETRES
130° W
60° N
120° W
55° N
50° N
130° W
120° W
55° N
50° N

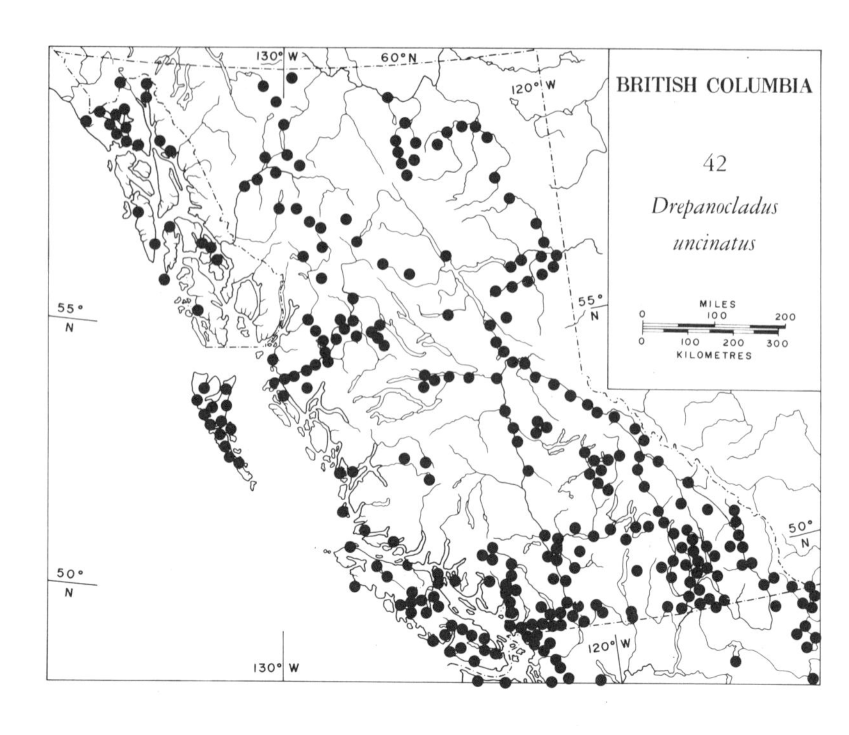
BRITISH COLUMBIA
42
Drepanocladus uncinatus
MILES
0
100
200
0
100
200
300
KILOMETRES
130° W
60° N
120° W
55° N
50° N

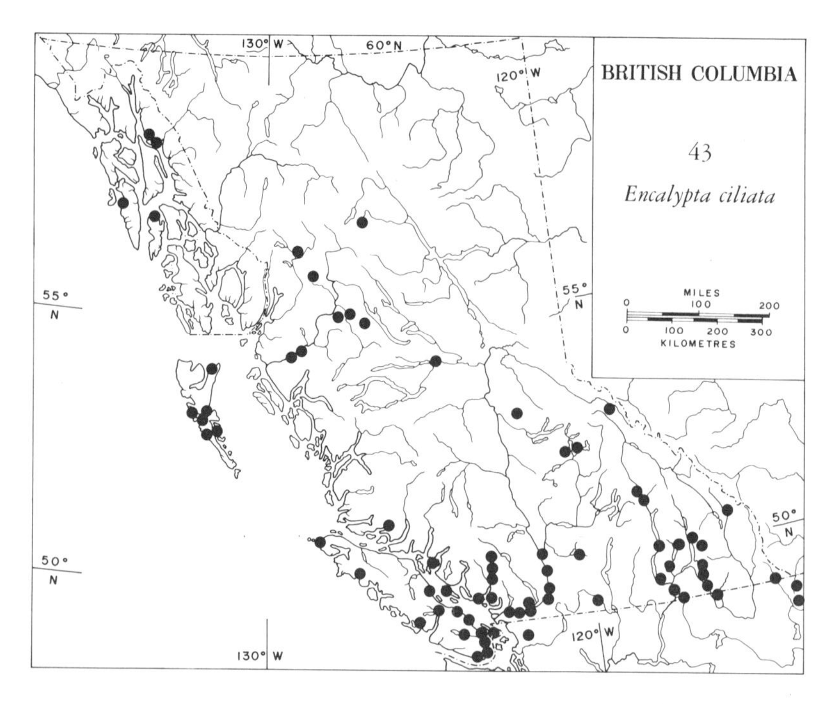
BRITISH COLUMBIA
43
Encalypta ciliata
MILES
0
100
200
0
100
200
300
KILOMETRES
130° W
60° N
120° W
55° N
50° N

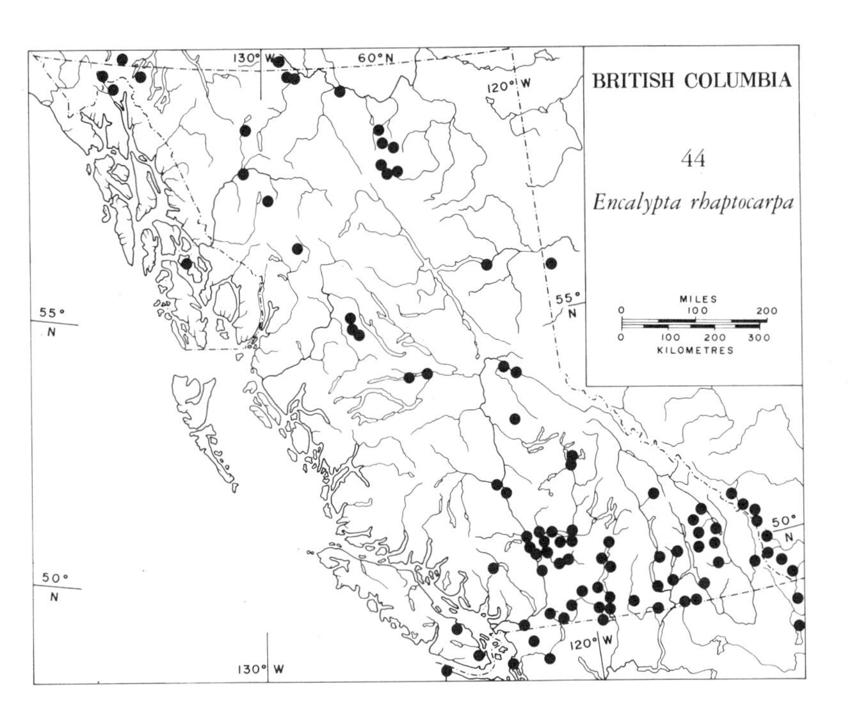
BRITISH COLUMBIA
44
Encalypta rhaptocarpa
MILES
0
100
200
0
100
200
300
KILOMETRES
130° W
60° N
120° W
55° N
50° N
120° W
130° W

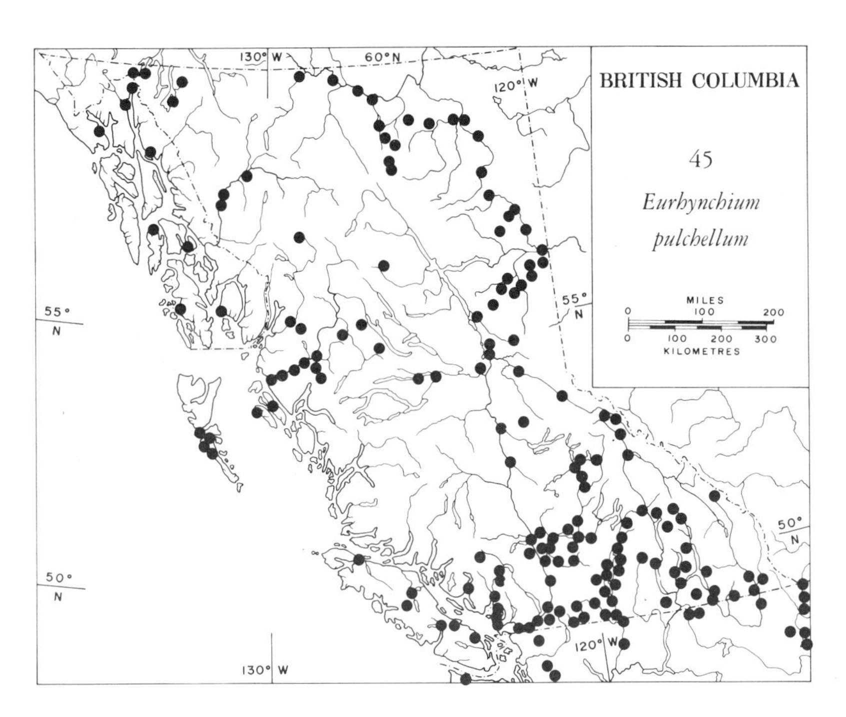
BRITISH COLUMBIA
45
Eurhynchium pulchellum
MILES
0
100
200
0
100
200
300
KILOMETRES
130° W
60° N
120° W
55° N
50° N
120° W
130° W

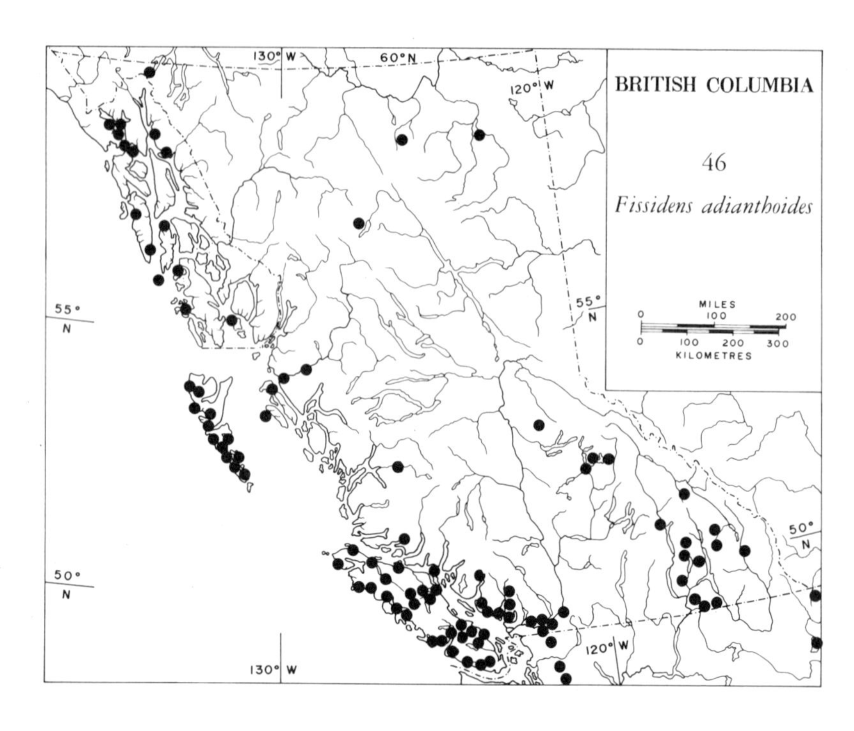

130° W
60° N
120° W
BRITISH COLUMBIA
46
Fissidens adiantboides
55° N
MILES
0 100 200
0 100 200 300
KILOMETRES
50° N
50° N
120° W
130° W

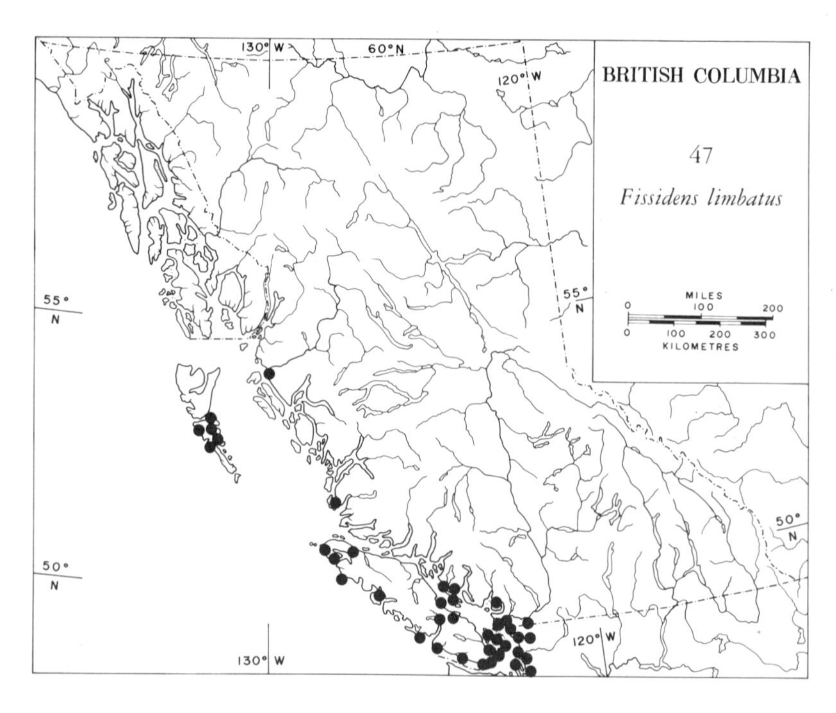

130° W
60° N
120° W
BRITISH COLUMBIA
47
Fissidens limbatus
55° N
MILES
0 100 200
0 100 200 300
KILOMETRES
50° N
50° N
120° W
130° W

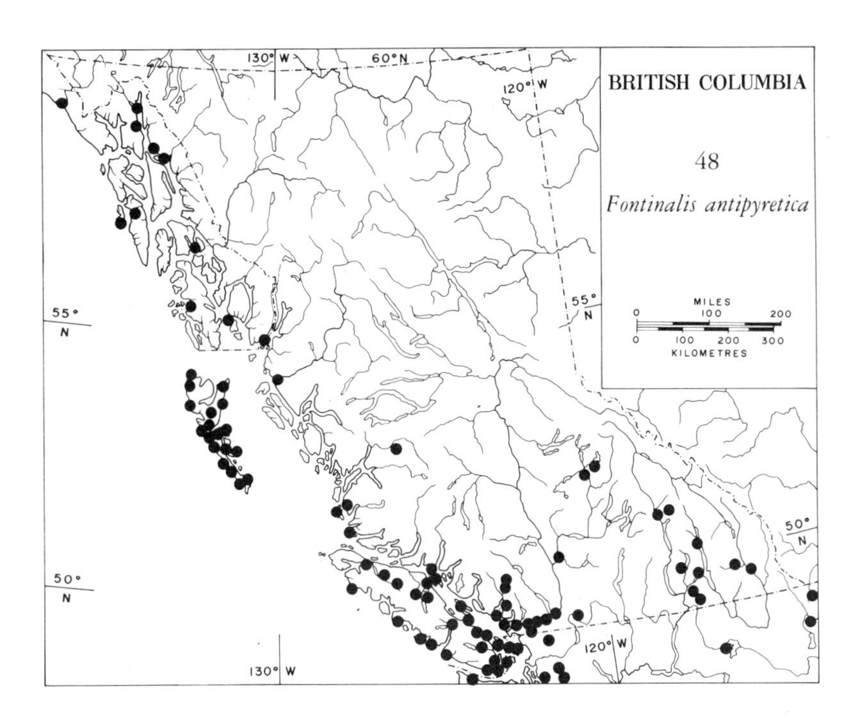
BRITISH COLUMBIA
48
Fontinalis antipyretica
MILES
0 100 200
0 100 200 300
KILOMETRES
130° W
60° N
120° W
55° N
50° N

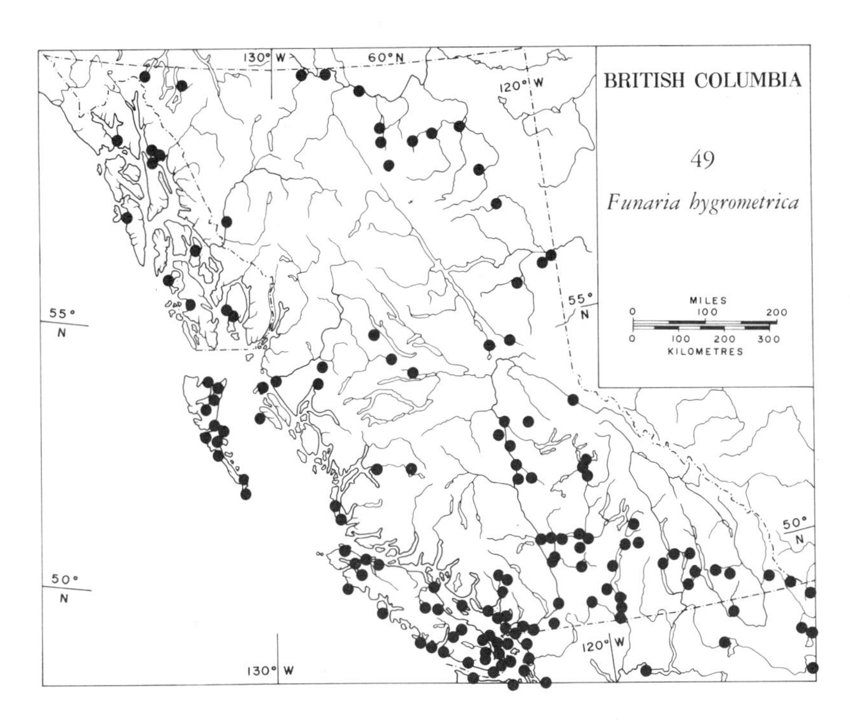
BRITISH COLUMBIA
49
Funaria hygrometrica
MILES
0 100 200
0 100 200 300
KILOMETRES
130° W
60° N
120° W
55° N
50° N

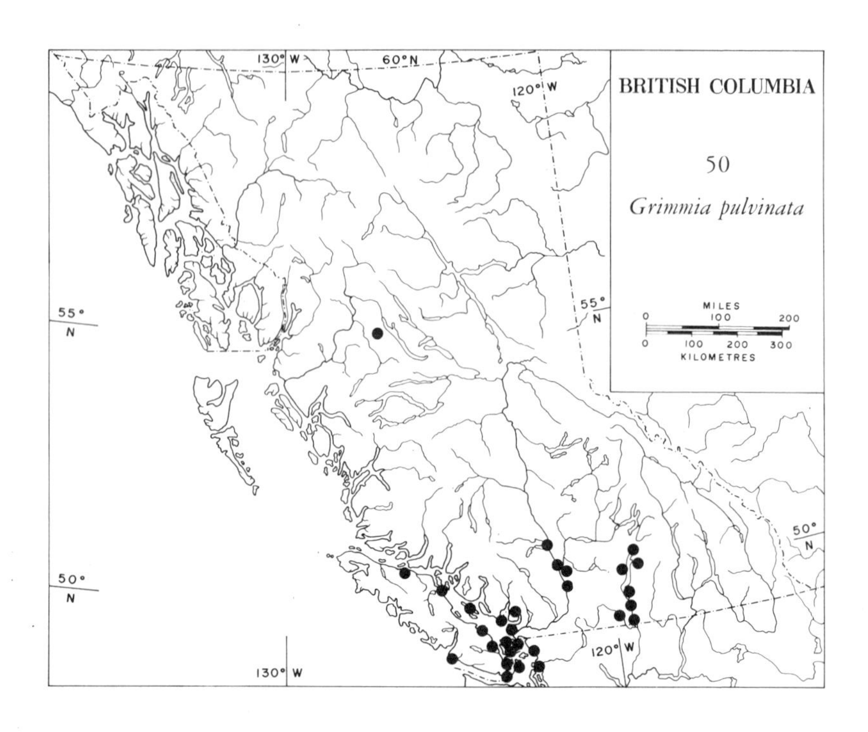
130° W
60° N
120° W
BRITISH COLUMBIA
50
Grimmia pulvinata
55° N
MILES
0 100 200
0 100 200 300
KILOMETRES
50° N
120° W
130° W

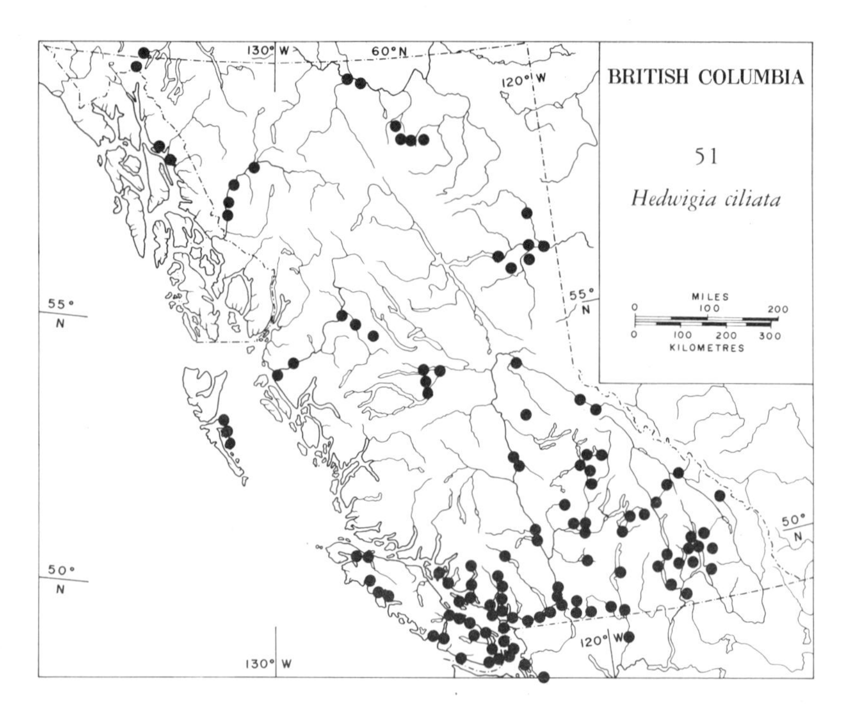
130° W
60° N
120° W
BRITISH COLUMBIA
51
Hedwigia ciliata
55° N
MILES
0 100 200
0 100 200 300
KILOMETRES
50° N
120° W
130° W

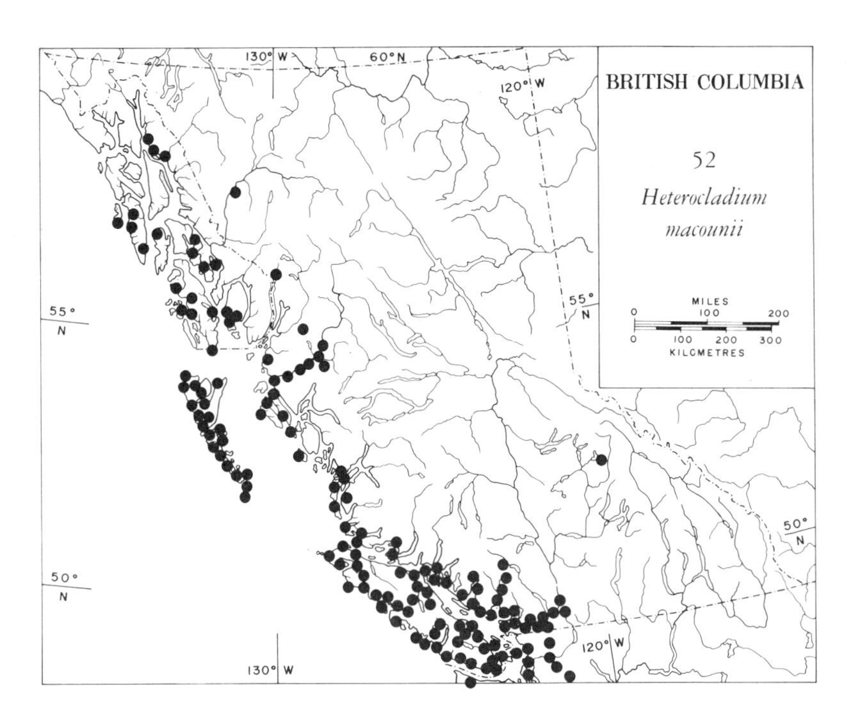
130° W
60° N
120° W
BRITISH COLUMBIA
52
Heterocladium macounii
MILES
0 100 200
0 100 200 300
KILOMETRES
55° N
50° N
120° W
130° W

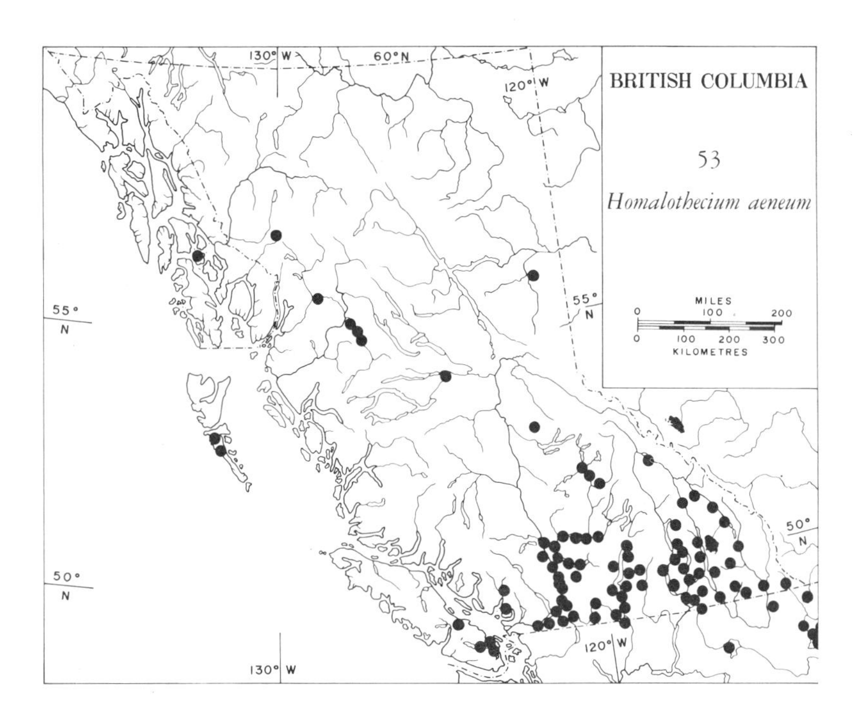
130° W
60° N
120° W
BRITISH COLUMBIA
53
Homalothecium aeneum
MILES
0 100 200
0 100 200 300
KILOMETRES
55° N
50° N
120° W
130° W

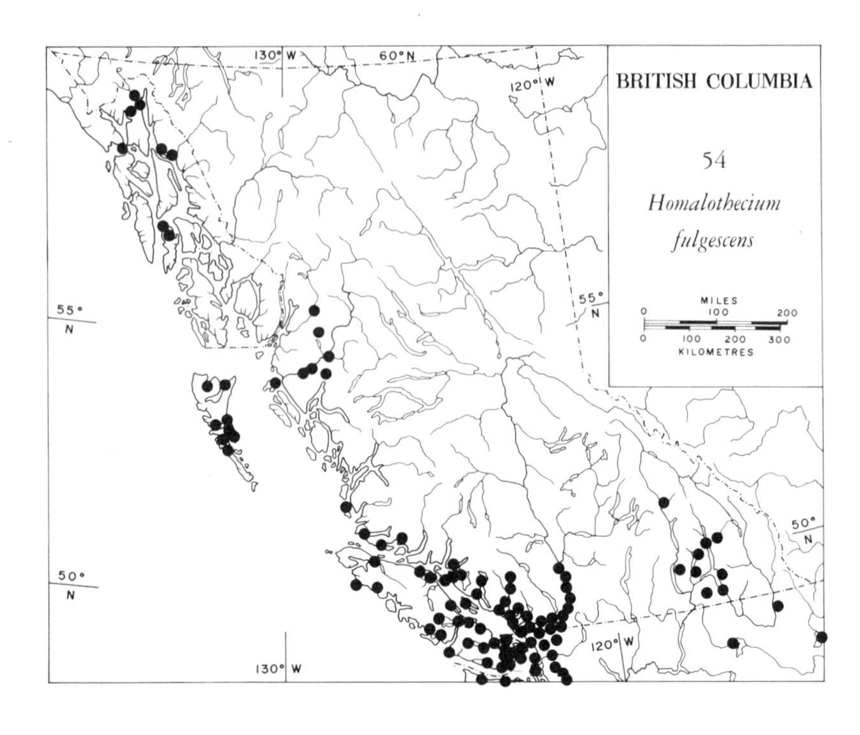
BRITISH COLUMBIA
54
Homalothecium
fulgescens
MILES
0
100
200
0
100
200
300
KILOMETRES
130° W
60° N
120° W
55° N
50° N
130° W
120° W
55° N
50° N

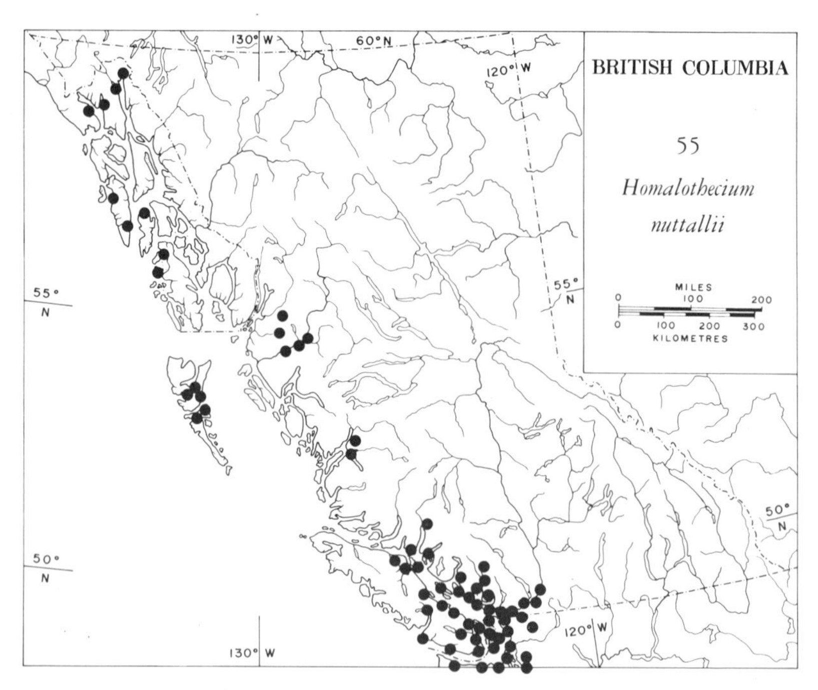
BRITISH COLUMBIA
55
Homalothecium
nuttallii
MILES
0
100
200
0
100
200
300
KILOMETRES
130° W
60° N
120° W
55° N
50° N
130° W
120° W
55° N
50° N

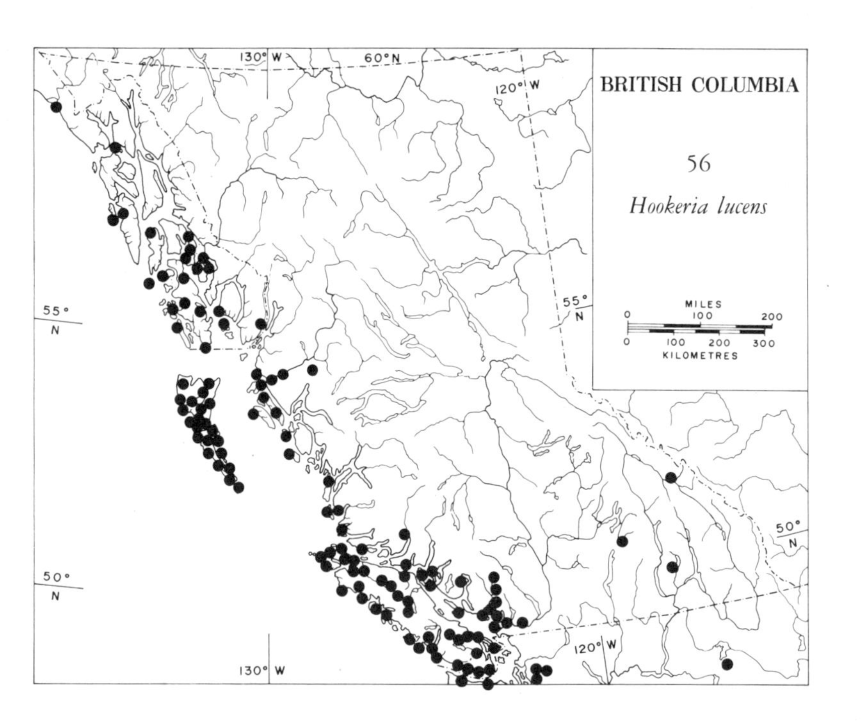
BRITISH COLUMBIA
56
Hookeria lucens
MILES
0 100 200
0 100 200 300
KILOMETRES
130° W
60° N
120° W
55° N
50° N
120° W
130° W

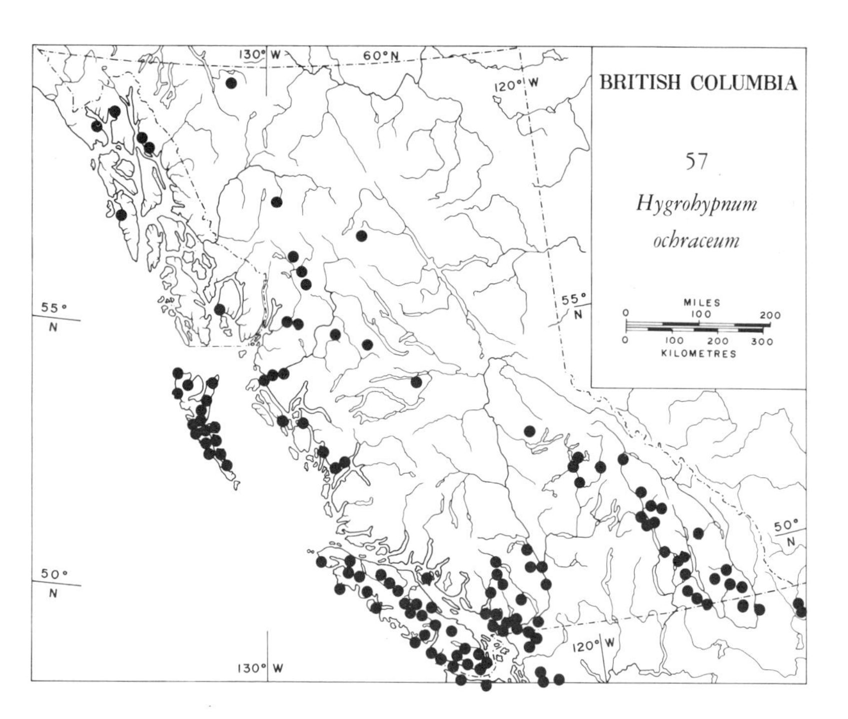
BRITISH COLUMBIA
57
Hygrohypnum ochraceum
MILES
0 100 200
0 100 200 300
KILOMETRES
130° W
60° N
120° W
55° N
50° N
120° W
130° W

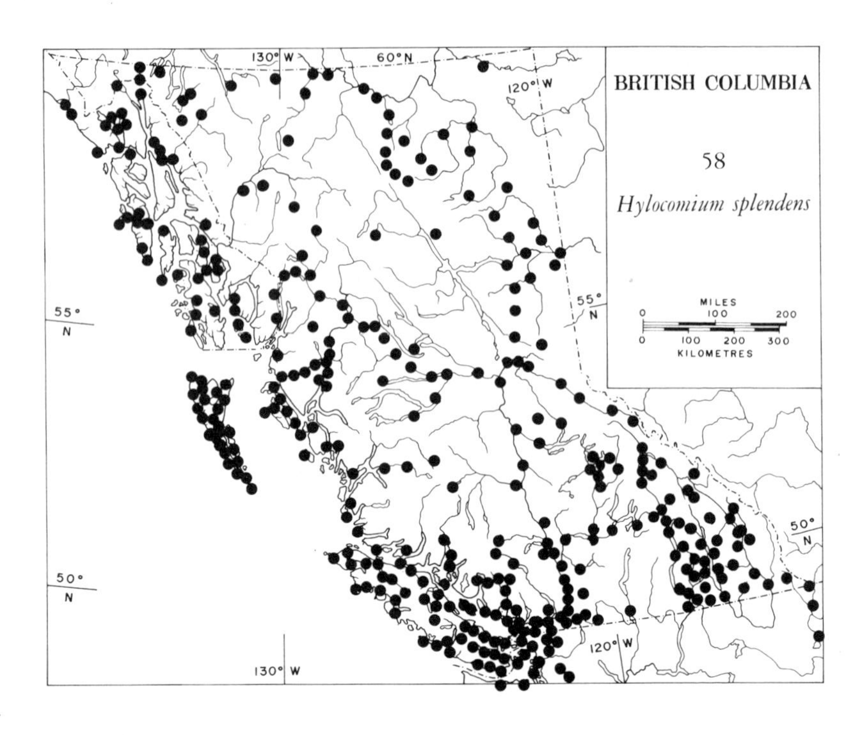
BRITISH COLUMBIA
58
Hylocomium splendens
MILES
0 100 200
0 100 200 300
KILOMETRES
130° W
60° N
120° W
55° N
50° N
130° W
120° W
55° N
50° N

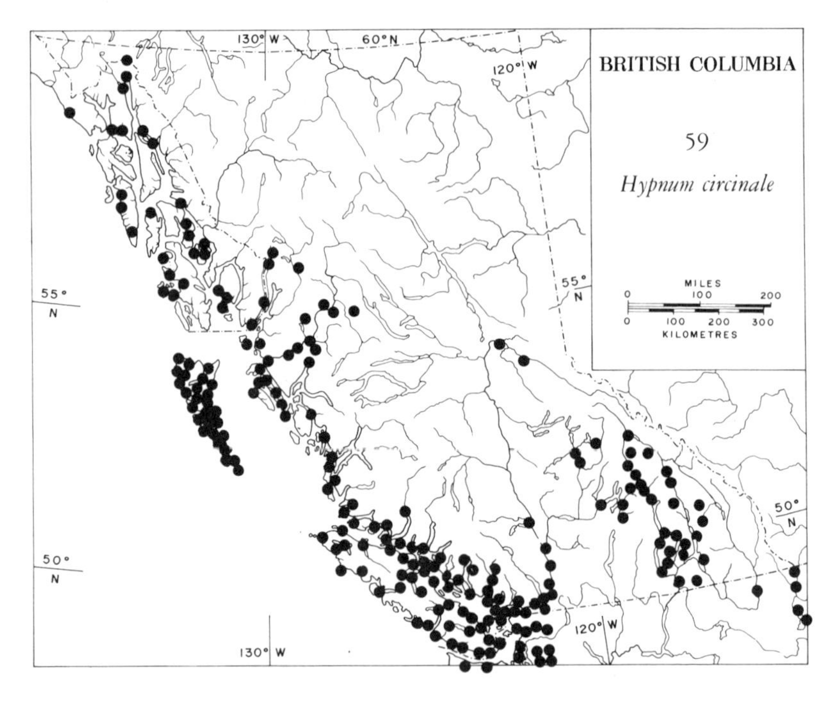
BRITISH COLUMBIA
59
Hypnum circinale
MILES
0 100 200
0 100 200 300
KILOMETRES
130° W
60° N
120° W
55° N
50° N
130° W
120° W
55° N
50° N

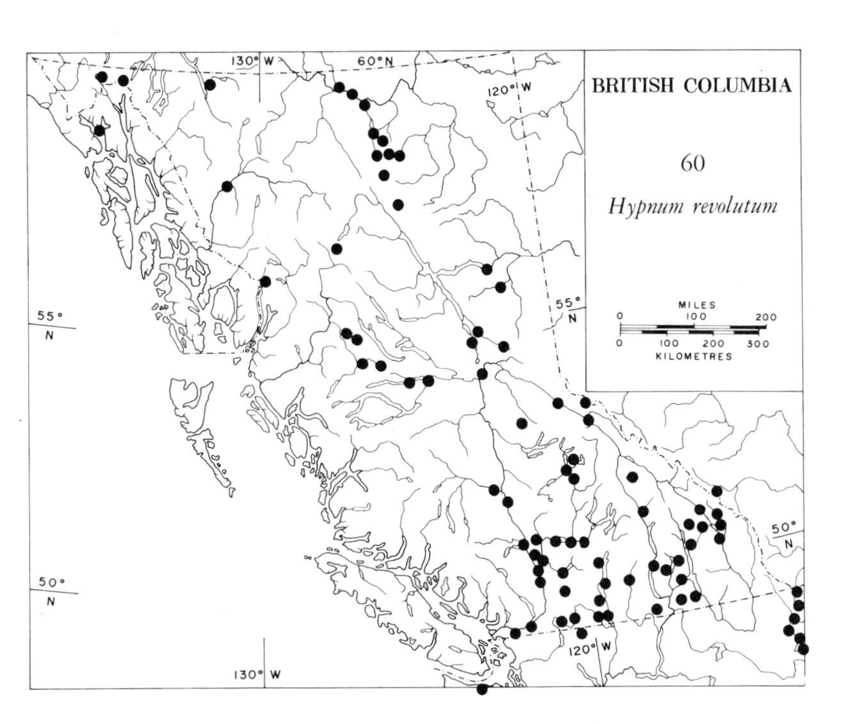
130° W
60° N
120° W
BRITISH COLUMBIA
60
Hypnum revolutum
55° N
MILES
0 100 200
0 100 200 300
KILOMETRES
50° N
120° W
130° W

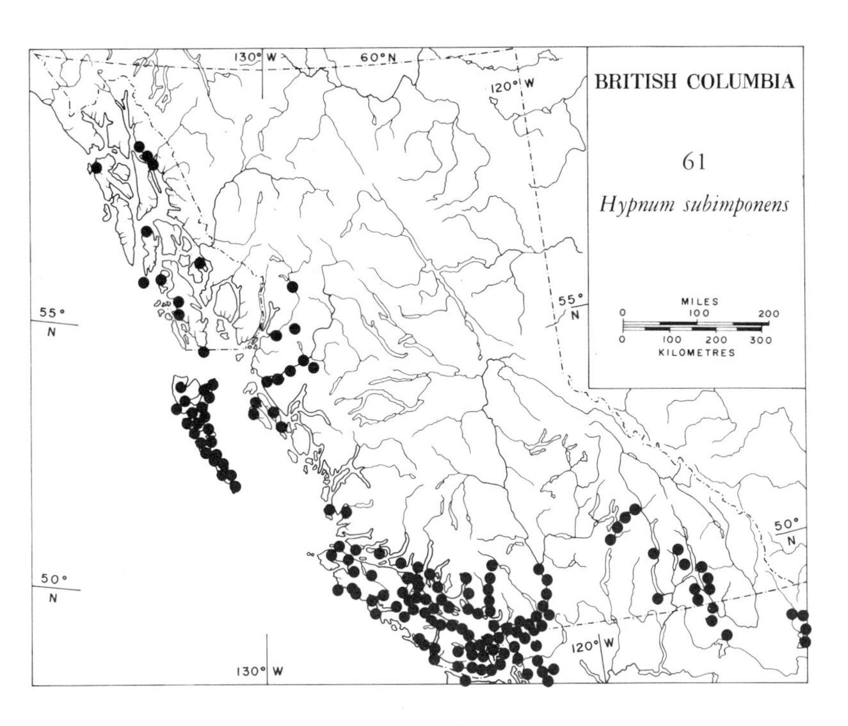
130° W
60° N
120° W
BRITISH COLUMBIA
61
Hypnum subimponens
55° N
MILES
0 100 200
0 100 200 300
KILOMETRES
50° N
120° W
130° W

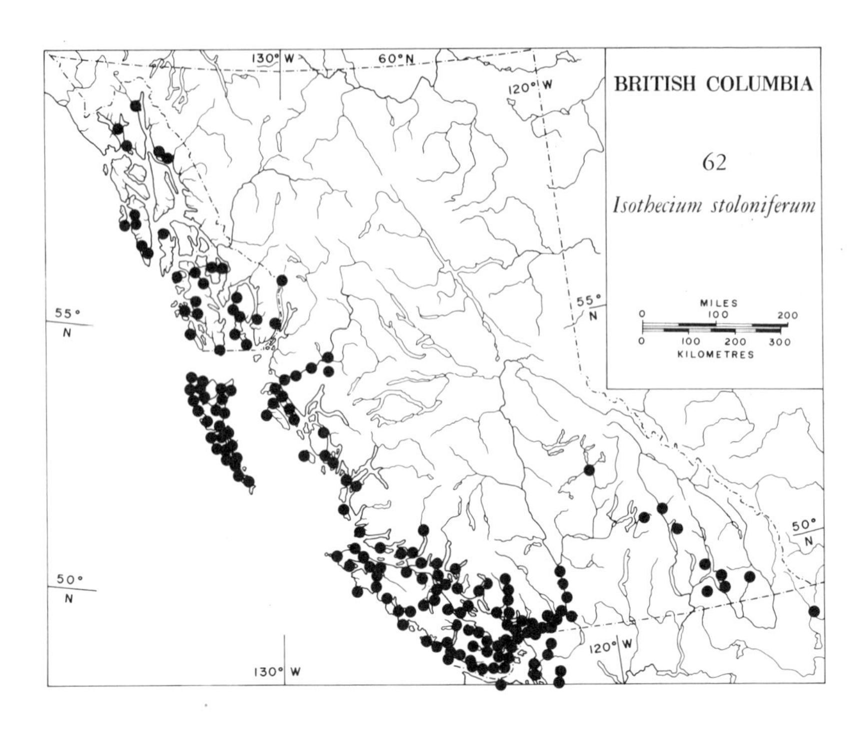
130° W
60° N
120° W
BRITISH COLUMBIA
62
Isothecium stoloniferum
MILES
0 100 200
0 100 200 300
KILOMETRES
55° N
50° N
130° W
120° W
50° N
55° N

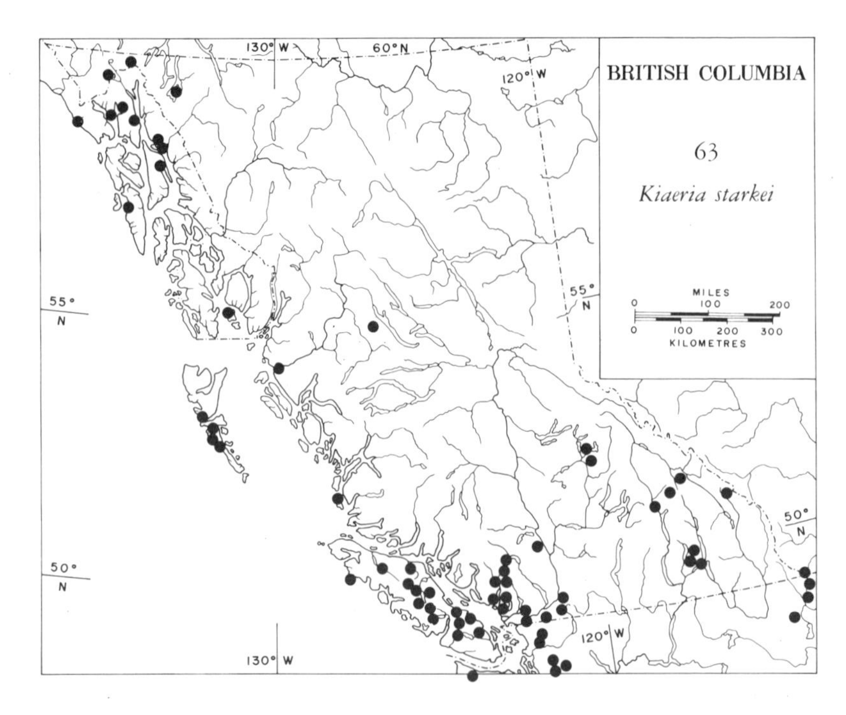
130° W
60° N
120° W
BRITISH COLUMBIA
63
Kiaeria starkei
MILES
0 100 200
0 100 200 300
KILOMETRES
55° N
50° N
130° W
120° W
50° N
55° N

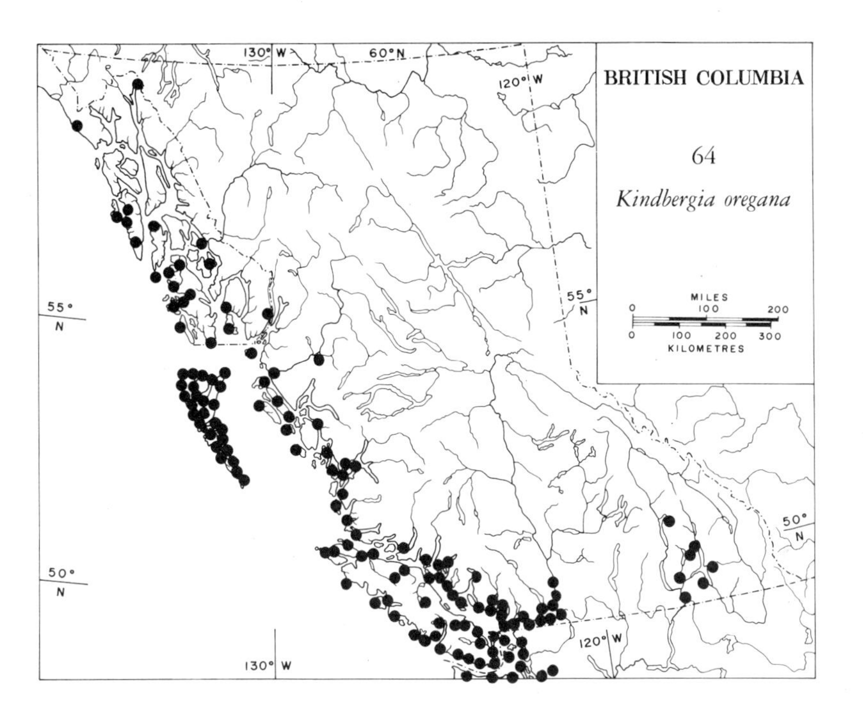
130° W
60° N
120° W
BRITISH COLUMBIA
64
Kindbergia oregana
55° N
MILES
0 100 200
0 100 200 300
KILOMETRES
50° N
130° W
120° W

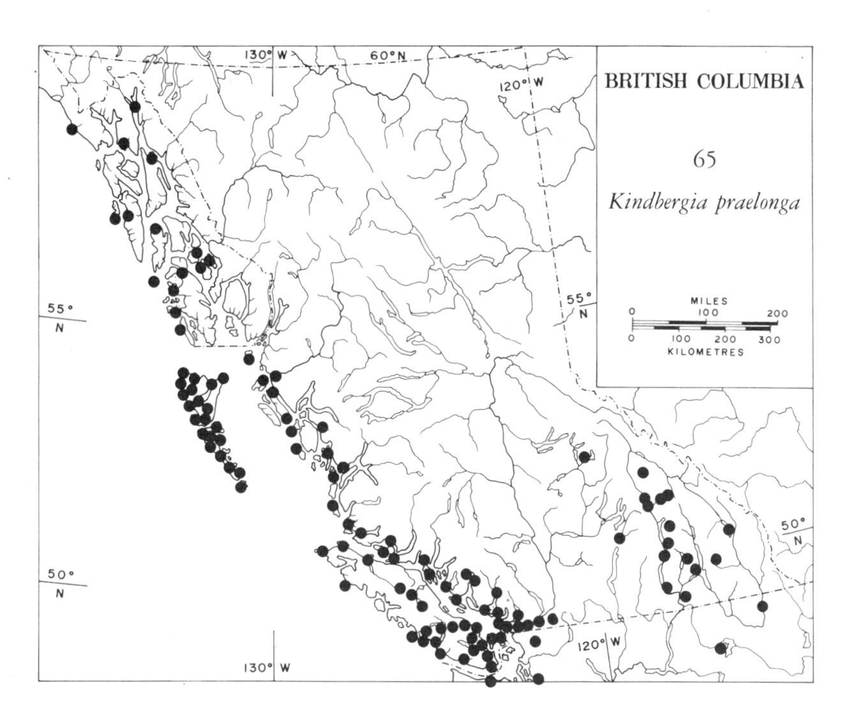
130° W
60° N
120° W
BRITISH COLUMBIA
65
Kindbergia praelonga
55° N
MILES
0 100 200
0 100 200 300
KILOMETRES
50° N
130° W
120° W

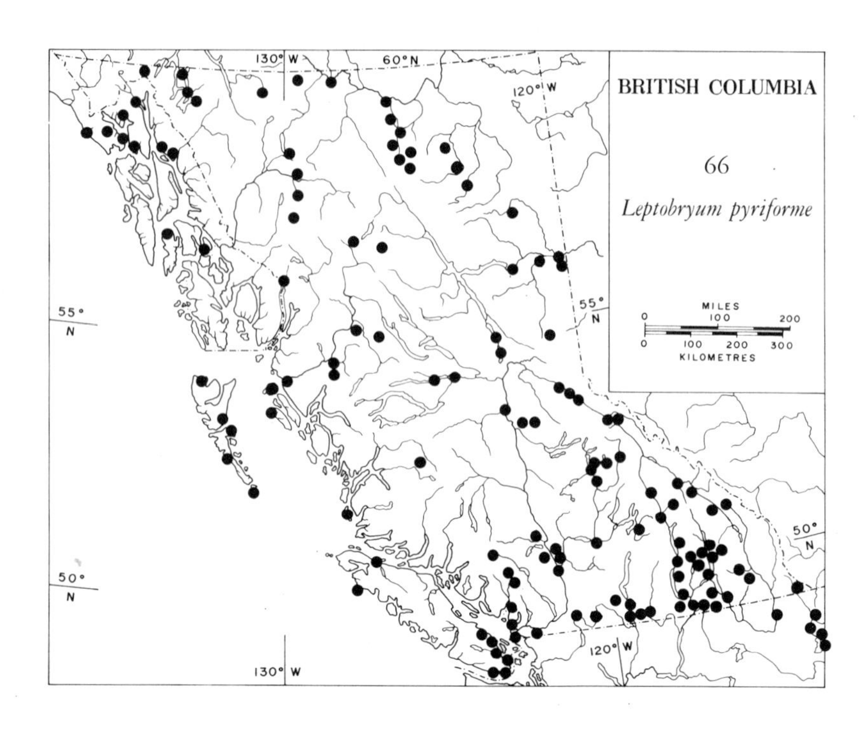

BRITISH COLUMBIA
66
Leptobryum pyriforme
MILES
0 100 200
0 100 200 300
KILOMETRES
130° W
60° N
120° W
55° N
50° N

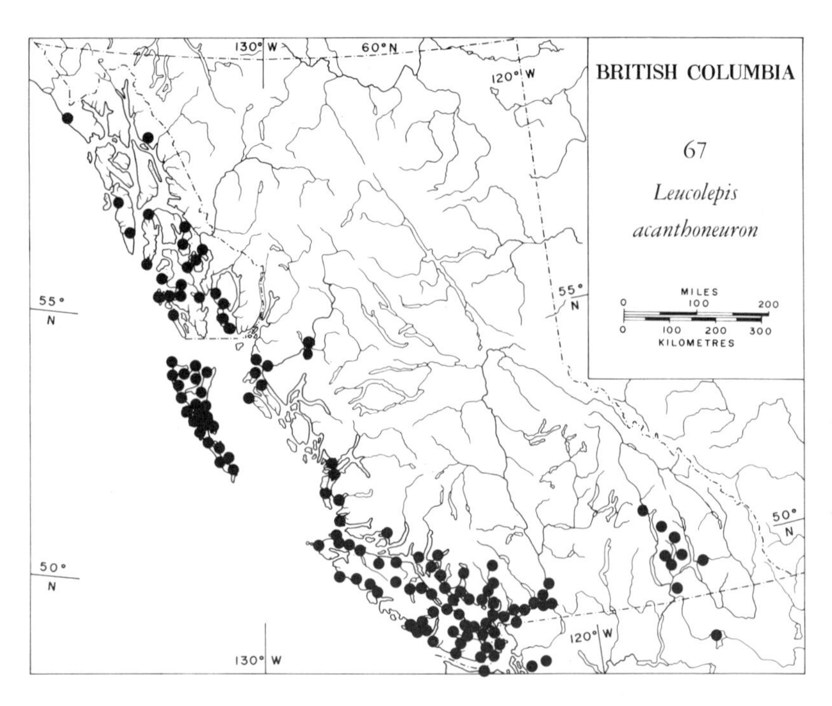

BRITISH COLUMBIA
67
Leucolepis acanthoneuron
MILES
0 100 200
0 100 200 300
KILOMETRES
130° W
60° N
120° W
55° N
50° N

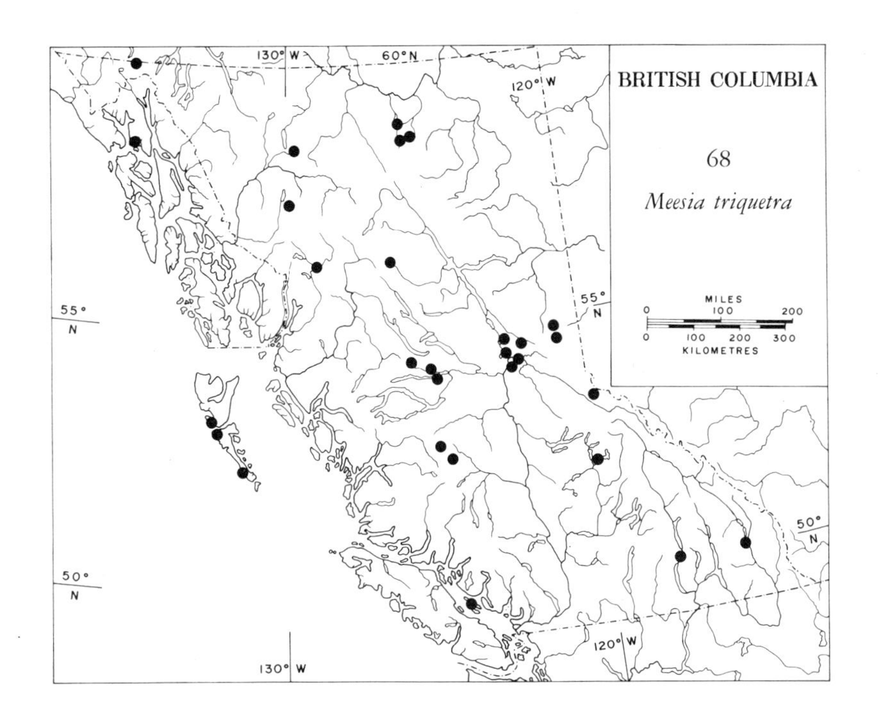
130° W
60° N
120° W
BRITISH COLUMBIA
68
Meesia triquetra
MILES
0
100
200
0
100
200
300
KILOMETRES
55°
N
55°
N
50°
N
50°
N
120° W
130° W

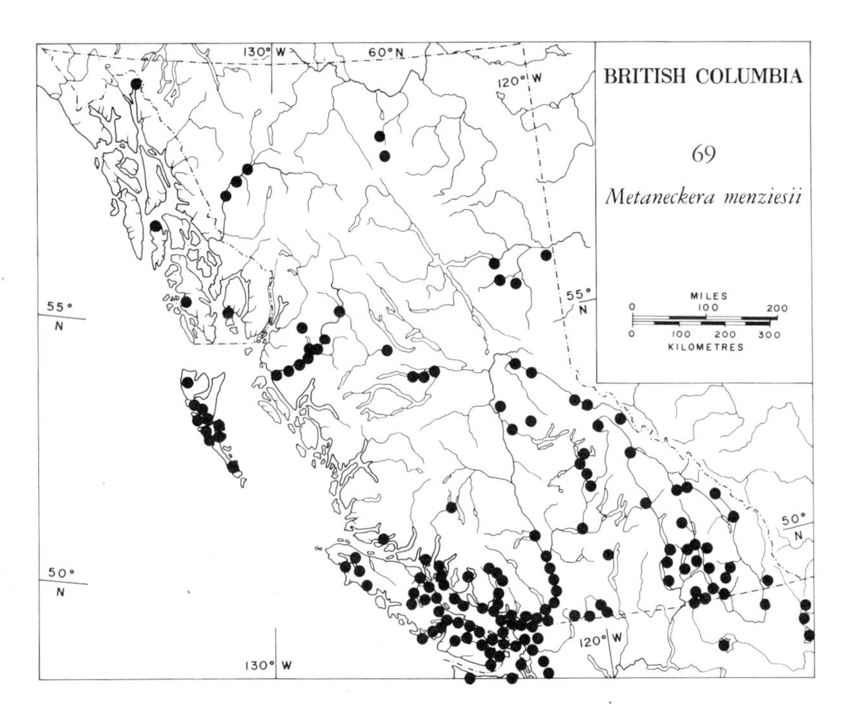
130° W
60° N
120° W
BRITISH COLUMBIA
69
Metaneckera menziesii
MILES
0
100
200
0
100
200
300
KILOMETRES
55°
N
55°
N
50°
N
50°
N
120° W
130° W

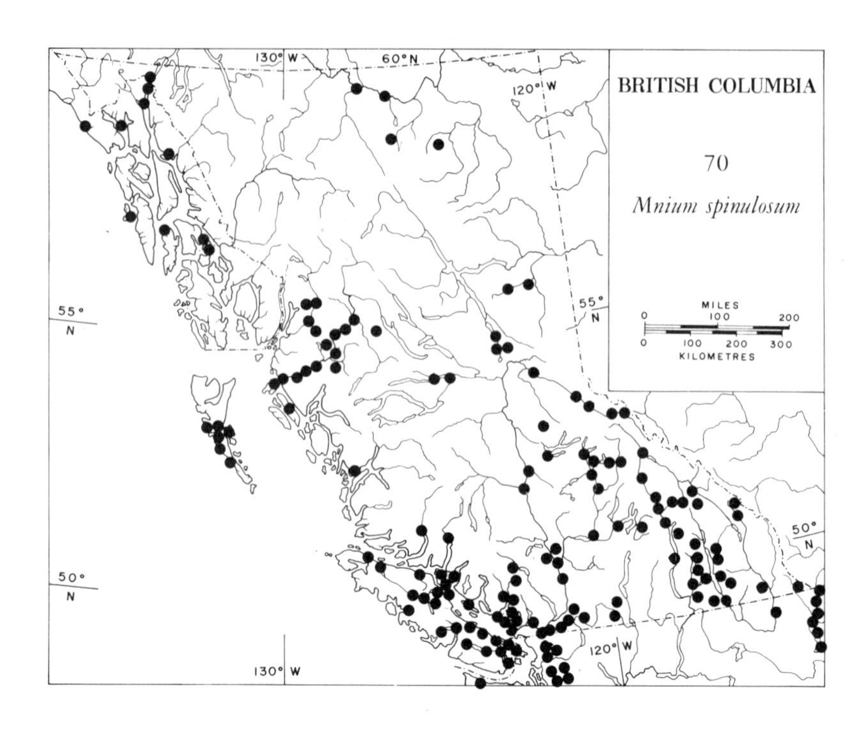
130° W
60° N
120° W
BRITISH COLUMBIA
70
Mniium spinulosum
55° N
MILES
0 100 200
0 100 200 300
KILOMETRES
50° N
120° W
130° W

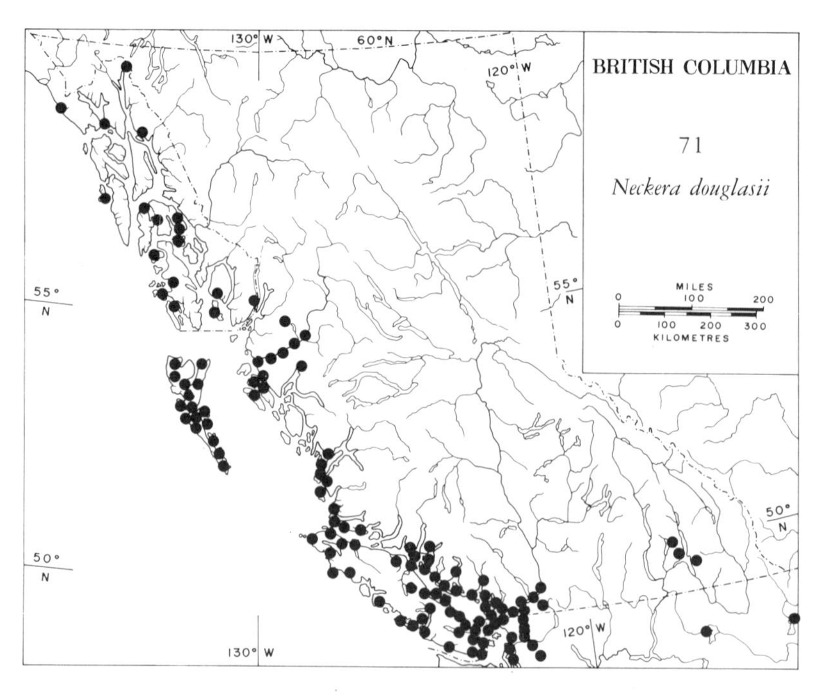
130° W
60° N
120° W
BRITISH COLUMBIA
71
Neckera douglasii
55° N
MILES
0 100 200
0 100 200 300
KILOMETRES
50° N
120° W
130° W

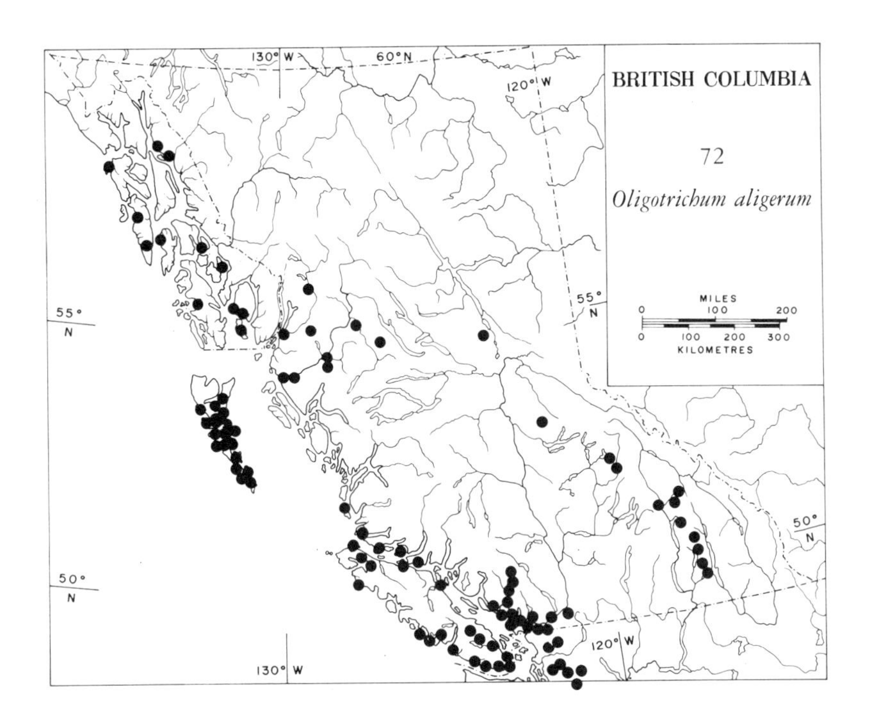
130° W
60° N
120° W
BRITISH COLUMBIA
72
Oligotrichum aligerum
55° N
MILES
0 100 200
0 100 200 300
KILOMETRES
50° N
120° W
130° W

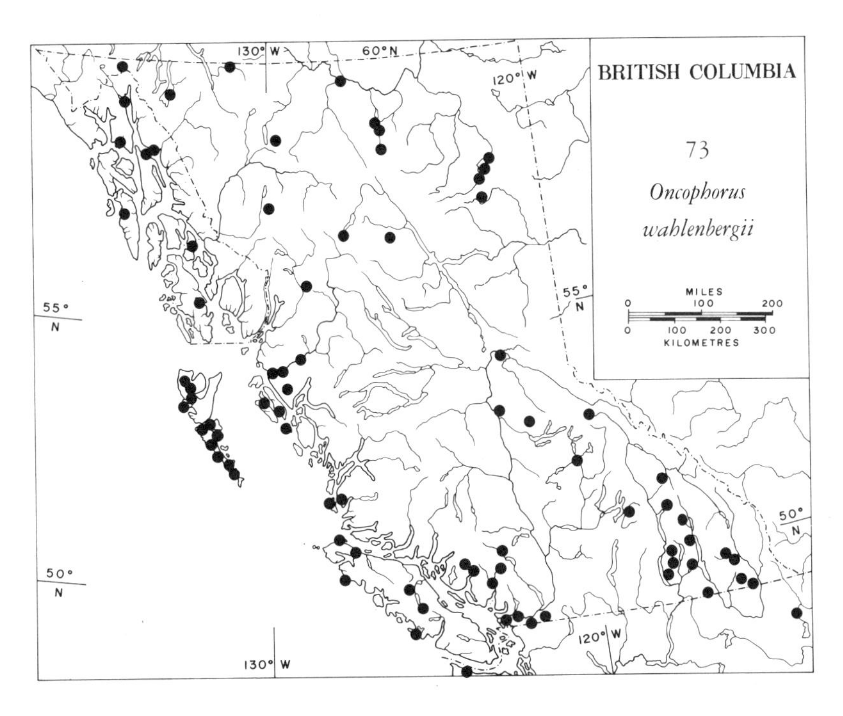
130° W
60° N
120° W
BRITISH COLUMBIA
73
Oncophorus wahlenbergii
55° N
MILES
0 100 200
0 100 200 300
KILOMETRES
50° N
120° W
130° W

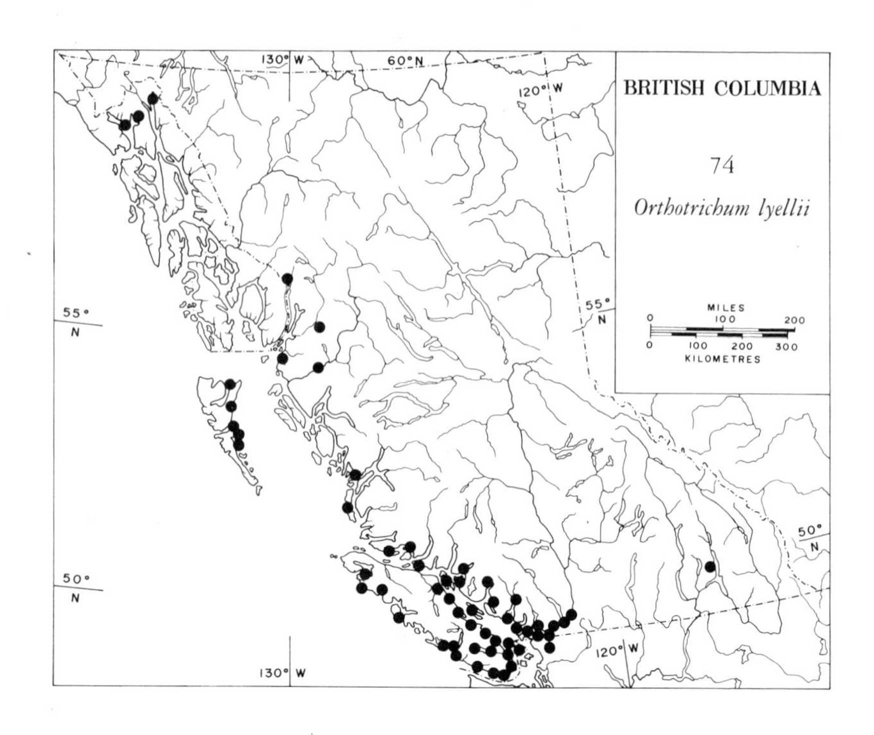
130° W
60° N
120° W
BRITISH COLUMBIA
74
Orthotrichum lyellii
55° N
MILES
0 100 200
0 100 200 300
KILOMETRES
50° N
50° N
120° W
130° W

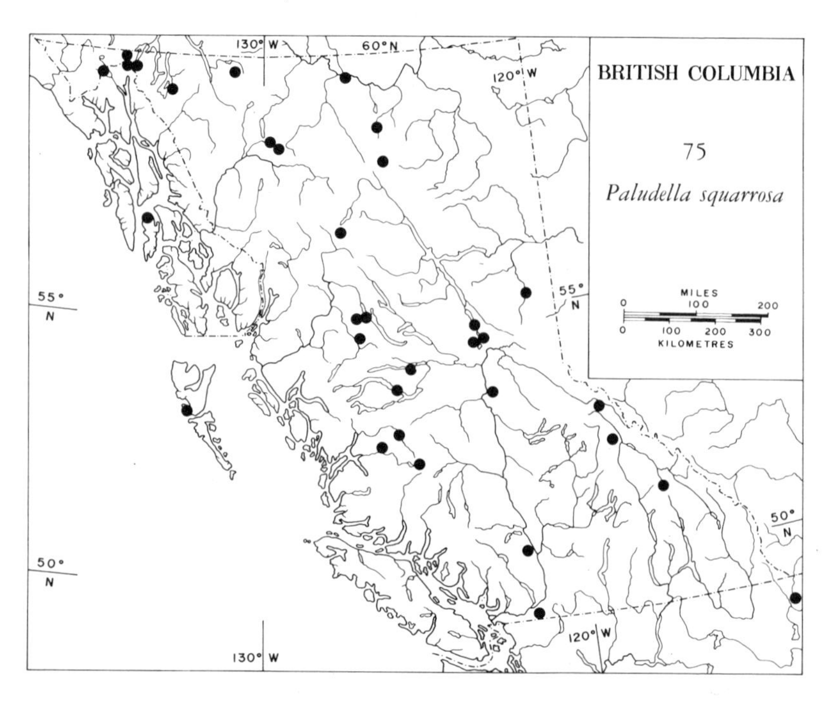
130° W
60° N
120° W
BRITISH COLUMBIA
75
Paludella squarrosa
55° N
MILES
0 100 200
0 100 200 300
KILOMETRES
50° N
50° N
120° W
130° W

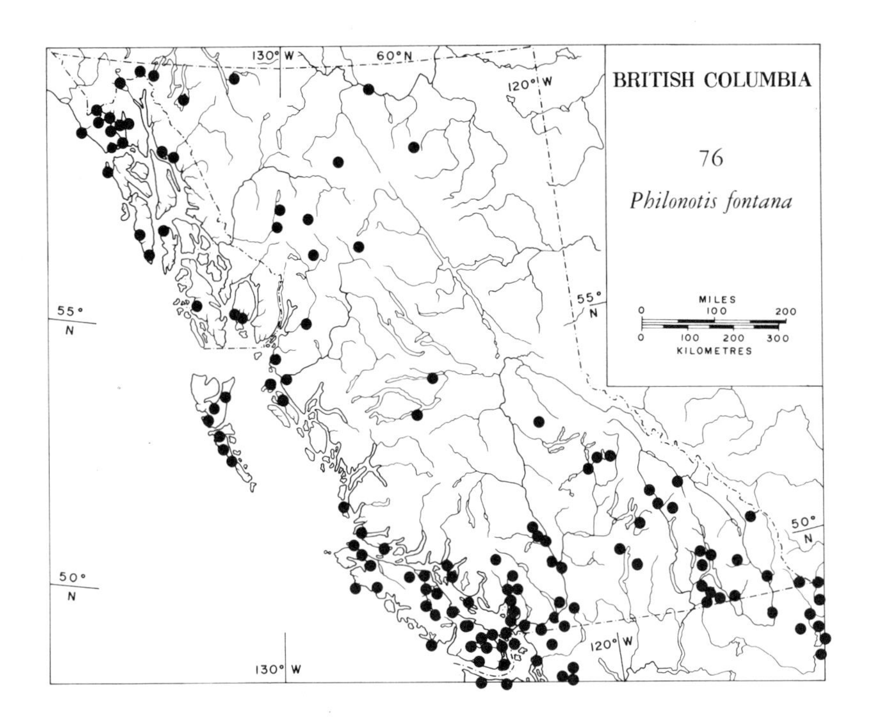
130° W
60° N
120° W
BRITISH COLUMBIA
76
Philonotis fontana
55° N
MILES
0 100 200
0 100 200 300
KILOMETRES
50° N
120° W
130° W

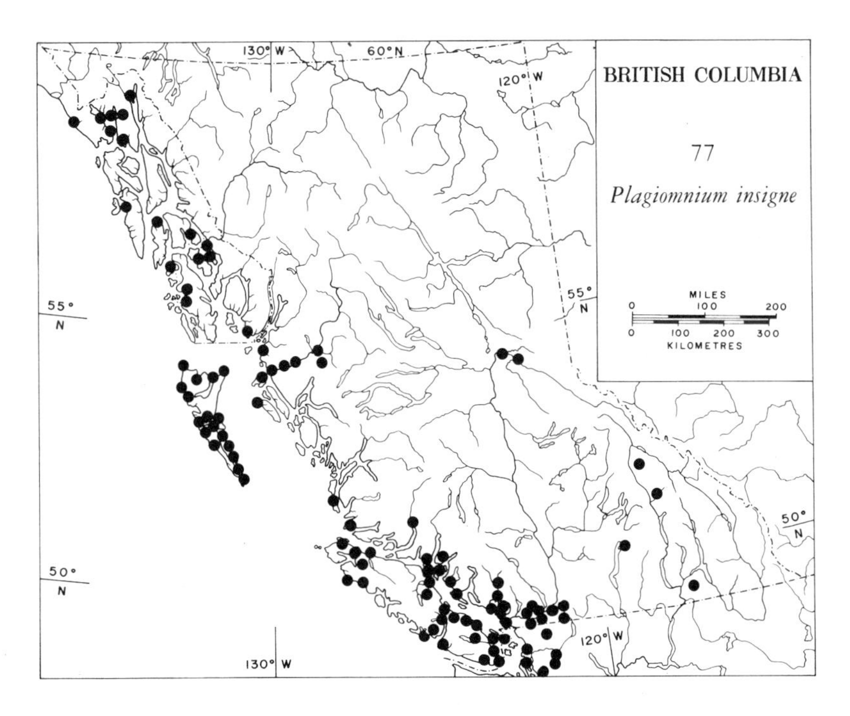
130° W
60° N
120° W
BRITISH COLUMBIA
77
Plagiomnium insigne
55° N
MILES
0 100 200
0 100 200 300
KILOMETRES
50° N
120° W
130° W

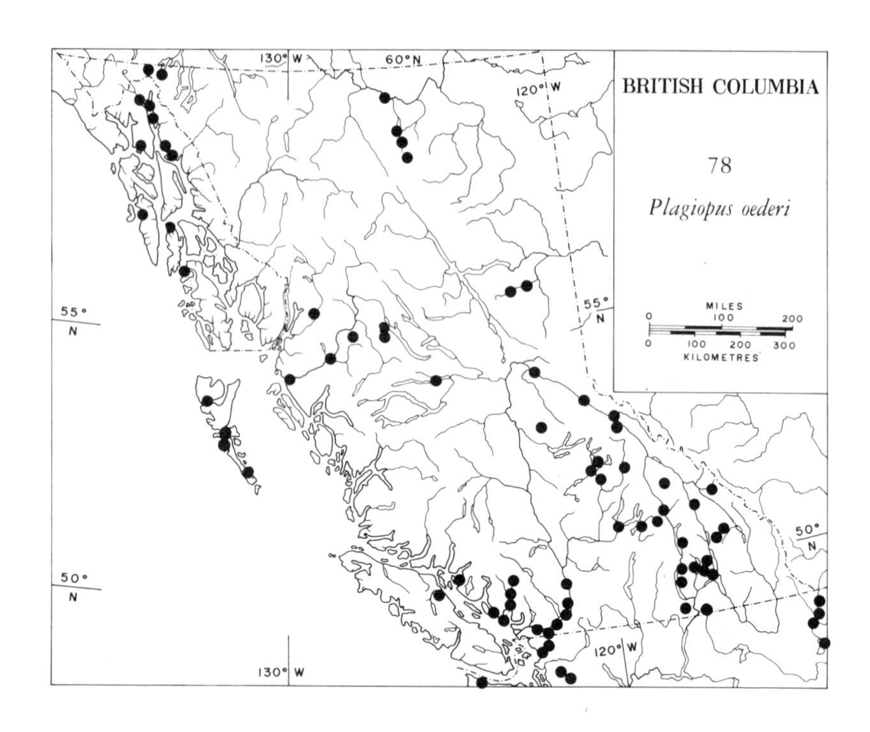
BRITISH COLUMBIA
78
Plagiopus oederi
MILES
0 100 200
0 100 200 300
KILOMETRES
130° W
60° N
120° W
55° N
50° N

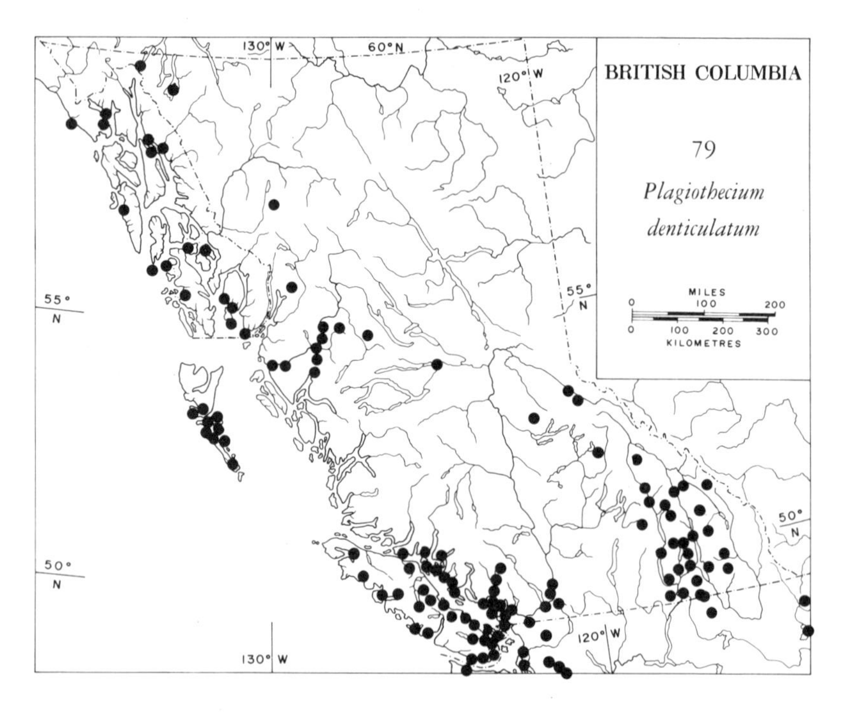
BRITISH COLUMBIA
79
Plagiothecium denticulatum
MILES
0 100 200
0 100 200 300
KILOMETRES
130° W
60° N
120° W
55° N
50° N

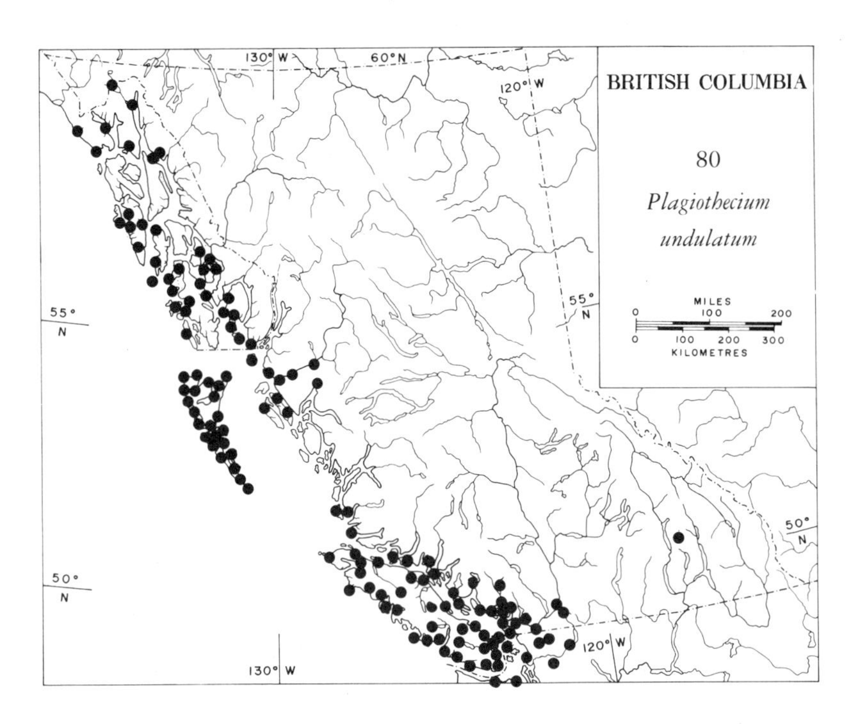
BRITISH COLUMBIA
80
Plagiothecium undulatum
MILES
0 100 200
0 100 200 300
KILOMETRES
130° W
60° N
120° W
55° N
50° N
130° W
120° W

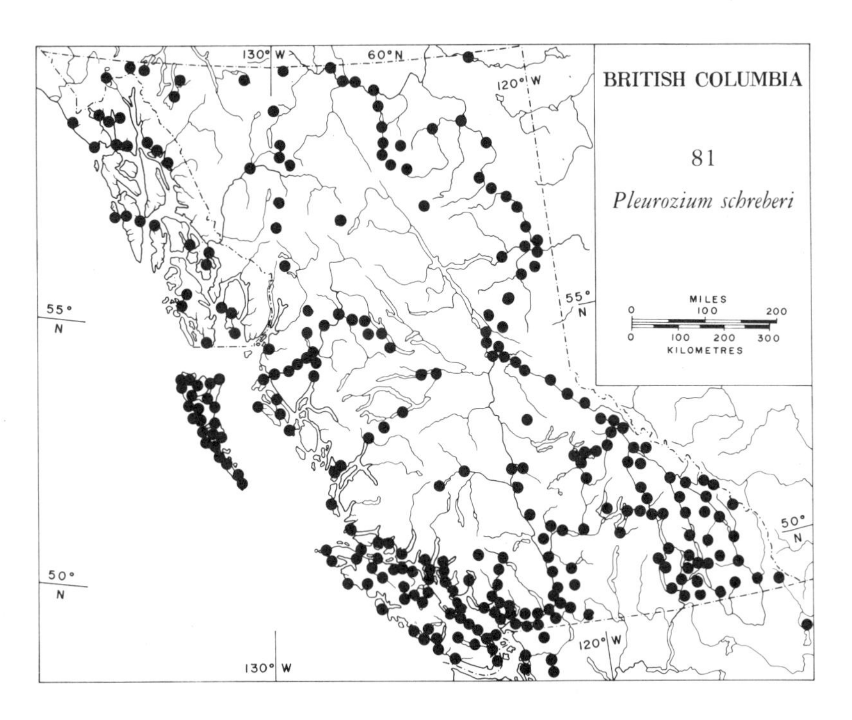
BRITISH COLUMBIA
81
Pleurozium schreberi
MILES
0 100 200
0 100 200 300
KILOMETRES
130° W
60° N
120° W
55° N
50° N
130° W
120° W

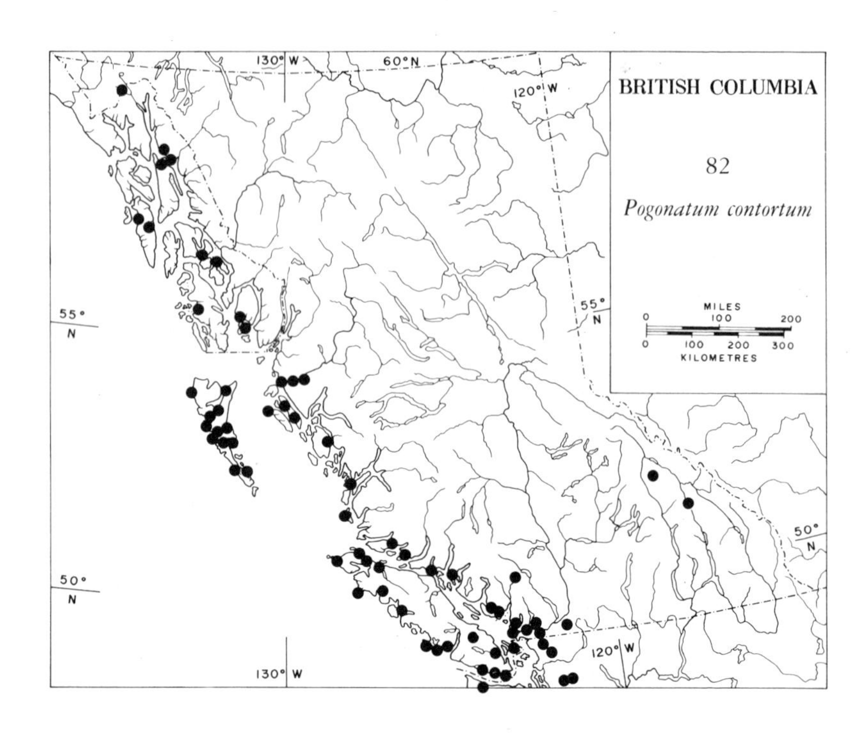
BRITISH COLUMBIA
82
Pogonatum contortum
MILES
0
100
200
0
100
200
300
KILOMETRES
130° W
60° N
120° W
55° N
50° N
130° W
120° W
55° N
50° N

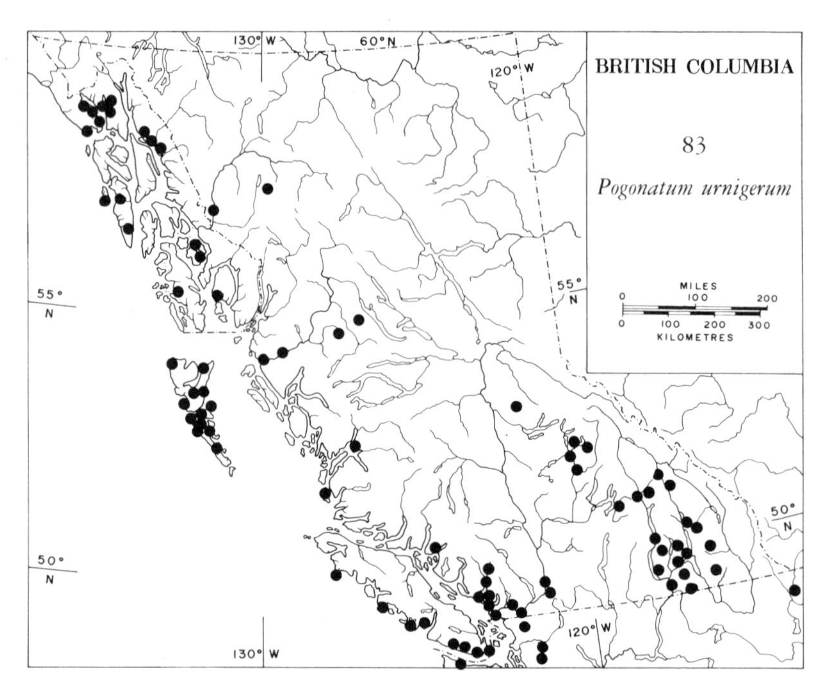
BRITISH COLUMBIA
83
Pogonatum urnigerum
MILES
0
100
200
0
100
200
300
KILOMETRES
130° W
60° N
120° W
55° N
50° N
130° W
120° W
55° N
50° N

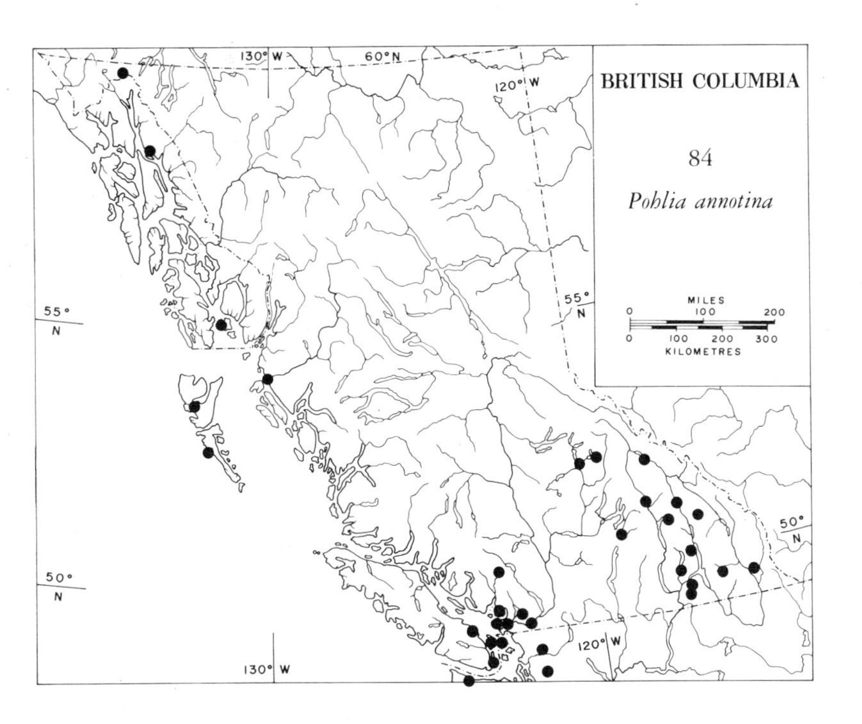
BRITISH COLUMBIA
84
Pohlia annotina
MILES
0 100 200
0 100 200 300
KILOMETRES
130° W
60° N
120° W
55° N
50° N

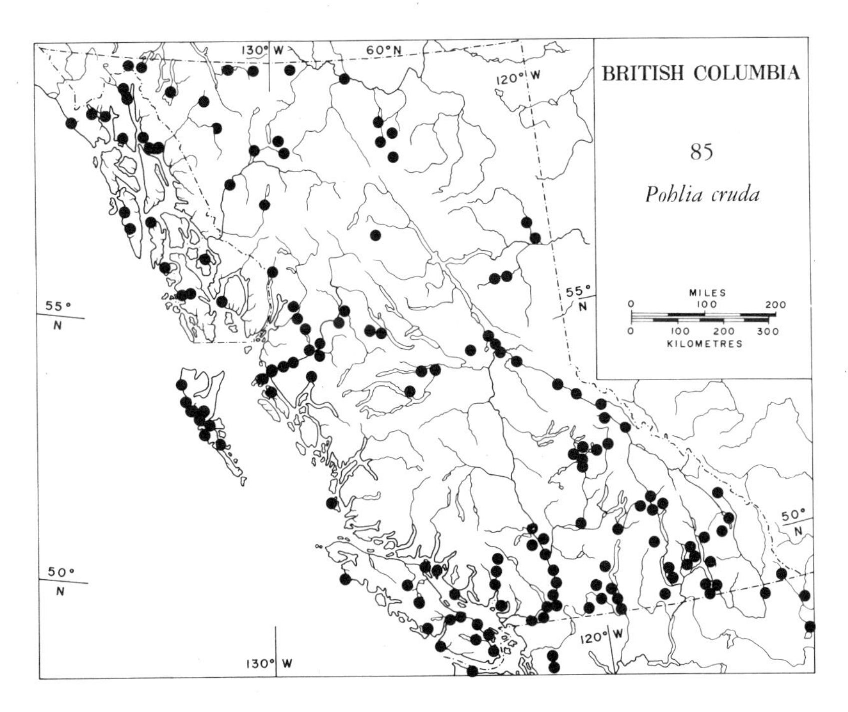
BRITISH COLUMBIA
85
Pohlia cruda
MILES
0 100 200
0 100 200 300
KILOMETRES
130° W
60° N
120° W
55° N
50° N

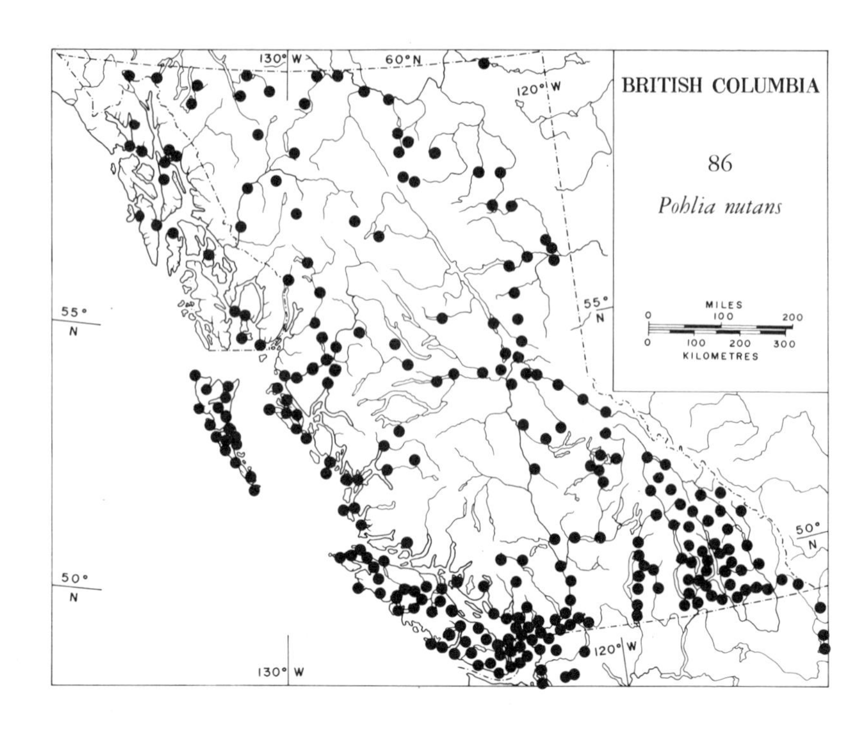
130° W
60° N
120° W
BRITISH COLUMBIA
86
Pohlia nutans
55° N
MILES
0 100 200
0 100 200 300
KILOMETRES
50° N
50° N
130° W
120° W

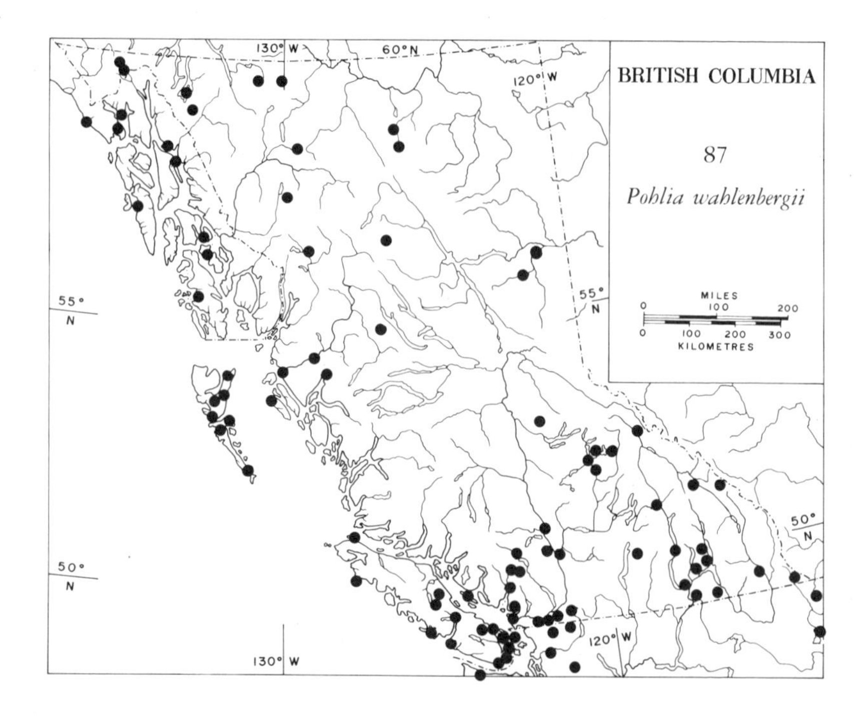
130° W
60° N
120° W
BRITISH COLUMBIA
87
Pohlia wahlenbergii
55° N
MILES
0 100 200
0 100 200 300
KILOMETRES
50° N
50° N
130° W
120° W

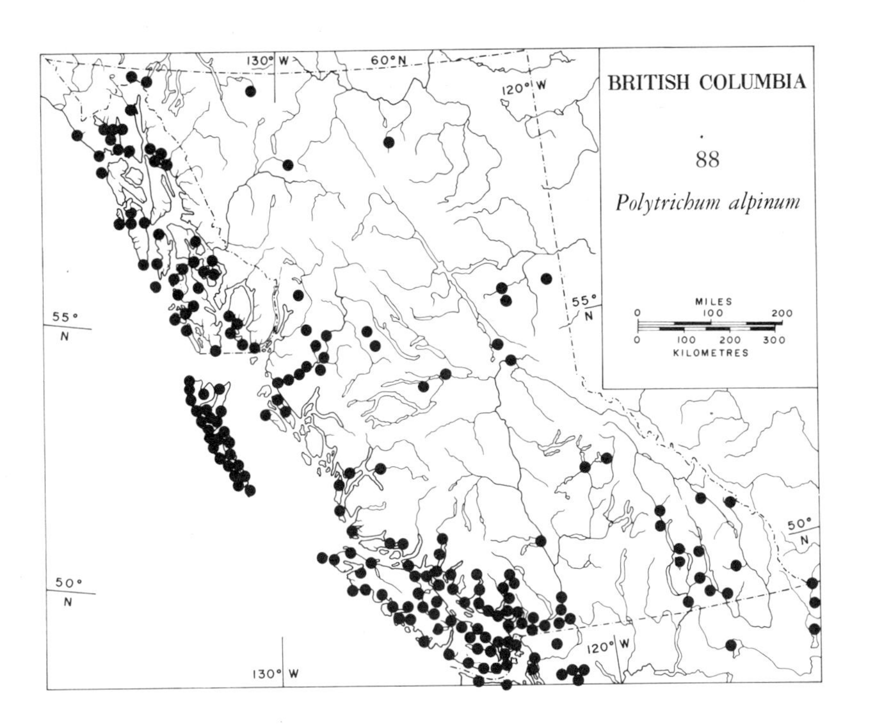
BRITISH COLUMBIA
88
Polytrichum alpinum
MILES
0 100 200
0 100 200 300
KILOMETRES
130° W
60° N
120° W
55° N
50° N

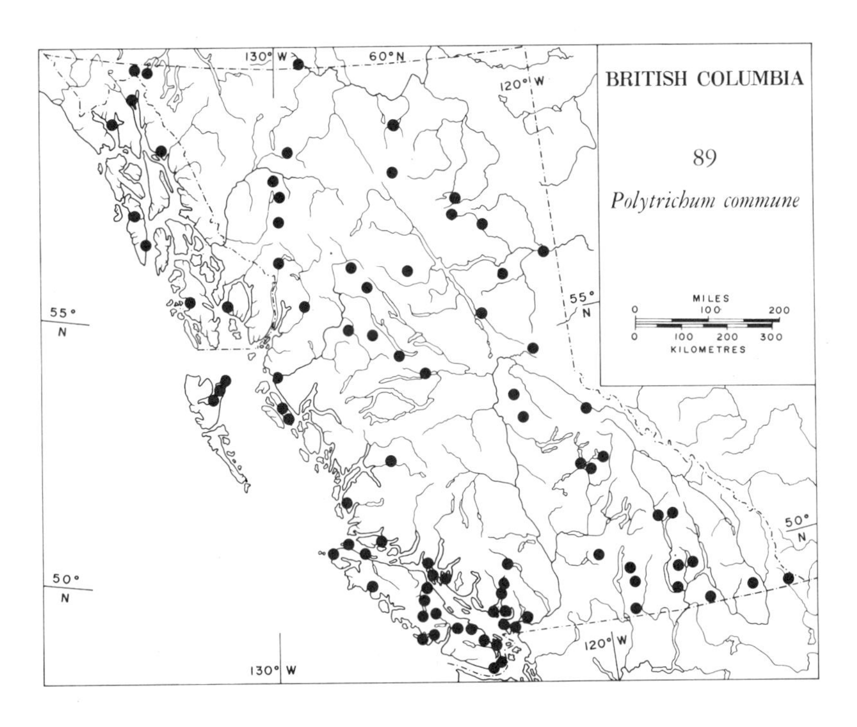
BRITISH COLUMBIA
89
Polytrichum commune
MILES
0 100 200
0 100 200 300
KILOMETRES
130° W
60° N
120° W
55° N
50° N

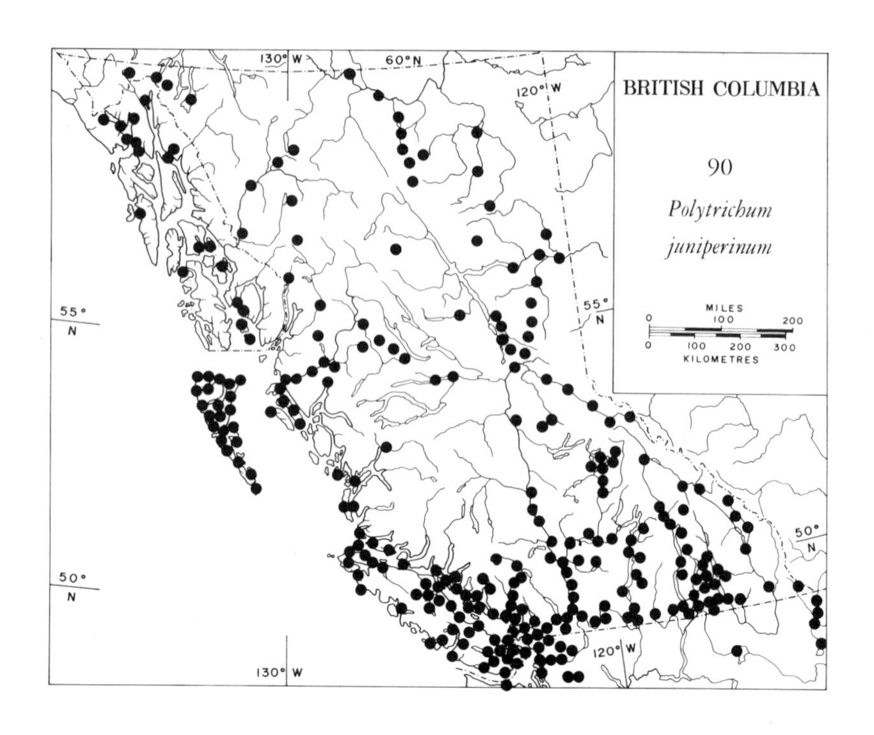
BRITISH COLUMBIA
90
Polytrichum juniperinum
MILES
0 100 200
0 100 200 300
KILOMETRES
130° W
60° N
120° W
55° N
50° N

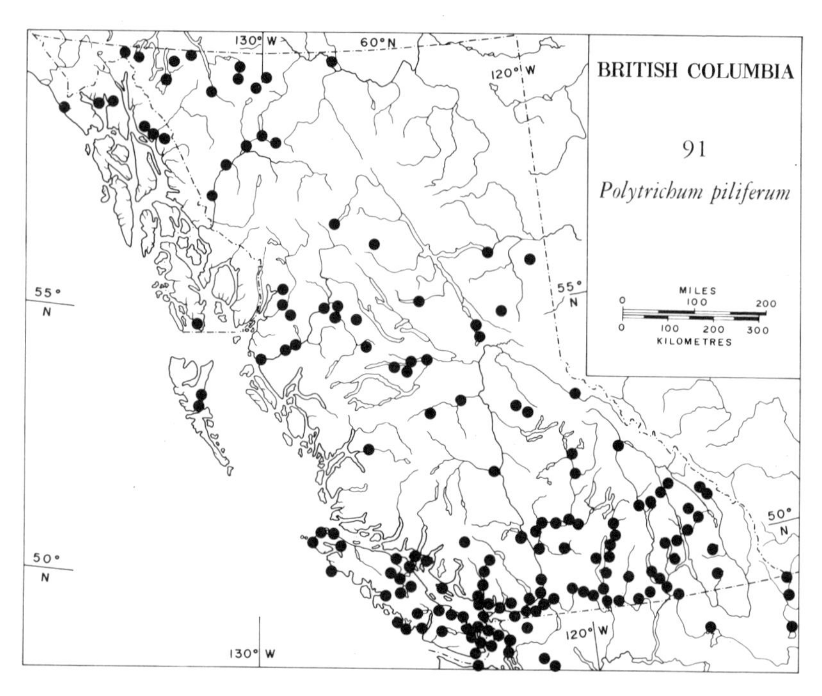
BRITISH COLUMBIA
91
Polytrichum piliferum
MILES
0 100 200
0 100 200 300
KILOMETRES
130° W
60° N
120° W
55° N
50° N

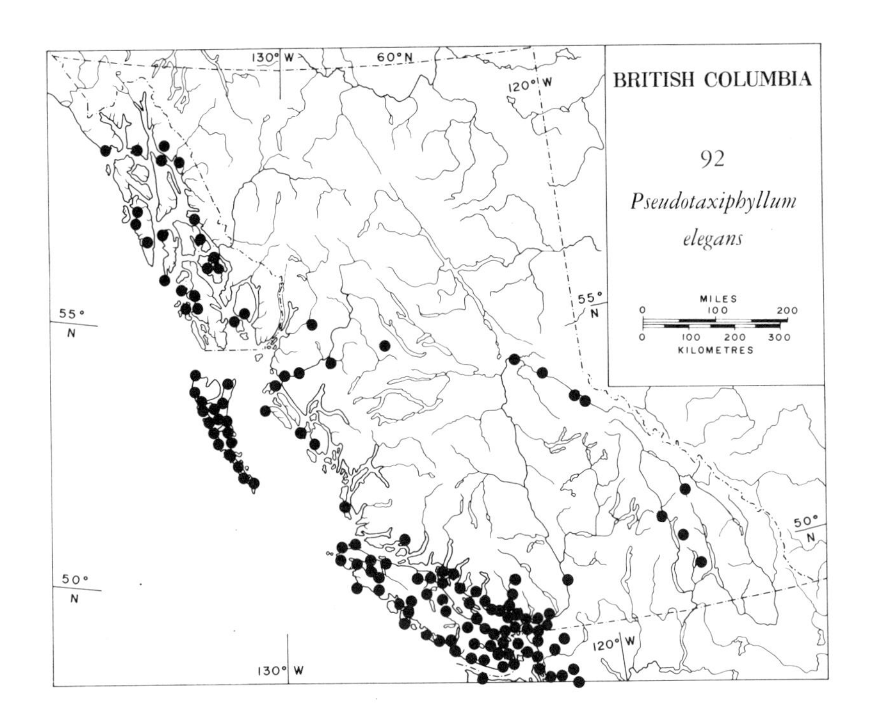
BRITISH COLUMBIA
92
Pseudotaxiphyllum elegans
MILES
0 100 200
0 100 200 300
KILOMETRES
130° W
60° N
120° W
55° N
50° N
130° W
120° W

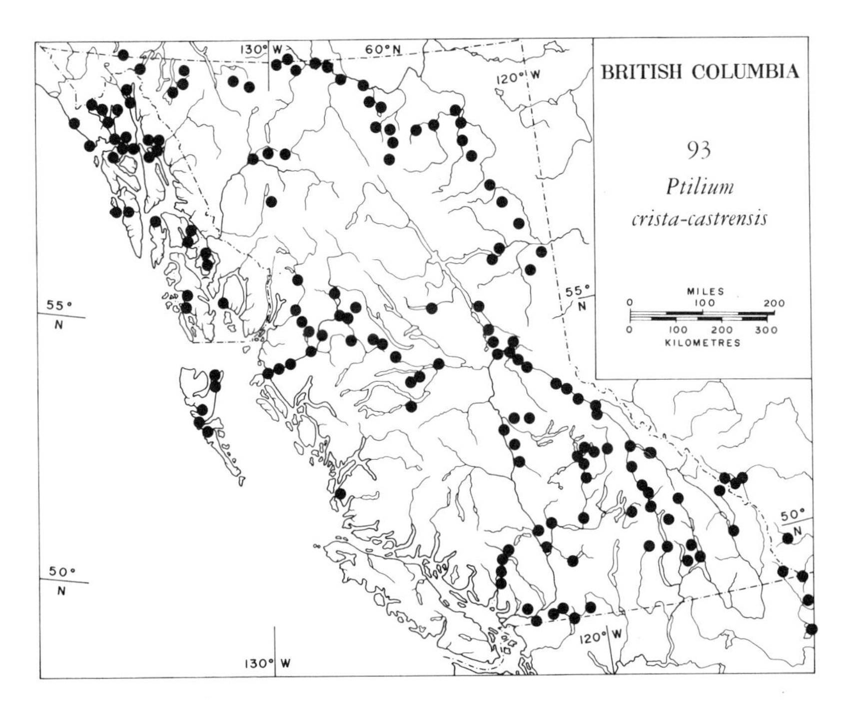
BRITISH COLUMBIA
93
Ptilium crista-castrensis
MILES
0 100 200
0 100 200 300
KILOMETRES
130° W
60° N
120° W
55° N
50° N
130° W
120° W

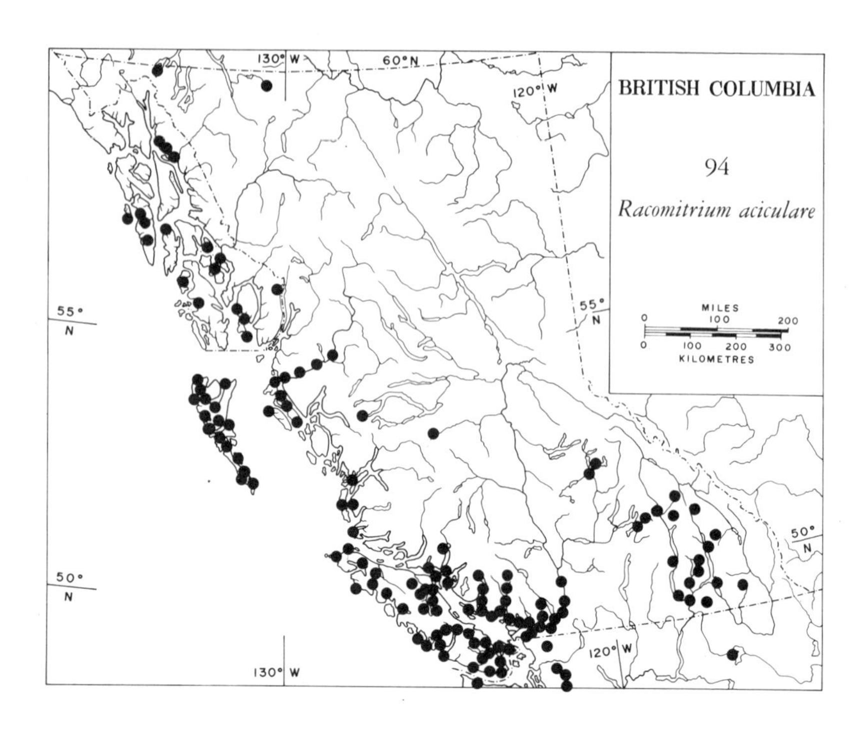
130° W
60° N
120° W
BRITISH COLUMBIA
94
Racomitrium aciculare
55° N
MILES
0 100 200
0 100 200 300
KILOMETRES
50° N
50° N
120° W
130° W

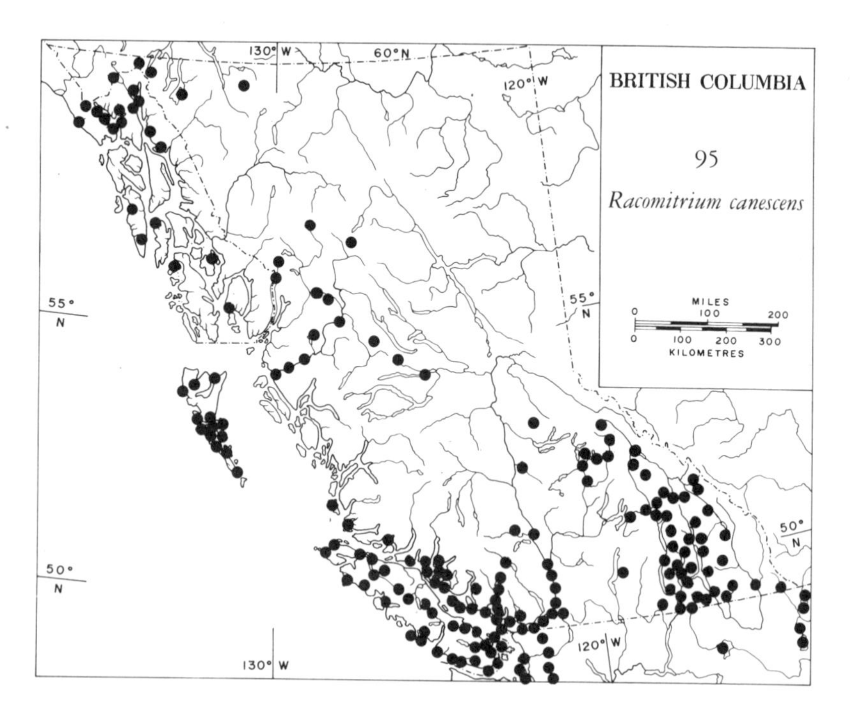
130° W
60° N
120° W
BRITISH COLUMBIA
95
Racomitrium canescens
55° N
MILES
0 100 200
0 100 200 300
KILOMETRES
50° N
50° N
120° W
130° W

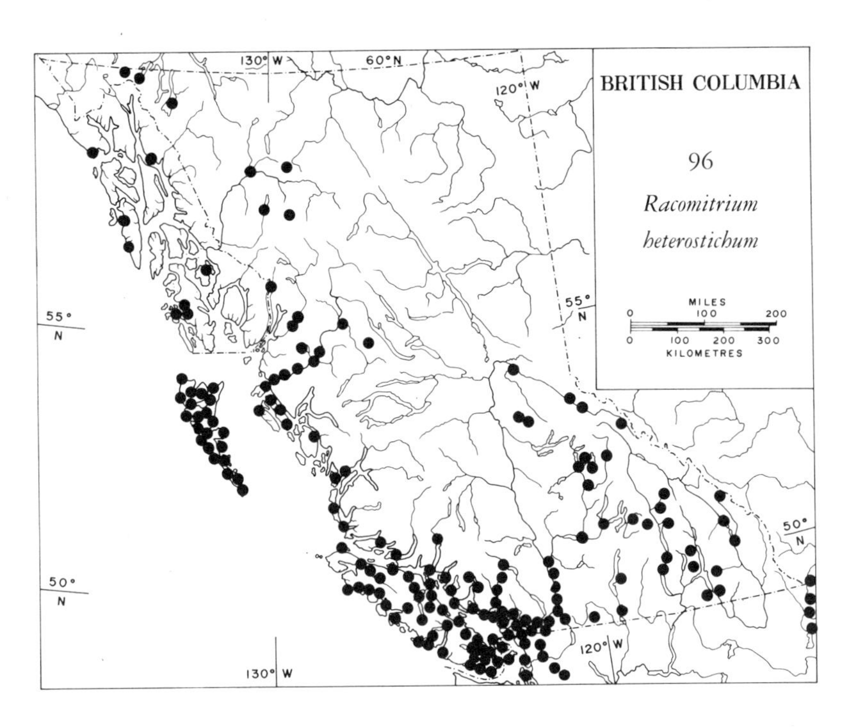
BRITISH COLUMBIA
96
Racomitrium heterostichum
MILES
0 100 200
0 100 200 300
KILOMETRES
130° W
60° N
120° W
55° N
50° N

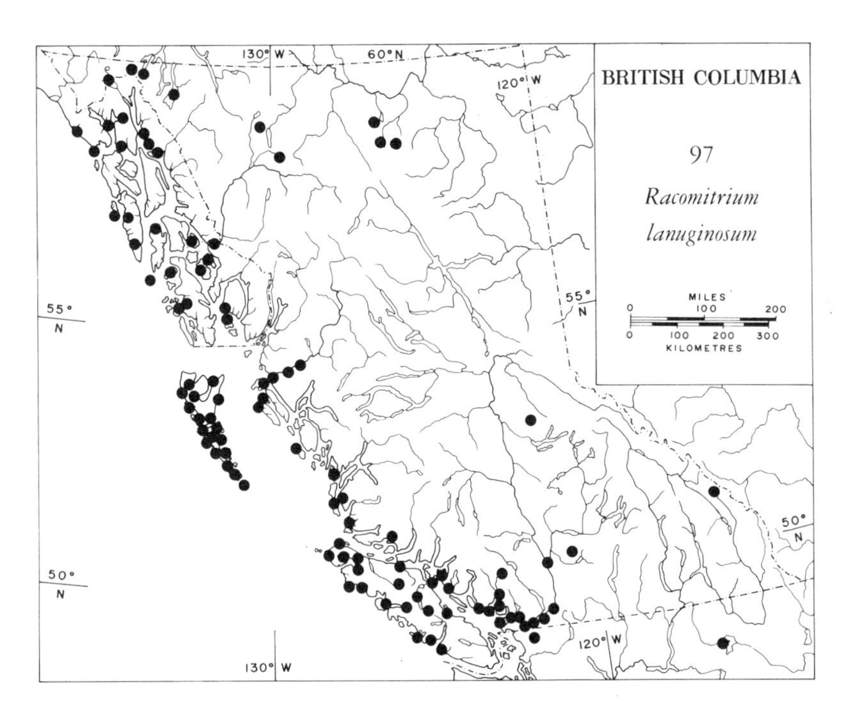
BRITISH COLUMBIA
97
Racomitrium lanuginosum
MILES
0 100 200
0 100 200 300
KILOMETRES
130° W
60° N
120° W
55° N
50° N

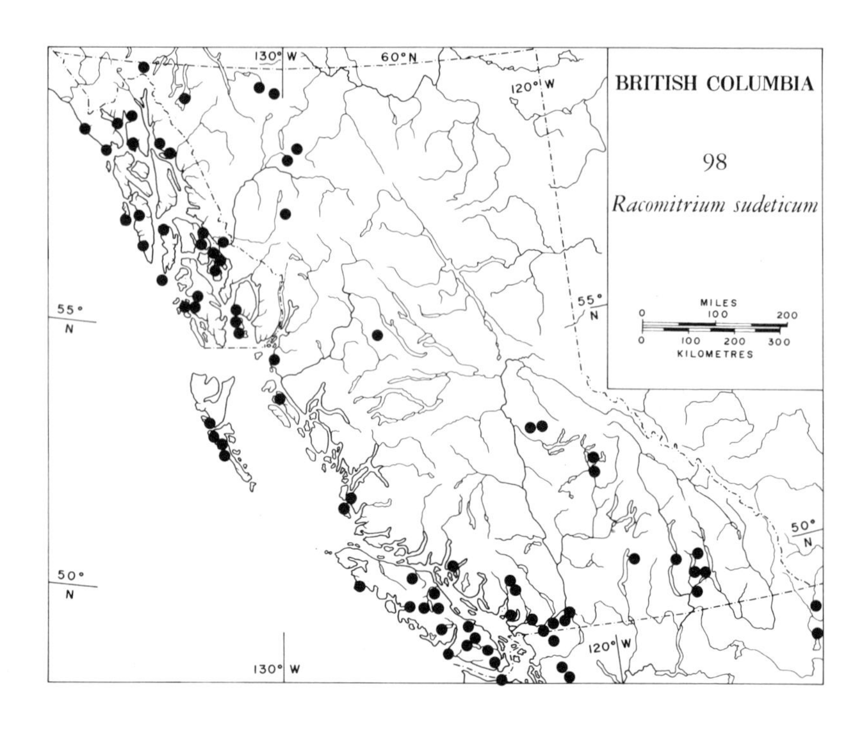

BRITISH COLUMBIA
98
Racomitrium sudeticum
MILES
0
100
200
0
100
200
300
KILOMETRES
130° W
60° N
120° W
55° N
50° N

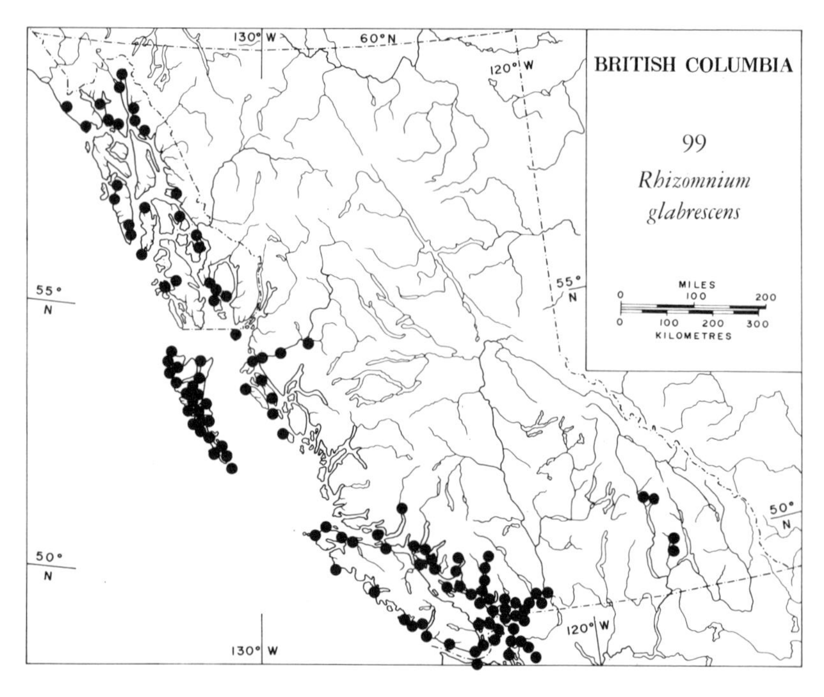

BRITISH COLUMBIA
99
Rhizomnium glabrescens
MILES
0
100
200
0
100
200
300
KILOMETRES
130° W
60° N
120° W
55° N
50° N

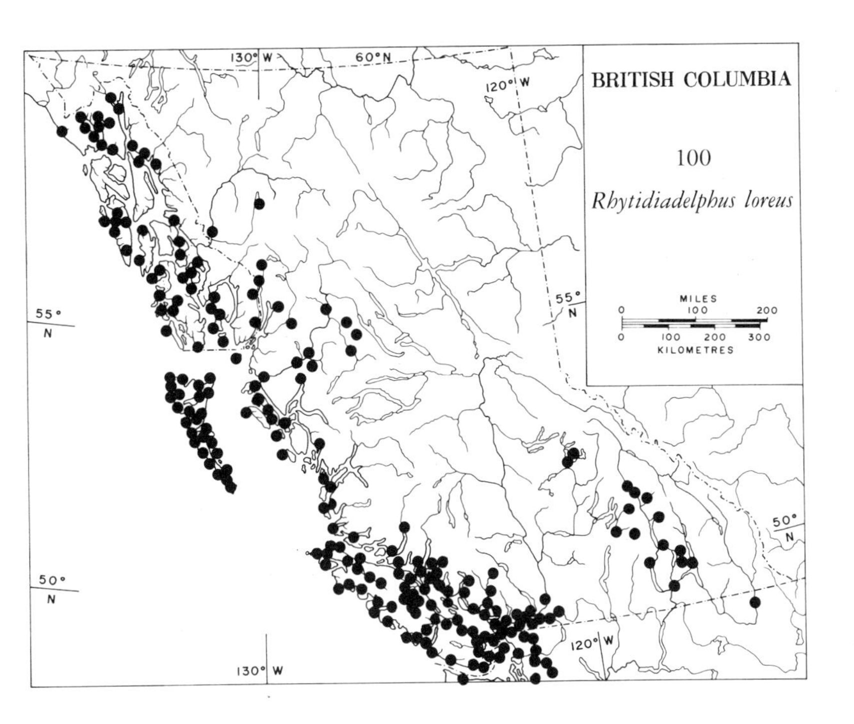

BRITISH COLUMBIA
100
Rhytidiadelphus loreus
130° W
60° N
120° W
55° N
50° N
MILES
0
100
200
300
KILOMETRES

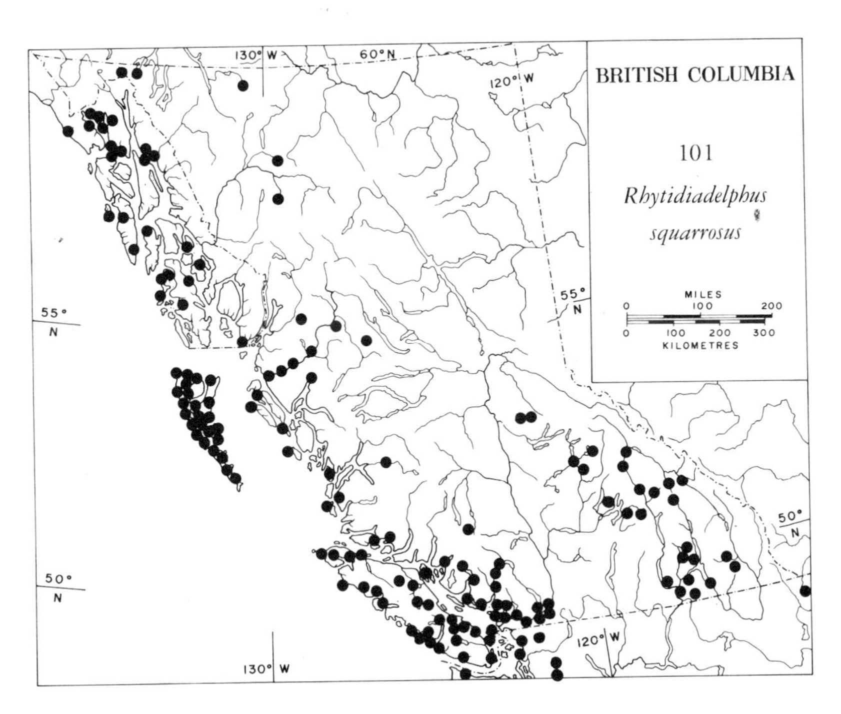

BRITISH COLUMBIA
101
Rhytidiadelphus squarrosus
130° W
60° N
120° W
55° N
50° N
MILES
0
100
200
300
KILOMETRES

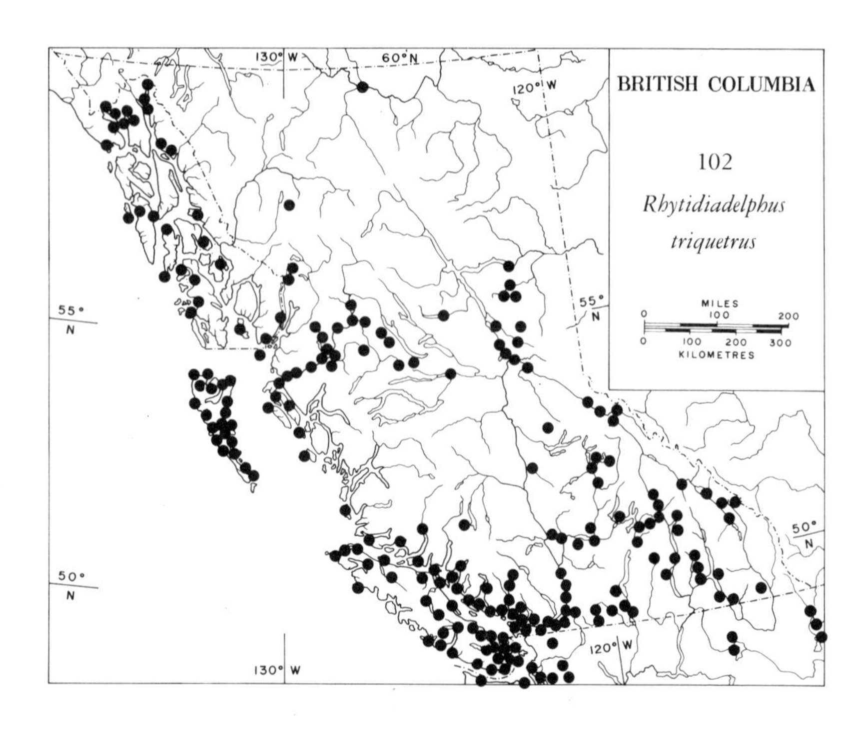
BRITISH COLUMBIA
102
Rhytidiadelphus triquetrus
MILES
0 100 200
0 100 200 300
KILOMETRES
130° W
60° N
120° W
55° N
50° N

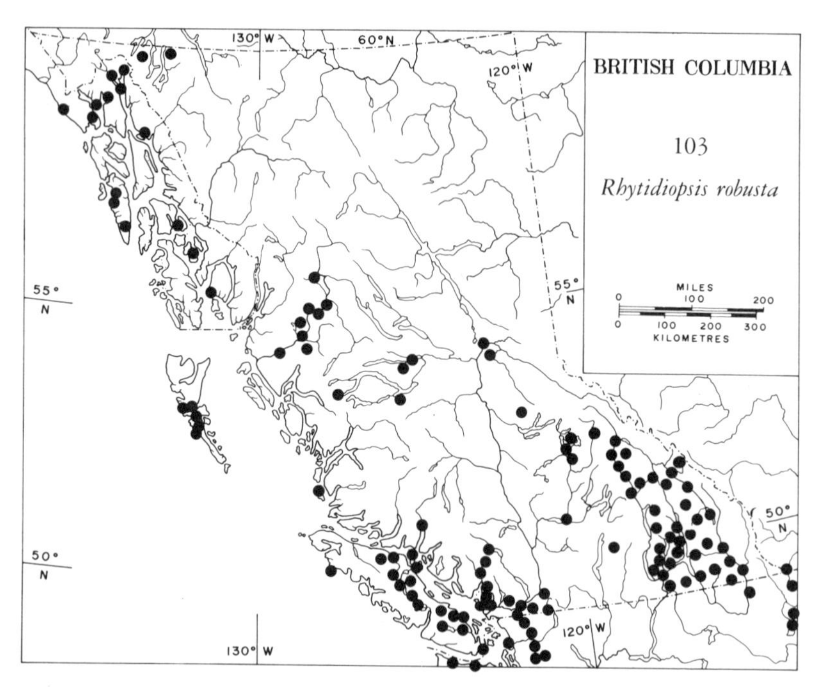
BRITISH COLUMBIA
103
Rhytidiopsis robusta
MILES
0 100 200
0 100 200 300
KILOMETRES
130° W
60° N
120° W
55° N
50° N

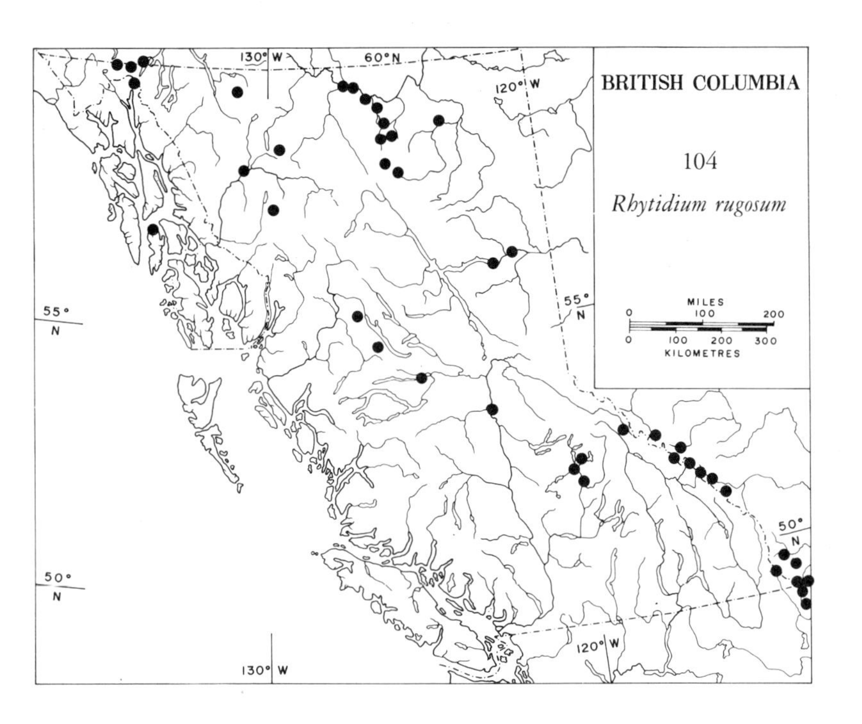
BRITISH COLUMBIA
104
Rhytidium rugosum
MILES
0
100
200
0
100
200
300
KILOMETRES
130° W
60° N
120° W
55° N
50° N
120° W
130° W

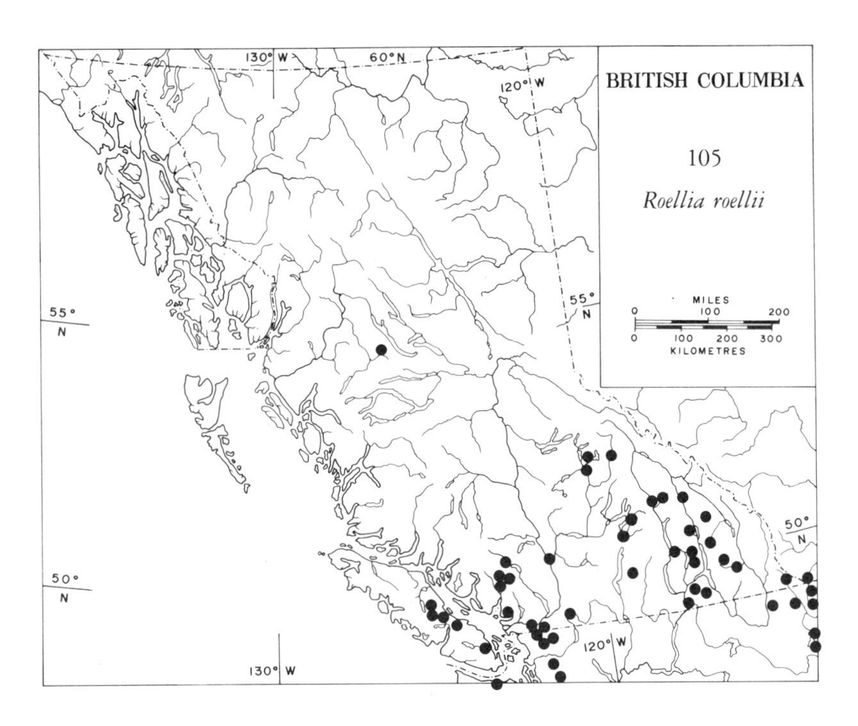
BRITISH COLUMBIA
105
Roellia roellii
MILES
0
100
200
0
100
200
300
KILOMETRES
130° W
60° N
120° W
55° N
50° N
120° W
130° W

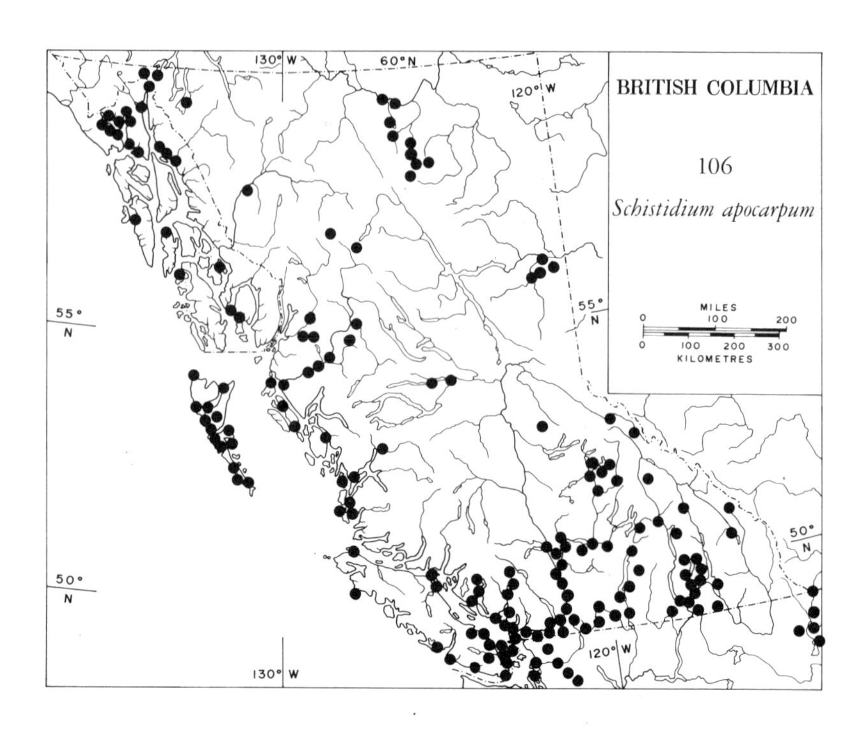
130° W
60° N
120° W
BRITISH COLUMBIA
106
Schistidium apocarpum
MILES
0
100
200
0
100
200
300
KILOMETRES
55°
N
55°
N
50°
N
50°
N
120° W
130° W

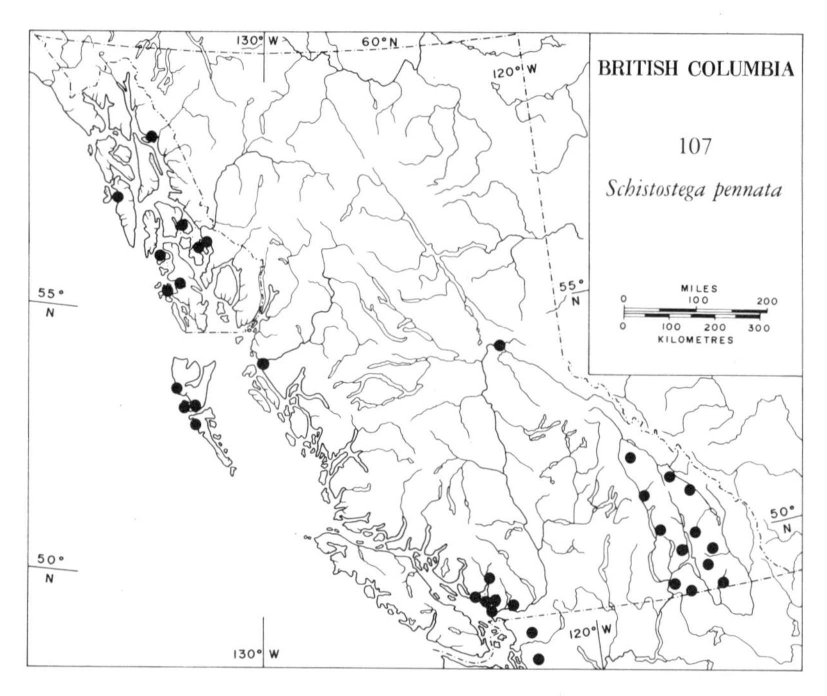
130° W
60° N
120° W
BRITISH COLUMBIA
107
Schistostega pennata
MILES
0
100
200
0
100
200
300
KILOMETRES
55°
N
55°
N
50°
N
50°
N
120° W
130° W

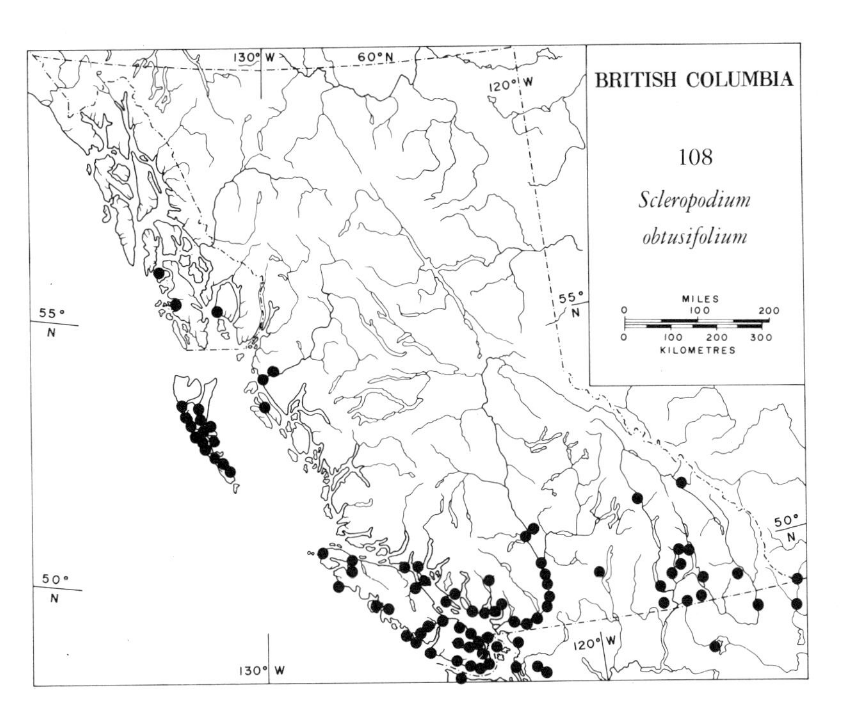
130° W
60° N
120° W
BRITISH COLUMBIA
108
Scleropodium obtusifolium
55° N
MILES
0 100 200
0 100 200 300
KILOMETRES
50° N
120° W
130° W

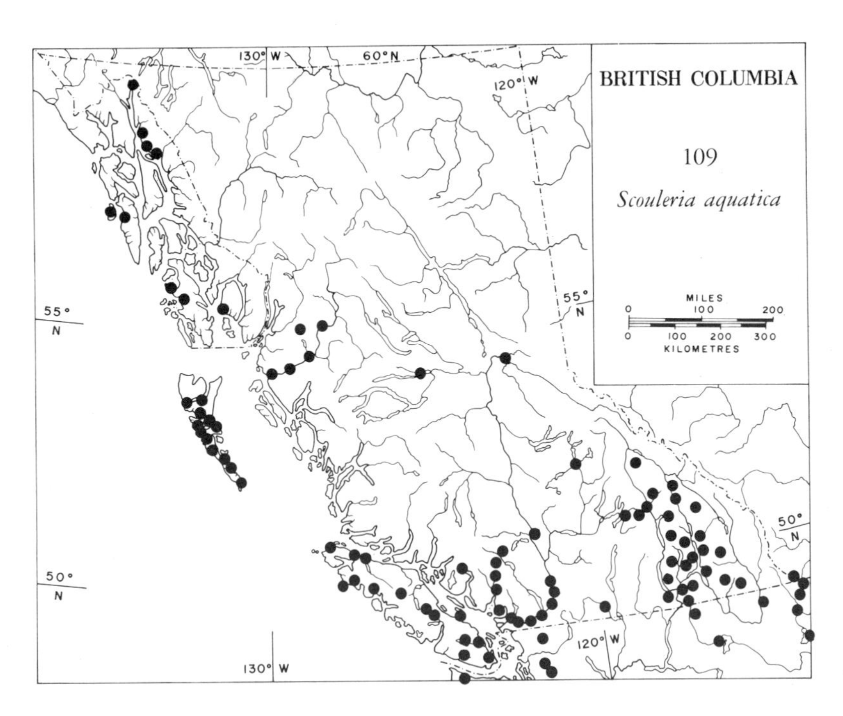
130° W
60° N
120° W
BRITISH COLUMBIA
109
Scouleria aquatica
55° N
MILES
0 100 200
0 100 200 300
KILOMETRES
50° N
120° W
130° W

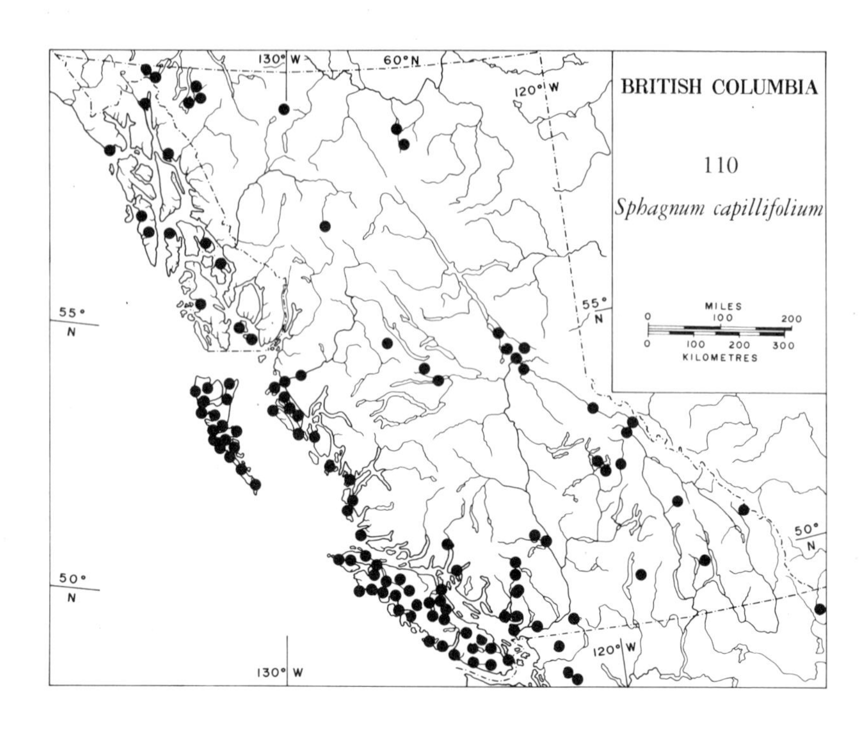
130° W
60° N
120° W
BRITISH COLUMBIA
110
Sphagnum capillifolium
55° N
MILES
0 100 200
0 100 200 300
KILOMETRES
50° N
50° N
130° W
120° W

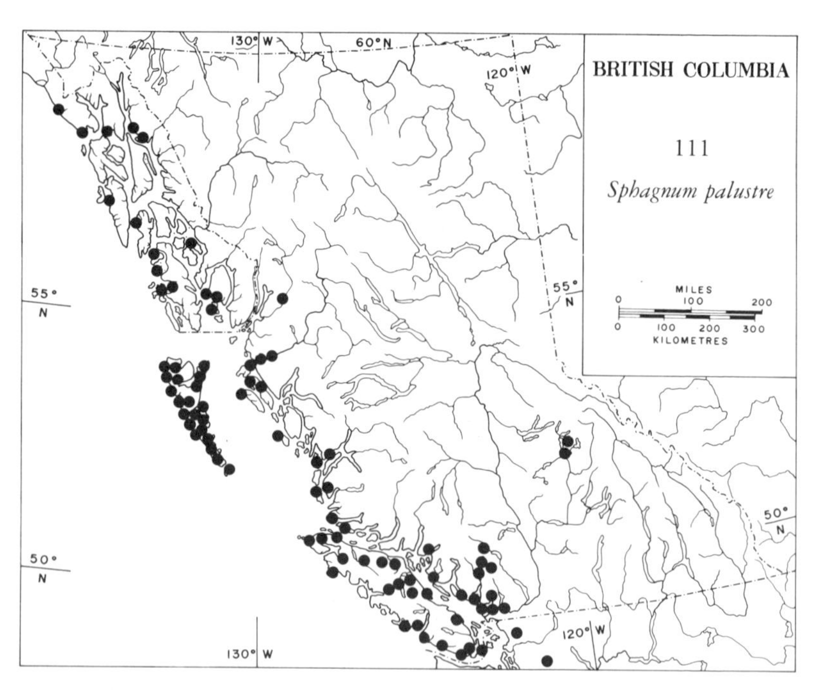
130° W
60° N
120° W
BRITISH COLUMBIA
111
Sphagnum palustre
55° N
MILES
0 100 200
0 100 200 300
KILOMETRES
50° N
50° N
130° W
120° W

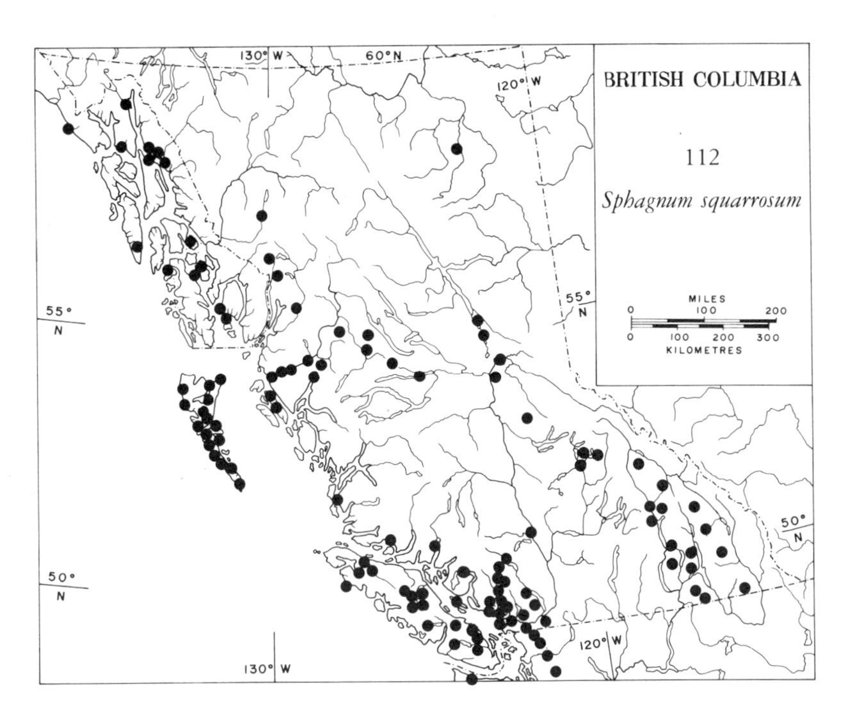
BRITISH COLUMBIA
112
Sphagnum squarrosum
130° W
60° N
120° W
55° N
50° N
MILES
0 100 200
0 100 200 300
KILOMETRES

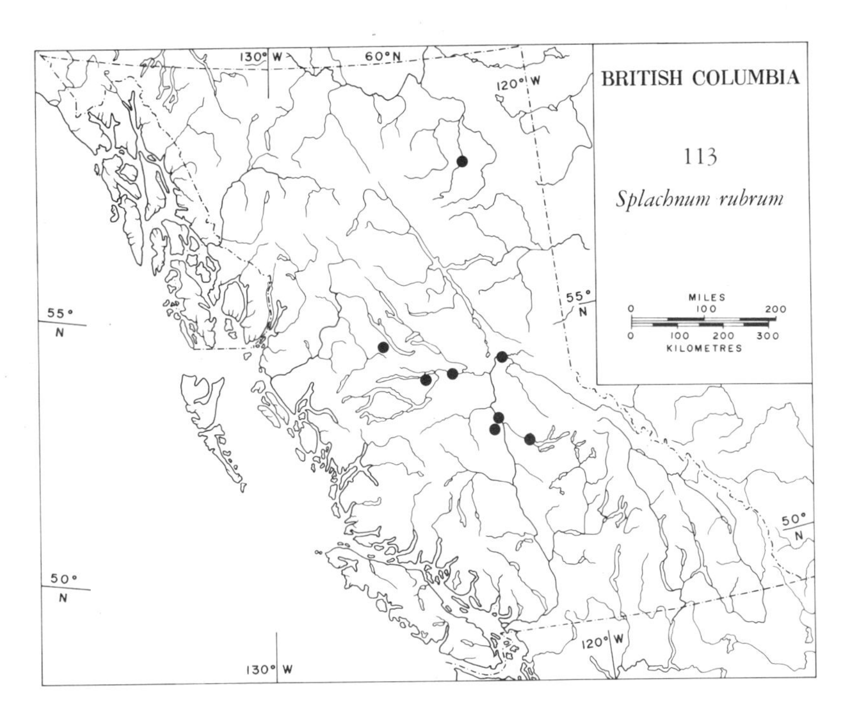
BRITISH COLUMBIA
113
Splachnum rubrum
130° W
60° N
120° W
55° N
50° N
MILES
0 100 200
0 100 200 300
KILOMETRES

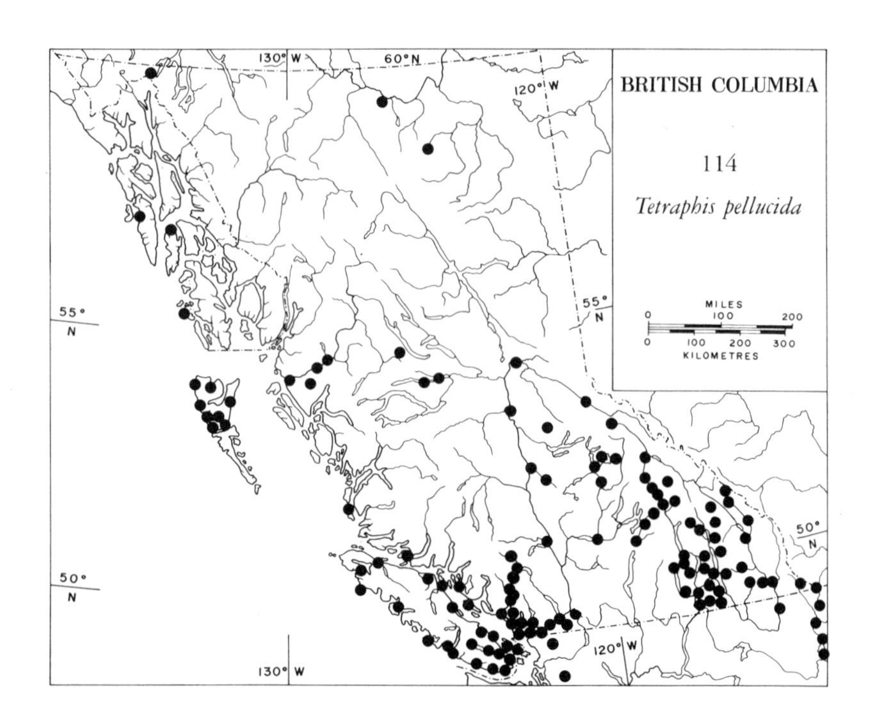
130° W
60° N
120° W
BRITISH COLUMBIA
114
Tetraphis pellucida
55° N
MILES
0 100 200
0 100 200 300
KILOMETRES
50° N
50° N
120° W
130° W

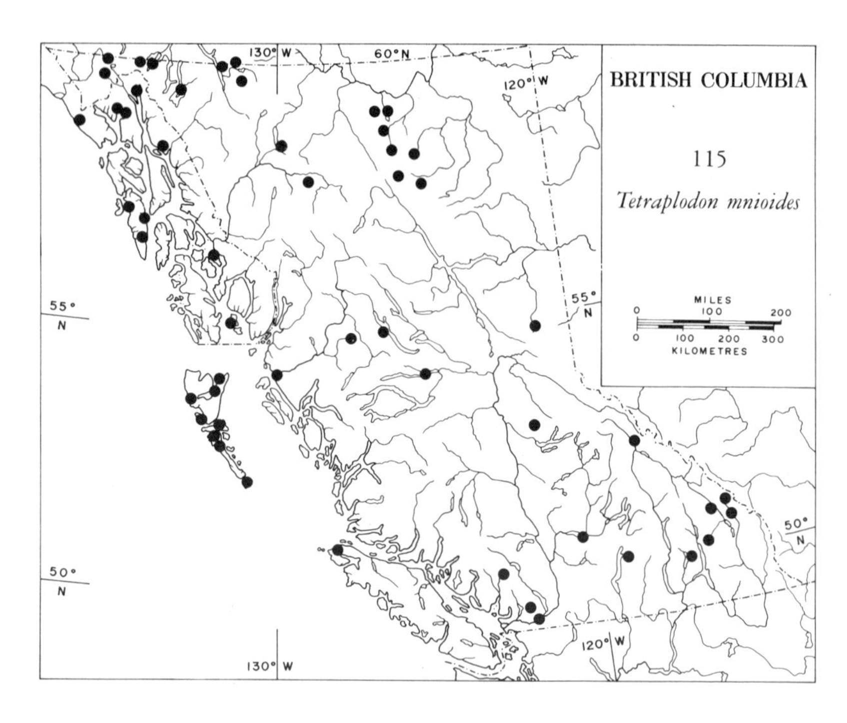
130° W
60° N
120° W
BRITISH COLUMBIA
115
Tetraplodon mnioides
55° N
MILES
0 100 200
0 100 200 300
KILOMETRES
50° N
50° N
120° W
130° W

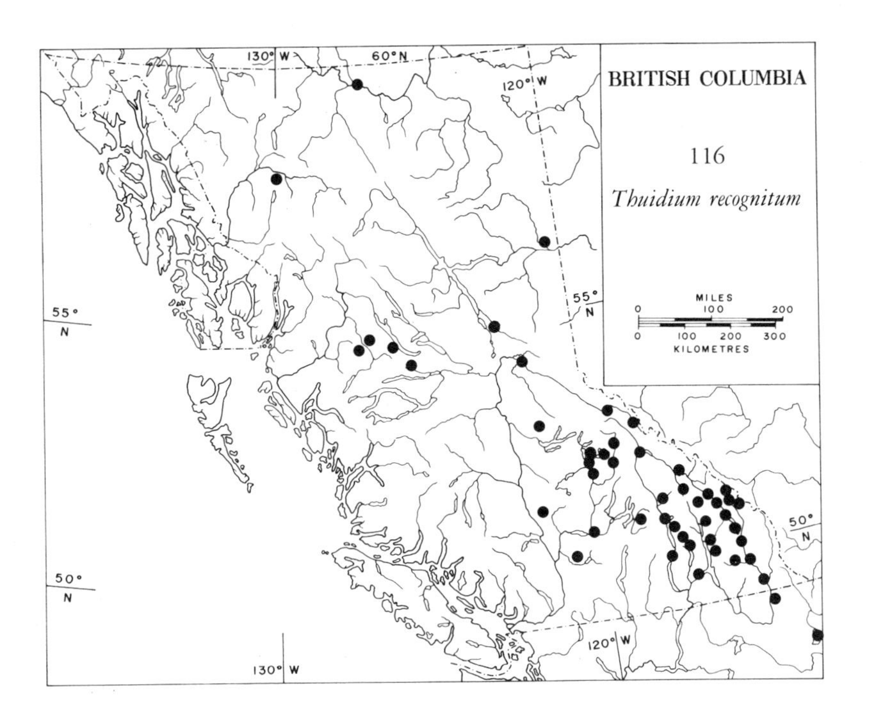
BRITISH COLUMBIA
116
Thuidium recognitum
MILES
0 100 200
0 100 200 300
KILOMETRES
130° W
60° N
120° W
55° N
50° N
130° W
120° W

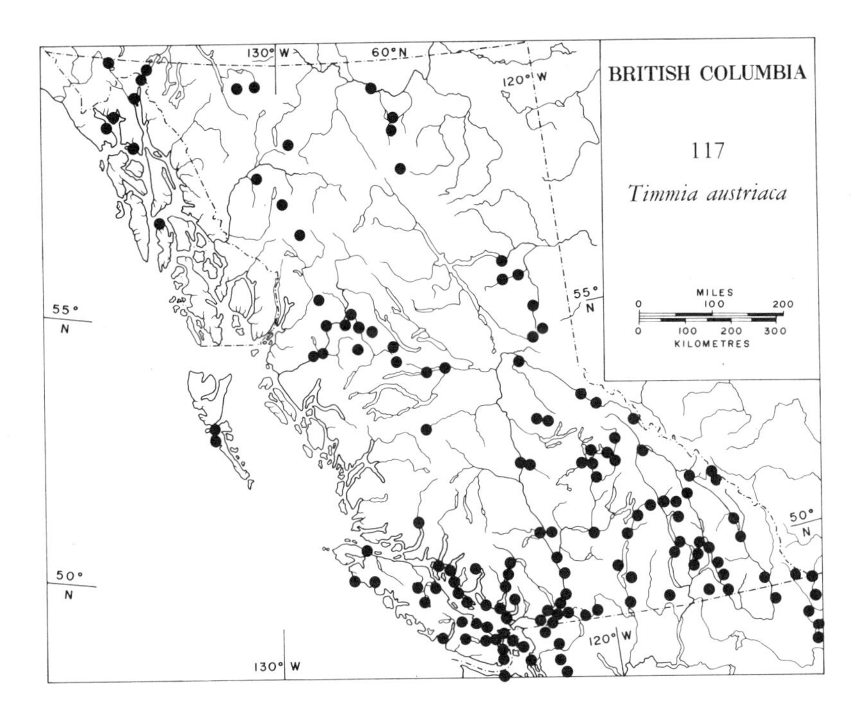
BRITISH COLUMBIA
117
Timmia austriaca
MILES
0 100 200
0 100 200 300
KILOMETRES
130° W
60° N
120° W
55° N
50° N
130° W
120° W

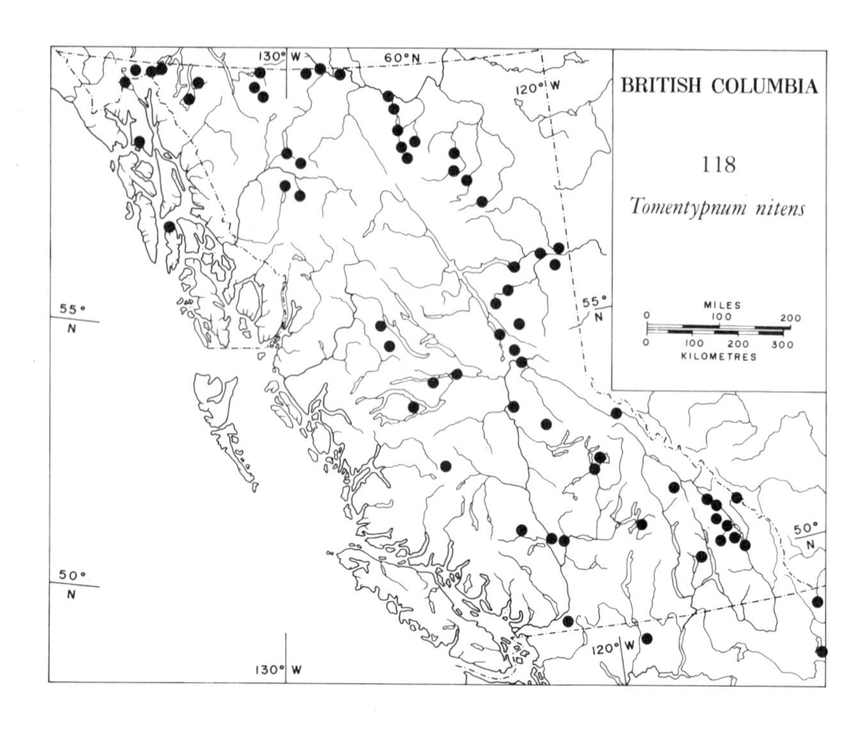
BRITISH COLUMBIA
118
Tomentypnum nitens
MILES
0
100
200
0
100
200
300
KILOMETRES
130° W
60° N
120° W
55° N
50° N
120° W
130° W
55° N
50° N

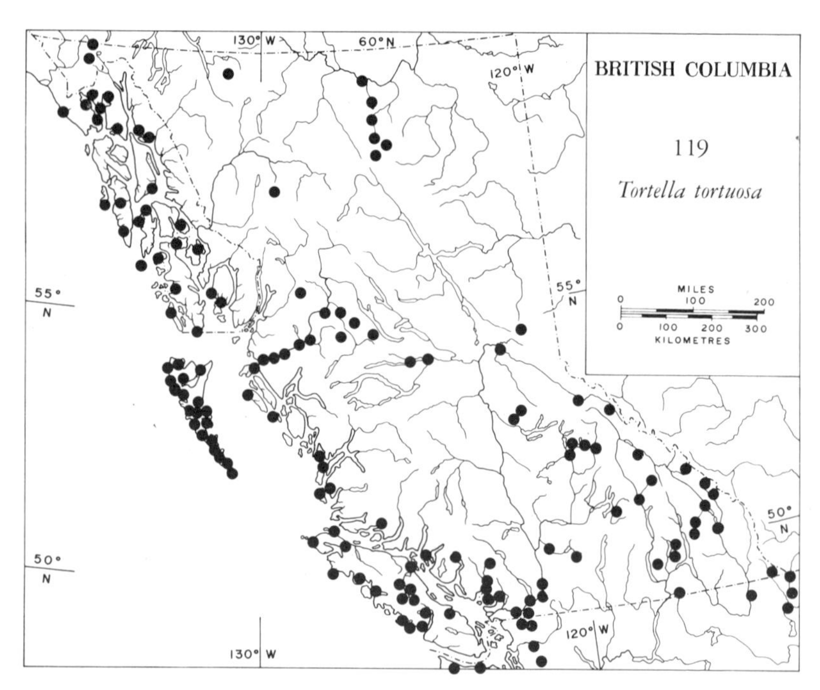
BRITISH COLUMBIA
119
Tortella tortuosa
MILES
0
100
200
0
100
200
300
KILOMETRES
130° W
60° N
120° W
55° N
50° N
120° W
130° W
55° N
50° N

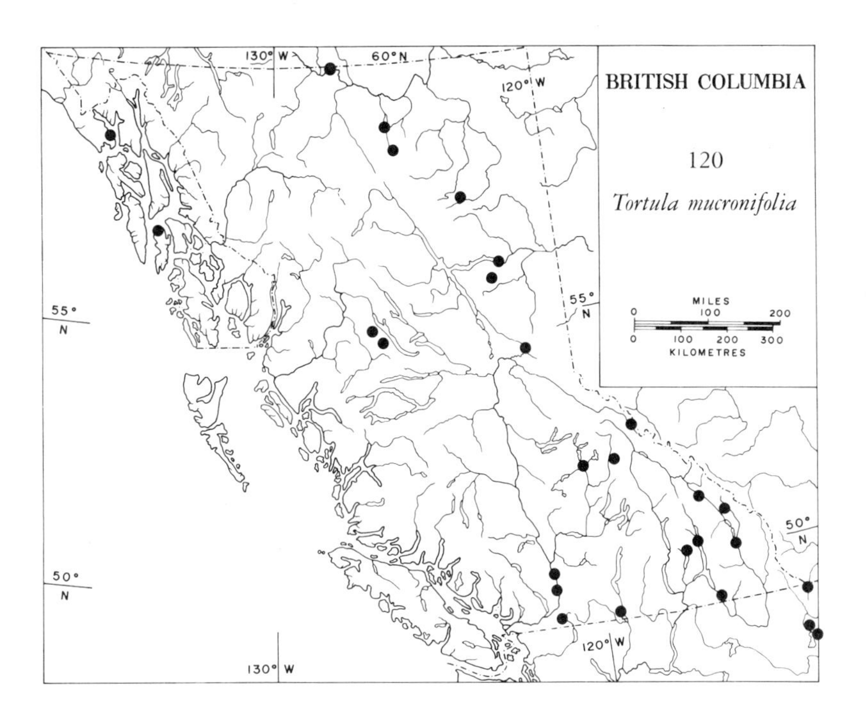
BRITISH COLUMBIA
120
Tortula mucronifolia
MILES
0 100 200
0 100 200 300
KILOMETRES
130° W
60°N
120° W
55° N
50° N
130° W
120° W

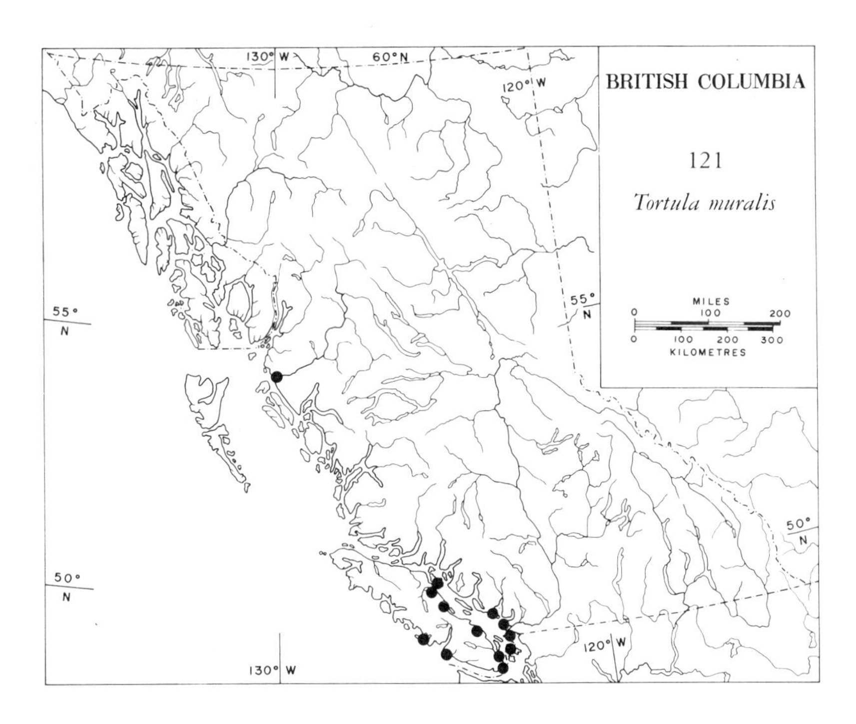
BRITISH COLUMBIA
121
Tortula muralis
MILES
0 100 200
0 100 200 300
KILOMETRES
130° W
60°N
120° W
55° N
50° N
130° W
120° W

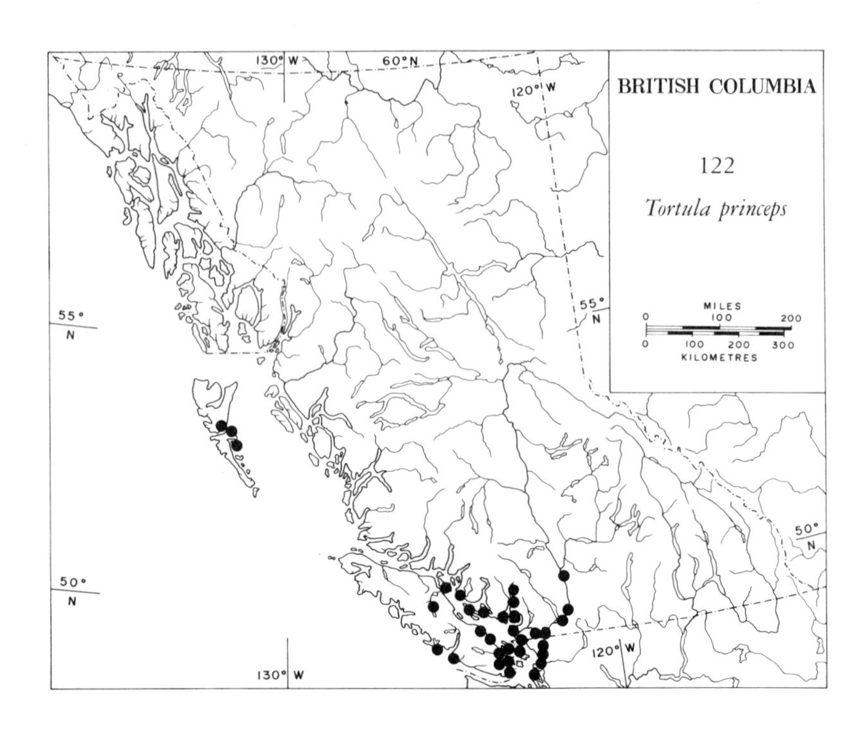
BRITISH COLUMBIA
122
Tortula princeps
MILES
0
100
200
0
100
200
300
KILOMETRES
130° W
60° N
120° W
55° N
50° N
130° W
120° W

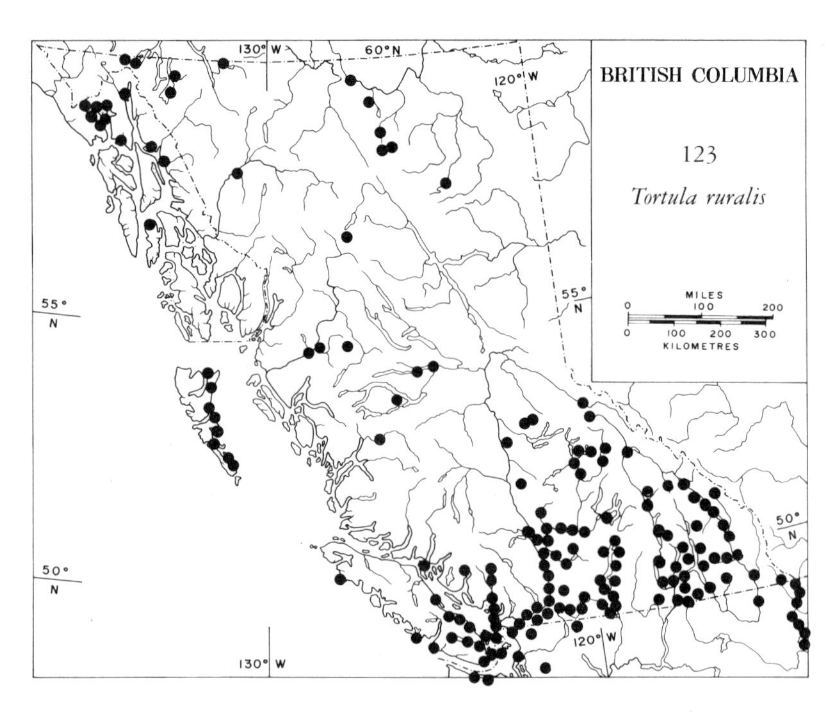
BRITISH COLUMBIA
123
Tortula ruralis
MILES
0
100
200
0
100
200
300
KILOMETRES
130° W
60° N
120° W
55° N
50° N
130° W
120° W

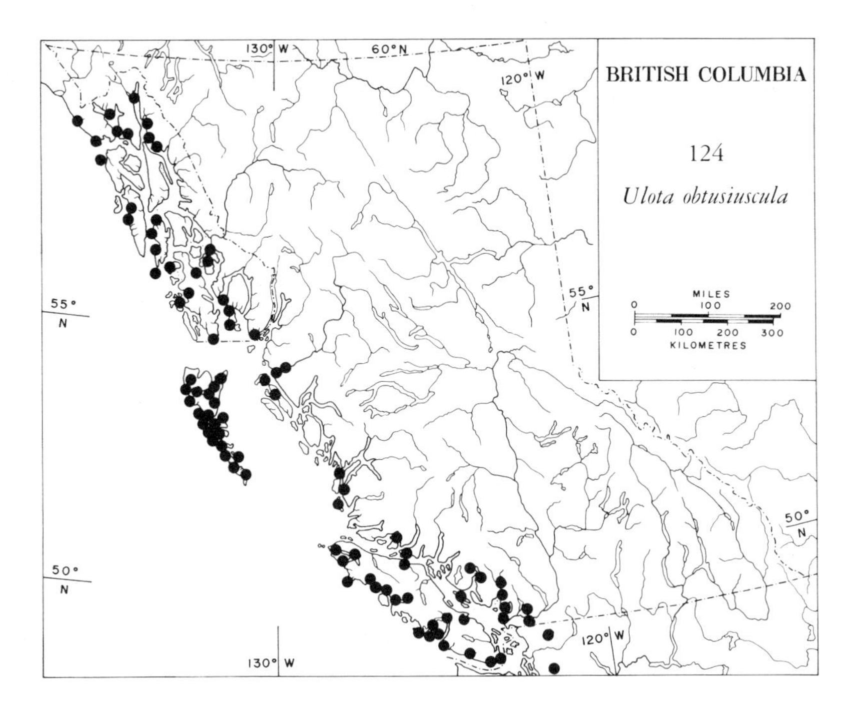
BRITISH COLUMBIA
124
Ulota obtusiuscula
MILES
0
100
200
0
100
200
300
KILOMETRES
130° W
60° N
120° W
55° N
55° N
50° N
50° N
130° W
120° W

GLOSSARY

acute—referring to the shape of the leaf apex, gradually narrowing to a point (see fig. 2, p. 6).
alar—a region of the leaf at the basal corners (see fig. 1, p. 5).
alpine—the elevation on a mountain above the limit of forest trees.
annulus—a ring of cells at the apex of a sporangium just at the base of the operculum, usually elastic and coiling outward from the sporangium before the operculum is shed.
apiculus—a short abrupt point.
attenuated—with a long, narrow apex.
bog—marshy land in which water and nutrients are added only from precipitation and in which there is accumulation of considerable undecomposed organic material; usually very acidic.
bordered—referring to a leaf margin different in colour and/or cell shape from the rest of the leaf (see p. 255).
boreal—referring to that portion of the northern hemisphere directly south of the arctic tundra and north of the sub-tropics, including much of Canada, Europe and northern Asia.
calyptra—a small sheath that covers the young sporangium and falls off before the operculum is shed (see fig. 1, p. 5).
capsule—a term often used instead of sporangium.
conic—having the shape of a cone.
coniferous—needle-leaved evergreen trees, usually producing cones.
cosmopolitan—of very wide geographic distribution.
costa—the midrib of a moss leaf (see fig. 1, p. 5).
crisped—referring to leaves that, when dry, become twisted and contorted.
cylindric—having the shape of a cylinder.

deciduous—referring to leaves that easily break off, from deciduous tre which lose their leaves for part of the year; in British Columbia, most broa leaved trees are deciduous.
decurrent—referring to the leaf that extends downward on both sides of t main attachment at the base (see fig. 2 G, p. 6).
disjunct—referring to geographic distribution, a species occurring in wide separated localities.
divergent—referring to leaves that angle outward from the stem, wi spaces between them.
elliptic—having the shape of an ellipse.
endemic—restricted in natural distribution to a limited geographic area.
entire—referring to a leaf's edge that is smooth or without any suggestion teeth.
exserted—referring to the sporangium that is slightly emergent beyond t perichaetial leaves.
falcate—a leaf that is curved, as the blade of a sickle (see p. 125).
fen—marshy land in which there is seepage of ground water adding nutrien from below, and in which there is an accumulation of an undecayed organ mat; usually alkaline or neutral.
flagelliferous—producing long slender branches with leaves much small than on the rest of the plant.
gemmae—special vegetative buds or clusters of cells that are deciduous ar serve as asexual reproductive structures (see fig. 1, p. 5).
glossy—with a somewhat shiny surface.
humus—decayed organic material, largely plant remains, producing a cha acteristic black to dark-brown surface layer, particularly in forest soils.
imbricate—leaves lying close to the stem, pointing toward the apex (s p. 161, habit sketch).
immersed—referring to the sporangium with the seta so short that it bare emerges from, or is sheathed by, the perichaetial leaves (see p. 269).
inclined—leaning over, referring to the sporangium oriented at an obtuse right angle to the seta (see p. 121).
incurved—referring to leaf margins that are curved upward and inward.
lamellae—flaps on the surface of a leaf (see *Polytrichum commune*, p. 235).
mats—mosses in which the shoots are interwoven and creeping over the su tratum.
montane—on mountains.
nodding—referring to the orientation of the sporangium, the seta curved an inverted J-shape at the apex and with the sporangium hanging downwa from the seta apex (see p. 255).
ovate—shaped like the outline of an egg (see p. 53).

ovoid—egg-shaped (sporangium shape).
papillae—small wart-like thickenings on the surface of a cell wall.
paraphyllia—small appendages on the stems of some creeping mosses, scattered among the leaf insertions, sometimes branched and green; best viewed in *Hylocomium splendens*.
perichaetium—the short special branch or the apex of a shoot containing the female sex organs and from which the sporophyte arises.
perichaetial leaves—the leaves that make up the perichaetium, often differing in size or structure from other leaves of the same plant (see fig. 1, p. 5).
peristome—the apical opening of the moss sporangium through which spores are shed, often ringed by peristome teeth (see fig. 1, p. 5).
pinnate—very regularly branched or feather-like in appearance, the branches all arising in the same horizontal plane (see p. 56).
pleated—referring to leaves with folds or creases parallel to the margins and midrib (see p. 257).
plumose—a branching pattern with many regular branches in one plane on both sides of the main stem, rather like a feather (see *Ptilium crista-castrensis*, p. 243).
recurved—referring to a leaf margin that is curved under.
revolute—the margins of a leaf that is strongly curled under.
rhizoids—filamentous, many-branched strands usually red-brown in colour, generally on the stems of mosses, often fixing the moss to its substratum.
rosette—an apical, somewhat flattened tuft of leaves at the apex of a shoot, giving the apex the appearance of a flower.
secund—leaves that have the points all oriented in the same direction, to one side of the shoot (see p. 125).
seta—the stalk of the sporophyte (see fig. 1, p. 5).
siliceous—substratum in which silica is a main component, making the chemical reaction acidic.
sporangium—the spore-producing structure (capsule) of a moss.
squarrose—leaves that spread out at right angles to the stem, often with the base somewhat clasping and the leaf tip curved downward (see *Paludella squarrosa*, p. 207).
stoloniferous—a moss that consists primarily of erect shoots, but in which there are also creeping shoots; the creeping shoots are said to be stoloniferous.
subalpine—the elevation on a mountain just below alpine, where the forest distribution is controlled by temperature, the nature of soil formation and snow persistence.
suberect—nearly erect, referring to the orientation of a sporangium on the seta apex.

swamp—wet, marshy land in which there is usually standing water at o
above the soil surface most of the year.
turfs—mosses with the shoots parallel and upright like the pile of a carpet
undulate—referring to leaves with wave-like, regular wrinkles running a
right angles to the leaf length (see *Metaneckera menziesii*, p. 195).

INDEX

Abietinella abietina 56–57, 288; map 319
Ahti, T. 13
alar regions 4–5, 76
Aloina 296
Alsia 108
Amblystegium compactum 58
Amblystegium serpens 58–59; map 320
Amphidium 118
Amphidium lapponicum 60–61; map 320
Amphidium mougeotii 60
Anacolia menziesii 74, 212
Anderson, L.E. 27
Andreaea blyttii 62
Andreaea megistospora 62
Andreaea nivalis 62
Andreaea rothii 62
Andreaea rupestris 62–63; map 321
Andreaea schofieldiana 62
annulus 136
Anoectangium aestiuum 60
antheridia 8
Antitrichia californica 64
Antitrichia curtipendula 64–65, 260; map 321
apophysis 4–5
apple moss 74
archegonium 8
Arctoa fulvella 76, 112
asexual reproduction 9
Atrichum selwynii 66–67; map 322
Atrichum tenellum 66
Atrichum undulatum 66
Aulacomnium acuminatum 70, 72
Aulacomnium androgynum 68, 69–70, 96, 284; map 322
Aulacomnium palustre 68, 70–71, 72; map 323
Aulacomnium turgidum 70, 72–73; map 323

Barbula 94
Barbula cylindrica 128
Bartramia 212
Bartramia ithyphylla 74
Bartramia pomiformis 74–75; map 324
Belland, R.J. 14
Blindia acuta 76–77; map 324
Boas, F.M. 14
bog 15, 22
Bovey, Robin B. 27
Brachythecium 90, 178, 186, 292
Brachythecium albicans 160
Brachythecium asperrimum 78–79; map 325
Brachythecium curtum 78

Brachythecium fendleri 144
Brachythecium frigidum 78
Brinkman, A.H. 13
broadly lanceolate 6
brown moss 124
Bryhnia hultenii 186
Bryoerythrophyllum 94
Bryoerythrophyllum recurvivostrum 128
Bryologist 27
Bryopsida 2
Bryum 188, 224
Bryum argenteum 80–81; map 325
Bryum bicolor 80
Bryum calobryoides 80
Bryum capillare 82–83; map 326
Bryum miniatum 84–85; map 326
Bryum muhlenbeckii 84
Bryum pseudotriquetrum 82
bug moss 32
Buxbaumia aphylla 86
Buxbaumia piperi 32, 86–87; map 327
Buxbaumia viridis 86

Calliergon cordifolium 88
Calliergon giganteum 88–89; map 327
Calliergon sarmentosum 84, 88
Calliergon stramineum 90–91; map 328
Calliergonella cuspidata 88, 150, 218
calyptra 4–5, 8
Campylium stellatum 92–93; map 328
Campylopus 76, 124
Catoscopium nigritum 94–95, 96; map 329
Ceratodon purpureus 68, 96–97, 128; map 329
Chuang, C-C. 14
Cirriphyllum cirrosum 272
Claopodium 158
Claopodium bolanderi 58, 98
Claopodium crispifolium 58, 98–99; map 330
cliff ledge mosses 20
Climacium dendroides 100–101, 190; map 330
club-mosses 2
coastal forest mosses 17–18
coastal mosses 19
collecting 10–14
columella 5, 8
complanate 3
Conostomum 208
Conostomum tetragonum 74
cord moss 152
Coscinodon calyptratus 102–3, 154, 248; map 331
costa 4–5
Cratoneuron commutatum 104–5, 106, 242; map 331
Cratoneuron filicinum 106–7, 136; map 332
Crossidium 296
Crum, H.A. 27
Crumia latifolia 296
Ctenidium schofieldii 242
cylindric 7

decurrent base 6
Dendroalsia abietina 108–9; map 332
Desmatodon latifolius 296
determination of mosses 26
Dichodontium flavescens 110
Dichodontium pellucidum 110–11, 192; map 333
Dicranella 76, 114
Dicranella heteromalla 112–13, 182; map 333
Dicranella palustris 92, 110
Dicranella rufescens 114–15; map 334
Dicranodontium 120
Dicranoweisia cirrata 116–17; map 334
Dicranoweisia crispula 118–19, 182; map 335
Dicranum affine 122
Dicranum fragilifolium 126

Dicranum fuscescens 120–21; map 335
Dicranum majus 122, 124
Dicranum pallidisetum 120
Dicranum polysetum 122–23; map 336
Dicranum scoparium 124–25; map 336
Dicranum tauricum 76, 126–27; map 337
Didymodon 94
Didymodon insulanus 128–29; map 337
Didymodon occidentalis 128
differentiated margin 6
Distichium capillaceum 130–31; map 338
Distichium inclinatum 130
disturbed site mosses 23
Ditrichum 76, 112, 114, 130
Ditrichum ambiguum 134
Ditrichum crispatissimum 120, 132
Ditrichum flexicaule 132–33; map 338
Ditrichum heteromallum 134–35; map 339
Ditrichum schimperi 134
Ditrichum zonatum 134
Djan-Chékar, N. 14
Drepanocladus 104, 168
Drepanocladus crassicostatus 136
Drepanocladus exannulatus 136–37; map 339
Drepanocladus fluitans 136
Drepanocladus pseudostramineus 90
Drepanocladus uncinatus 138–39, 172, 174, 242; map 340
Drummond, Thomas 13
Dryptodon patens 248, 252

ecology 14–16
egg 8
elaters 3
electrified cat tail moss 260
elliptic 7
Encalypta affinis 140
Encalypta alpina 140
Encalypta brevicolla 140
Encalypta brevipes 140
Encalypta ciliata 140–41; map 340
Encalypta intermedia 142
Encalypta procera 140, 142
Encalypta rhaptocarpa 142–43; map 341
Encalypta spathulata 142
Encalypta vulgaris 142
Eurhynchium 179
Eurhynchium oreganum 184
Eurhynchium praelongum 186
Eurhynchium pulchellum 144–45, 184, 186; map 341
Eurhynchium pulchellum var. *barnesiae* 184
Eurhynchium stokesii 186
extinguisher moss 140

fairy gold 270
falcate 6
falcate-secund 44
feather mosses 28
Fissidens 270
Fissidens adianthoides 146–47; map 342
Fissidens bryoides 148
Fissidens grandifrons 146
Fissidens limbatus 148–49; map 342
Fissidens osmundioides 146
Fissidens ventricosus 148
Flowers, Seville 27
Fontinalis antipyretica 150–51; map 343
Fontinalis howellii 150
Fontinalis neomexicana 150
forest as habitat 16, 17–18
forest floor mosses 22
Funaria hygrometrica 152–53; map 343
Funaria muhlenbergii 152

gametophyte 2, 4–5, 8
garden weed mosses 23
Gebeebia gigantea 10, 294
gemmae 4–5, 10

Grimmia apocarpa 268
Grimmia pulvinata 102, 154–55, 248; map 344
Grimmia torquata 60
Grimmia trichophylla 154
Gymnostomum 60

habitat 10–11, 16, 19–26
hair cap mosses 28, 234
Halbert, R. 14
hand lens 31
heart-shaped 6
Hedwigia ciliata 156–57, 248; map 344
Helodium blandonii 57, 106
Hepaticae 3
herbarium 26
Heterocladium dimorphum 158
Heterocladium heteropteroides 158
Heterocladium macounii 58, 158–59; map 345
history of collectors 13–14
Homalothecium 179, 292
Homalothecium aeneum 160–61, 162, 164; map 345
Homalothecium arenarium 160
Homalothecium fulgescens 160, 162–63; map 346
Homalothecium lutescens 162
Homalothecium nevadense 160, 162, 164
Homalothecium nuttallii 162, 164–65; map 346
Homalothecium pinnatifidum 164
Hookeria acutifolia 166, 214
Hookeria lucens 166–67, 214; map 347
Horton, D.G. 14
Hübschmann, A. von 14
Hygrohypnum 272
Hygrohypnum ochraceum 168–69; map 347
Hylocomium splendens 10
Hylocomium splendens 10, 170–71, 186, 218, 288; map 348
Hylocomium umbratum 170
Hymenostylium 60
Hypnum 104, 138, 168
Hypnum callichroum 242
Hypnum circinale 172–73, 176; map 348
Hypnum crista-castrensis 242
Hypnum cupressiforme 174
Hypnum imponens 176
Hypnum procerrimum 242
Hypnum recurvatum 174
Hypnum revolutum 174–75; map 349
Hypnum subimponens 172, 176–77, 242; map 349
Hypnum vaucheri 174
Hypopterygium fauriei 190

inclined sporangium 7
Interior humid forest mosses 17–18
Ireland, R.R. 14, 27
Isopterygium borrerianum 240
Isopterygium elegans 240
Isothecium cristatum 179
Isothecium spiculiferum 179
Isothecium stoloniferum 58, 108, 178–81, 270; map 350

Jamieson, D.W. 14

key separating mosses from other organisms 3–4
key to common mosses 35–55
Kiaeria 76, 112, 118
Kiaeria blyttii 182
Kiaeria falcata 182
Kiaeria starkei 182–83; map 350
Kindbergia 144
Kindbergia oregana 106, 184–85, 186; map 351
Kindbergia praelonga 58, 106, 170, 184, 186–87; map 351
knight's plume moss 242

Krajina, V.J. 14
Kujala, V. 13

lamellae 6
lanceolate 6
lawn mosses 23–24
Lawton, Elva 26
leaf shapes of mosses 6
Leptobryum pyriforme 188–89; map 352
Leptodontium recurvifolium 10, 294
Leskea polycarpa 158
Leskeella 158
Leucolepis acanthoneuron 100, 190–91; map 352
Leucolepis menziesii 190
lichens 2
life cycle of moss 8–10
logs as moss habitat 25–26
luminous moss 32, 270
Lyall, David 13

MacFadden, Faye 13
MacKenzie, Elizabeth 13
Macoun, John 13
Marsh, Janet E. 27
McIntosh, T.T. 14
Meesia longiseta 94
Meesia triquetra 94, 192–93; map 353
Meesia uliginosa 94
Menzies, Archibald 13
Metaneckera menziesii 194–95; map 353
Mnium ambiguum 196
Mnium arizonicum 196
Mnium glabrescens 254
Mnium insigne 210
Mnium marginatum 196
Mnium menziesii 190
Mnium spinulosum 196–97; map 354
Mnium thomsonii 196
mortar as moss habitat 24
moss gardens 29–30
mountain fern moss 170
Musci 2
Myurella julacea 80

names of mosses 27–29
Neckera douglasii 194, 198–99; map 354
Neckera menziesii 194
Neckera pennata 194
northern mosses 19

oblong 6
oblong-ovate 6
Oligotrichum aligerum 200–201; map 355
Oligotrichum hercynicum 200
Oligotrichum parallelum 66
Oncophorus virens 192, 202
Oncophorus wahlenbergii 182, 202–3; map 355
operculum 2, 4, 5, 8
Orthothecium chryseum 292
Orthotrichum consimile 116, 304
Orthotrichum lyellii 204–5; map 356
Orthotrichum rivulare 244
Orthotrichum striatum 204
ovate 6
ovate-lanceolate 6
ovoid 7
Oxystegus tenuirostris 192, 294

packet construction 11
palm tree moss 190
Paludella squarrosa 206–7; map 356
Palustriella commutata 104
Paraleucobryum 120, 124
paraphyllia 47, 104
peat moss 28, 30, 276, 278, 280
perianth 3
perichaetial leaves 4–5, 8
perichaetium 8
perigonia 8
perigonial leaves 8

peristome 4–5, 9
peristome teeth 4–5, 9
Persson, H. 14
Philonotis 94, 230
Philonotis capillaris 208
Philonotis fontana 208–9; map 357
pigeon wheat 234
Plagiobryum zierii 80
Plagiomnium ciliare 210
Plagiomnium cuspidatum 210
Plagiomnium drummondii 210
Plagiomnium ellipticum 210
Plagiomnium insigne 210–11; map 357
Plagiomnium medium 210
Plagiomnium venustum 210
Plagiopus oederi 74, 212–13; map 358
Plagiothecium 194, 240
Plagiothecium cavifolium 214
Plagiothecium denticulatum 214–15; map 358
Plagiothecium elegans 240
Plagiothecium laetum 214
Plagiothecium undulatum 216–17; map 359
Pleuroziopsis ruthenica 190
Pleurozium schreberi 10, 88, 218–19; map 359
Pogonatum alpinum 232
Pogonatum alpinum var. *sylvaticum* 232
Pogonatum contortum 220–21; map 360
Pogonatum dentatum 222
Pogonatum macounii 232
Pogonatum urnigerum 222–23; map 360
Pohlia 94, 188
Pohlia annotina 224–25; map 361
Pohlia columbica 230
Pohlia cruda 226–27; map 361
Pohlia longibracteata 230
Pohlia nutans 82, 228–29; map 362
Pohlia wahlenbergii 230–31; map 362
Polytrichastrum alpinum 232
Polytrichum 290
Polytrichum alpinum 222, 232–33, 234; map 363
Polytrichum commune 234–35; map 363
Polytrichum formosum 234
Polytrichum juniperinum 234, 236–37, 238; map 364
Polytrichum lyallii 234
Polytrichum piliferum 234, 236, 238–39; map 364
Polytrichum sexangulare 222
Polytrichum strictum 236
Porotrichum bigelovii 214
protonema 8, 9
Pseudoscleropodium purum 24, 218
Pseudotaxiphyllum elegans 172, 240–41; map 365
Psilopilum cavifolium 200
Ptilium crista-castrensis 138, 176, 242–43; map 365

Racomitrium 156
Racomitrium aciculare 244–45, 252; map 366
Racomitrium affine 248
Racomitrium aquaticum 244
Racomitrium brevipes 252
Racomitrium canescens 246–47; map 366
Racomitrium elongatum 246
Racomitrium ericoides 246
Racomitrium heterostichum 248–49; map 367
Racomitrium lanuginosum 246, 248, 250–51; map 367
Racomitrium macounii 252
Racomitrium microcarpon 248
Racomitrium muticum 246
Racomitrium obesum 248
Racomitrium occidentale 248
Racomitrium pacificum 244
Racomitrium sudeticum 252–53; map 368
Racomitrium varium 248

red stem moss 218
reindeer mosses 2
revolute margins 6
rhizoids 3, 4–5, 8
Rhizomnium glabrescens 254–55; map 368
Rhizomnium gracile 254
Rhizomnium magnifolium 254
Rhizomnium nudum 254
Rhizomnium pseudopunctatum 254
Rhizomnium punctatum 254
Rhodobryum roseum 266
Rhynchostegium riparioides 272
Rhynchostegium serrulatum 214
Rhytidiadelphus loreus 256–57; map 369
Rhytidiadelphus loreus, uses 30
Rhytidiadelphus squarrosus 92, 256, 258–59; map 369
Rhytidiadelphus subpinnatus 258
Rhytidiadelphus triquetrus 64, 260–61; map 370
Rhytidiopsis robusta 260, 262–63; map 370
Rhytidium 256, 262
Rhytidium rugosum 262, 264–65; map 371
rock faces as moss habitat 15–16, 19–21
Roellia roellii 266–67; map 371
roots of overturned trees as moss habitat 25
rough neck moss 260

Sanionia uncinata 138
scale mosses 3
Schistidium 156
Schistidium apocarpum 268–69; map 372
Schistidium apocarpum var. *strictum* 268
Schistidium maritimum 268
Schistidium rivulare 244, 268
Schistostega pennata 32, 270–71; map 372
Scleropodium 179
Scleropodium obtusifolium 272–73; map 373
Scleropodium touretii 272
Scouleria aquatica 244, 274–75; map 373
sea mosses 2
seepage area mosses 20
semi-arid Interior mosses 19
seta 3, 4–5, 8
sexual reproduction 9
shaggy moss 260
silver moss 80, 81
sinuous apex 6
Spanish moss 2
Spence, J.R. 14
sperms 8
Sphagnum austinii 278
Sphagnum capillifolium 276–77; map 374
Sphagnum compactum 278
Sphagnum *see* peat moss
Sphagnum fuscum 276
Sphagnum girgensohnii 276
Sphagnum henryense 278
Sphagnum magellanicum 278
Sphagnum palustre 278–79; map 374
Sphagnum papillosum 278
Sphagnum rubellum 276
Sphagnum squarrosum 280–81; map 375
Sphagnum warnstorfii 276
Splachnaceae 14
Splachnum ampullaceum 282
Splachnum luteum 282
Splachnum rubrum 15, 32, 282–83; map 375
Splachnum sphaericum 286
Splachnum vasculosum 286
sporangium 2, 4–5
sporangium shapes 7
spore 8, 10
spore mother cells 8

spore release in mosses 9
sporophyte 2, 4–5, 8
squarrose 38
squarrose leaves 92, 93
stair-step moss 170
Stokesiella oregana 184
Stokesiella praelonga 186
storage of specimens 11
storks-bill moss 124
structure of a moss 4–5
stumps as moss habitat 25–26
subspherical 7
swamp mosses 22

Tan, B.C. 14
Taylor, T. 14
Tayloria 286
terrestrial mosses 21–22
Tetraphis geniculata 284
Tetraphis pellucida 68, 284–85; map 376
Tetraplodon angustatus 286
Tetraplodon mnioides 15, 286–87; map 376
Tetraplodon pallidus 286
Tetraplodon urceolatus 286
Thamnobryum neckeroides 190
Thuidium 170
Thuidium abietinum 58
Thuidium delicatulum 288
Thuidium philibertii 288
Thuidium recognitum 288–89; map 377
Timmia austriaca 290–91; map 377
Tomentypnum falcifolium 138, 292
Tomentypnum nitens 106, 292–93; map 378
tongue-shaped 6
Tortella fragilis 126, 294
Tortella humilis 294
Tortella inclinata 294
Tortella tortuosa 294–95, 304; map 378
Tortula amplexa 296
Tortula bolanderi 296
Tortula brevipes 296, 298
Tortula caninervis 302
Tortula laevipila 302
Tortula latifolia 296
Tortula mucronifolia 296–97; map 279
Tortula muralis 154, 296, 298–99; map 379
Tortula norvegica 302
Tortula princeps 298, 300–301; map 380
Tortula ruralis 300, 302–3; map 380
Tortula subulata 296
tree trunks as moss habitat 24–25
Trichostomum arcticum 294
tundra mosses 26
type specimen 28

Ulota drummondii 304
Ulota megalospora 304
Ulota obtusiuscula 294, 304–6; map 381
Ulota phyllantha 304
umbrella moss 32
undulate margins 6
undulate surface 6
uses of mosses 30–31

Vancouver, George 13
Vitt, D.H. 14, 27

Weissia 116
wet site mosses 22–23
wetland mosses 15
widespread mosses in B.C. 17, 18–19
Wijkia carlottae 10
Williams, R.S. 13

Zygodon viridissimus 60, 116
zygote 8